BUILDING CHURCHES IN NORTHERN CHINA.
A 1926 HANDBOOK IN CONTEXT

1926年法国传教士所撰中国北方教堂
营造手册的翻译和研究

舶来与本土

中英对照

著 [比] 高曼士 Thomas Coomans
　　　徐怡涛 Xu Yitao
译 　　吴美萍 Wu Meiping

图书在版编目（CIP）数据

舶来与本土：1926年法国传教士所撰中国北方教堂营造手册的翻译和研究：汉英对照 /（比）高曼士，徐怡涛著．—北京：知识产权出版社，2016.5

ISBN 978-7-5130-4144-7

Ⅰ.①舶… Ⅱ.①高… ②徐… Ⅲ.①教堂–建筑设计–研究–中国–汉、英 Ⅳ.①TU252

中国版本图书馆CIP数据核字（2016）第069800号

责任编辑：陆彩云　张　珑　　责任印制：卢运霞　　内文排版、设计：贺天作坊

舶来与本土：1926年法国传教士所撰中国北方教堂营造手册的翻译和研究

BOLAI YU BENTU:1926NIAN FAGUO CHUANJIAOSHI SUOZHUAN ZHONGGUO BEIFANG JIAOTANG YINGZAO SHOUCE DE FANYI HE YANJIU

［比］高曼士　徐怡涛　著

出版发行：知识产权出版社有限责任公司	网　　址：http://www.ipph.cn
电　　话：010-82004826	http://www.laichushu.com
社　　址：北京市海淀区西外太平庄55号	邮　　编：100081
责编电话：010-82000860转8574	责编邮箱：riantjade@sina.com
发行电话：010-82000860转8101/8573	发行传真：010-82000893/82003279
印　　刷：北京画中画印刷有限公司	经　　销：各大网上书店、新华书店及相关专业书店
开　　本：720mm×1000mm　1/16	印　　张：30.25
版　　次：2016年5月第1版	印　　次：2016年5月第1次印刷
字　　数：470千字	定　　价：180.00元

ISBN 978-7-5130-4144-7

出版权专有　侵权必究

如有印装质量问题，本社负责调换。

出版说明

1926年，法国传教士撰写并出版了手册《传教士建造者：建议 - 方案》。该手册介绍如何在中国北方建造教堂，内容涉及方方面面，如教堂选址、建筑材料选用、砖砌工艺、屋顶形式、建筑装饰等。该手册不仅提供了一份详尽的关于教堂建造的技术资料，也从另一角度帮助我们了解当时西方传教士是如何看待中国传统建筑的，以及他们是如何将西方建造技术传授给当地中国工人的。

《舶来与本土：1926年法国传教士所撰中国北方教堂营造手册的翻译和研究》是对该手册进行语境化分析和研究的集成，并全文再现了该手册。该手册撰写的年代是社会变革转型期的民国，当时有着大量关于"传统与现代建筑技术"以及"教堂建筑应该采用哥特样式还是中国本土化样式"的讨论。手册中提到的一处案例——河北省大名天主堂，现已成为中国"全国重点文物保护单位"。在对其进行深入研究的过程中，本书两位作者结合手册内容、现场勘测及其他西方文献进行了综合研究分析，从而产生了这本有着丰富插图的中英双语出版物。

本书提供了欧洲建筑如何转移并在中国实地建造的珍贵史料，展现了当时中、欧两大建筑传统之间的交融过程，具有非常独特的意义，对从事建筑历史、建筑考古、宗教历史和遗产保护等方面研究的中外学者具有重要的学术价值。

| 著者简介

高曼士

高曼士,考古和艺术史方向博士。现就职于比利时鲁汶大学工学院建筑系,也是鲁汶大学雷蒙德·勒麦尔国际保护中心的教授。他负责的专业课程主要有遗产保护历史和理论、建筑考古和建筑历史。现今他的研究主要集中于建筑考古、传统建筑材料和技术史、教堂建筑、1840—1940年西方建筑在中国的建设及其过程中产生的中西方建筑文化交融等。发表著作15部、学术论文百余篇。目前他还兼任加拿大魁北克大学城市遗产研究中心的副研究员以及香港中文大学的副教授,他也是比利时弗兰芒皇家科学院、国际古迹遗址理事会(ICOMOS)、中国文物保护基金会等学术机构的重要成员,同时还是国际古迹遗址理事会之项目——"共有遗产"的科学委员会的专家成员。

Thomas Coomans, doctor in archaeology and art history, is a professor at the Department of Architecture, Faculty of Engineering Science, KU Leuven (Belgium), where he teaches history and theory of conservation, building archaeology, and architectural history. He is also a staff member of the Raymond Lemaire International Centre for Conservation, an associate researcher at the Canada Research Chair on Urban Heritage (UQAM), and has been an adjunct assistant professor at the School of Architecture of The Chinese University of Hong Kong. His current research and publications focus on church architecture and East-West architectural transfers in China in the 1840s-1940s. He has published 15 books and more than hundred book chapters and articles in international publications. He is a member of the Royal Flemish Academy of Belgium, of ICOMOS International Scientific Committee on Shared Built Heritage, and of China Cultural Relics Protection Foundation.

著者简介

徐怡涛,历史学博士。现任北京大学考古文博学院副教授、文物建筑教研室主任。主要教学与科研方向为中国建筑史、中国建筑考古和中国文化遗产保护。历年发表学术论文、著作40余篇(部)。2006年当选北京大学第十一届十佳教师,2009年获北京大学教学优秀奖,2010年、2013年、2015年分别获北京大学人文杰出青年学者奖。2014年入选中国国家文物局专家库。

Xu Yitao, doctor in history, is an associate professor at the School of Archaeology and Museology, Peking University, and director of the Department of Architectural Heritage. His teaching and current research are mainly on the history of Chinese architecture, historic building archaeology and heritage conservation in China. He has published more than 40 papers and articles. In 2006, he was elected as one of the 11th Top Ten Teachers of Peking University, and in 2009, won the Teaching Excellence Award of Peking University. In 2010, 2013 and 2015, he was awarded as Outstanding Young Scholar of Peking University. Since 2014 he is a member of the expert panel of the State Administration of Cultural Heritage, China.

徐怡涛

| 译者简介

吴美萍

吴美萍,建筑遗产保护和管理方向博士。现居比利时,从事建筑遗产保护历史和理论、建筑遗产的预防性保护、城市建筑遗产可持续利用、世界文化遗产监测管理等方面的研究工作。目前已发表学术论文20余篇,出版论著2部,参与完成国际合作、国家级和省级科研项目近10项,参与完成各类地方保护工程10余项。

Wu Meiping, holds a doctorate in architectural heritage and conservation from Southeast University, and is now an independent scholar living in Belgium. Her research is mainly on the history and theory of architectural heritage conservation, preventive conservation and monitoring of architectural heritage, adaptive reuse of urban historic buildings, conservation planning of historic cities/towns/villages in China. She has published two books and more than 20 papers.

特别鸣谢

鲁汶大学
University of Leuven

北京大学
Peking University

上海天华建筑设计有限公司
Tianhua Architecture Planning & Engineering Ltd.

鸢园
Iris Garden

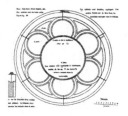

本书的意义和分工（代序）
Meaning and Division of Labor of the Book

本书是对 1926 出版的一本法语手册《传教士建造者：建议-方案》的翻译及相关研究成果的集成。其学术意义主要体现在以下三个方面。

（1）这部手册的研究和翻译，让我们得以从异域的角度审视自身历史文化轨迹，丰富了我们对中国近代建筑文化的认知。手册的意义并不仅仅在于传授建造教堂的技法，而是隐喻出建筑风格背后的文化立场。面对中国近代建筑，我们应该清楚地意识到，它们不仅仅代表着中国人的思想，同时也映射着西方人的思想。从这个意义上说，中国的近代建筑所蕴含的文化意向必然是多解的，该手册就让我们了解到西方传教士所兴建的中国近代哥特式教堂所蕴含的外来意向。那么，遍布中国各地、数以万计的近代建筑遗存，它们又蕴含着多少我们尚未认知的意向呢？

This book contextualizes and translates a French handbook *The Missionary-Builder: Adivce-Plans* from 1926 about how to build a church in China. Its academic significance is mainly reflected in the following aspects.

Studying and translating this handbook revealed unknown historical and cultural aspects of the cross-influences between China and the western world in the specific context of the society's modernization at the time of the Chinese Republic. Therefore, it enriches our knowledge about the intention of Chinese modern architectural culture. This means that the handbook is much more than a compilation of technical hints for building a church; it is a metaphor of a cultural world vision behind architecture. Research on China's modern architecture never may forget to pay interest to the Chinese people's thoughts as well as to the Westerners' views on the Chinese society and their projects in and for China. The analysis of such crossed perception and reception reveals the complexity of the path to modernity and the multiple

（2）这部手册的发现、翻译、注释等研究工作和所获得的学术成果，树立了国际合作开展中国近代建筑遗产研究的范式，即由中欧学者基于共同的建筑考古学术理念，发挥各自所长，合作完成的研究，在揭示中国近代建筑遗产价值的高度和文化背景的广度上，皆有所突破。这种突破源于历史文献和建筑实例研究的有机结合，这一研究范式为全面发掘中国近代建筑的历史文化内涵指明了前进的方向。

（3）这部手册的研究与翻译，为中国本土建筑的研究提供了重要史料，这是本次研究的另一个重要的收获和启示。历史研究的核心永远是史料，而西方人在中国近百年的活动历史，积累了大量关于中国建筑的史料，尤其是影像和地图资料，更加无可替代，弥足珍

solutions architecture offered. The handbook from 1926 is a unique source about the intentions of foreign Christian missionaries, who built Gothic churches together with Chinese people. Furthermore, this approach suggests that today a large number of intentions still remains unperceived behind the tens of thousands Chinese modern architectural relics.

Discovering the French handbook from 1926 was at the origin of translations and research based on a common academic philosophy in the fields of architectural history and building-archaeology. The research collaboration between Chinese and European scholars achieved results that individual research never could have expected. It may, therefore, be considered as a breakthrough in the field of Chinese modern cultural heritage studies, in the dynamic perspective of shared built heritage. This breakthrough results from the creative combination of research questions based on historical documents and architectural case studies, which has set up a research paradigm for international cooperation.

Contextualizing the French handbook of 1926 provided important unknown historical material for the study of Chinese local architecture, which is another important achievement and inspiration of this study. New archival and visual sources are essential for all historical studies. The presence and activities of

本书的意义和分工（代序）
Meaning and Division of Labor of the Book

贵。史料所蕴含的能量，永远超出研究者的想象，对欧洲所藏中国建筑史料的收集、整理和研究，势必对未来的学术发展产生巨大推动。从这个意义上说，2014 年的大名天主堂宛如 1900 年的敦煌藏经洞，破壁的那一刻，一个学术的时代已经悄然开启。

本书参与者的具体分工如下：

本书的引言、第一部分和第三部分的英文内容由高曼士教授和徐怡涛副教授两位作者共同完成，第二部分的英文内容由高曼士教授完成。附录部分由两位作者共同完成。全书的中文翻译由吴美萍完成。

翻译和出版本书得到了很多人的帮助和支持，我们在这里向他们表示最诚挚的感谢。

我们要向吴美萍博士表示感谢，她

westerners in China from the 1840s to the 1940s generated a huge amount of varied archival, literary, and visual sources. The information contained in historical material is often beyond the expectations of the researchers. Important sources about historical buildings are spread over the world in public and private collections, while the buildings are standing in China. Unravelling such sources, therefore, will generate important progress in future academic research. We could, to a certain extent, compare the research on the handbook from 1926 and the fieldwork in and around the Catholic church of Daming in 2014, with the discovering in 1900 of the Dunhuang Caves and their precious manuscripts, which opened a new era for academic research.

The specific contributions of the authors are:

Professor Thomas Coomans and Professor Xu Yitao authored together the introduction, the first and third pars in English, and Appendix 2. The English translation and comments of the second part as well as the French transcription of Appendix 1 have been done by Thomas Coomans. Dr. Wu Meiping has done the Chinese translation of the whole book.

Translating and publishing the handbook would not have been possible without the support of many people we sincerely thank.

The authors want to express their

舶来与本土：1926 年法国传教士所撰中国北方教堂营造手册的翻译和研究
Building Churches in Northern China. A 1926 Handbook in Context

负责本书的中文翻译并全身心投入了这个项目。她毕业于东南大学建筑学院。在鲁汶与这样一位优秀学者合作，每周就建筑与中国方面的讨论丰富了这个项目的内容。没有她的努力及她与知识产权出版社的联系，本书的出版永远不会成为现实。

感谢两位毕业于北京大学考古文博学院，现就读于鲁汶大学建筑系的年轻学者：崔金泽，他从一开始就对这个项目给予了支持；谭镭，她在后期给予了极大的帮助。感谢鲁汶南怀仁研究中心主任陈聪铭博士，他在后期参与了部分内容的翻译工作。

感谢鲁汶大学和北京大学这两所学校为促进双方研究、教学和合作提供的巨大的精神和物质支持。感谢鲁汶大学的工学院院长 Michiel Steyaert 教授，雷蒙德·勒麦尔国际保护中心主任 Koen Van Balen 教授、建筑系主任 Hilde Heynen 教授和 Krista de Jonge 教授。感谢在本项合作研究进行期间给予支持的北京大学考古文博学院前后两任院

gratitude to Dr Wu Meiping who made the Chinese translation of the handbook and most other text. She was fully devoted to the project. Working in Leuven with such a fine scholar, educated at the School of Architecture of Southeast University, enriched the project thanks to weekly discussions on architecture and China. Without her commitment and her contacts with academic publishers in China, this book never would have been realised.

Two other young Chinese scholars graduated from the School of Archaeology and Museology of Peking University and pursuing their studies at the Department of Architecture of the University of Leuven deserve a special mention: Cui Jinze supported the project from its very beginnings, and Tam Lui helped with great efficiency during the last phase of the editorial work. Dr Chen Tsung-ming, director of research at Ferdinand Verbiest Institute, University of Leuven, deserves special thanks too.

The support from our both universities for developing a structural scientific collaboration based on research, education and friendship, was of great moral and material help. At Leuven's side, Prof. Michiel Steyaert, dean of the faculty of Engineering Science, Prof. Koen Van Balen, director of the Raymond Lemaire International Centre for Conservation, Prof. Hilde Heynen and Prof. Krista de

本书的意义和分工（代序）
Meaning and Division of Labor of the Book

长——赵辉教授和杭侃教授。

感谢以下杰出的建筑历史学家、汉学家和历史学家，他们帮助我们理解这本手册的重要性并鼓励我们对手册进行研究和翻译，他们是 Jeffrey Cody 博士（盖蒂研究院）、何培斌教授（香港中文大学建筑学院）、朱光亚教授（东南大学建筑学院）、钟鸣旦(Nicolas Standaert)教授（鲁汶大学汉学系）、刘亦师博士（清华大学建筑学院）、梅谦立(Thierry Meynard)教授（中山大学哲学系）、Jan de Maeyer 教授（鲁汶大学历史系）、徐苏斌教授（天津大学建筑学院）、卢永毅教授（同济大学建筑学院）、郑扬文教授（曼彻斯特大学历史系）。

2014 年的暑期培训班由北京大学考古文博学院组织，感谢北京大学的张剑葳博士和王书林女士，大名县文广新体旅局的康玉娥局长、陈振山副局长、

Jonge, past and present presidents of the Department of Architecture, deserve a special mention. At Peking's side, all our gratitude goes to Prof. Zhao Hui, former dean of the School of Archaeology and Museology, and Prof. Hang Kan, the present dean.

Several outstanding architectural historians, sinologists and historians helped us to understand the importance of the handbook and encouraged our project to analyze and translate it: Dr Jeffrey Cody (The Getty Institute), Prof. Ho Puay-peng (The Chinese University of Hong Kong, School of Architecture), Prof. Zhu Guangya (Southeast University, School of Architecture), Prof. Nicolas Standaert (University of Leuven, Department of Sinology), Dr Liu Yishi (Tsinghua University, School of Architecture), Prof. Thierry Meynard (Sun Yatsen University, Department of Philosophy), Prof. Jan de Maeyer (University of Leuven, Department of History), Prof. Xu Subin (Tianjin University, School of Architecture), Prof. Lu Yongyi (Tongji University, School of Architecture), and Prof. Zheng Yangwen (University of Manchester, Department of History).

The workshop organised by the School of Archaeology and Museology of Peking University in Daming in June 2014 would not have been possible without the involvement of Dr Zhang Jianwei and

舶来与本土：1926年法国传教士所撰中国北方教堂营造手册的翻译和研究
Building Churches in Northern China. A 1926 Handbook in Context

文物科长王建平和文保所副所长任志强，大名天主教会的李雍绍会长和张神父等人，他们对暑期培训班的组织工作给予了大力支持。培训班学生的学习热情是对我们双方工作的最大激励。

我们还要感谢 Jean-Luc de Moerloose 和 Loup Browaeys，感谢胡新宇、罗薇博士、Martine de Meulemeester、Friquette Smets、Chantal Browaeys、冯澜、吕舟教授、姚安、张光玮博士、徐桐博士、王颖博士、谭永亮教授、朱益宜教授、Dirk Van Overmeire、Odile Compagnon、魏扬波博士、卢悦、冯源、宁华、舒畅雪博士。

感谢 Whitney de Courcel 为英文翻译进行的修改和润色。几家档案馆欢迎我们并允许我们重新出版他们的收藏，在这里我们要感谢：魁北克拉瓦尔大学的 James Harold Lambert、Marie-Claude Bouchard 和 Dave Anderson，他们为我们提供了原始的手册和图纸；鲁汶大学 KADOC 档案中心和宗教、文化和社会研究中心的 Luc Vints、Patricia Quaghebeur 和 Greet de Neef，鲁汶南怀仁研究中心的韩德力和 Pieter Ackerman，法国巴黎-旺弗耶稣会档案馆的 Fr Robert Bonfils 和 Barbara Baudry，巴黎外方传教会档案馆的 Annie Salavert 和 Ghislaine Olive，鲁汶大学东方图书馆的华贝妮。

Wang Shulin, and the welcome of the city heritage officials Tang Yu'e, Chen Zhenshan, Wang Jianping and Ren Zhiqiang, as well as Li Yongshao and father Zhang from the Catholic church of Daming.

For their support and interest, we would like to thank warmly Jean-Luc de Moerloose and Loup Browaeys, as well as Hu Xinyu, Dr Luo Wei, Martine de Meulemeester, Friquette Smets, Chantal Browaeys, Anke Van Lancker, Yao An, Prof. Lyu Zhou, Dr Zhang Guangwei, Dr Xu Tong, Dr Wang Ying, Prof. Patrick Taveirne, Prof. Cindy Chu Yik-yi, Dirk Van Overmeire, Odile Compagnon, Dr Jean-Paul Wiest, Lu Yue, Feng Yuan, Ning Hua and Dr Shu Changxue.

Whitney de Courcel kindly revised the English translation. Several archives welcomed us and authorized us to reproduce documents from their collections. We would like to thank: James Harold Lambert, Marie-Claude Bouchard and Dave Ander son, for the scanning of the original illustration of the handbook (Université Laval, Québec), Luc Vints, Patricia Quaghebeur and Greet de Neef (KADOC Documentation and Research Centre for Religion, Culture and Society, University of Leuven), father Jeroom Heyndrickx and Pieter Ackerman (Ferdinand Verbiest Institute, Leuven), father Robert Bonfils and Barbara Baudry (Archives Jésuites, Compagnie de Jésus,

最后，感谢2014年参加大名建筑考古培训班的北京大学考古文博学院的学生。他们最早体验了手册的文本资料和大名天主堂的实物资料之间的互动：韩爽、尚劲宇、王一臻、季宇、罗登科、王宇、章亿安、马青龙。

Province de France, Vanves), Annie Salavert and Ghislaine Olive (Missions Étrangères de Paris, bibliothèque, iconothèque et archives, Paris), Dr Isabelle Lecocq (Royal Institute for Cultural Heritage, Brussels), and Benedicte Vaerman (Eastern Collections, Central Library, University of Leuven).

At last, we thank the students of the School of Archaeology and Museology of Peking University who participated with enthusiasm in the building archaeological workshop of Daming in June 2014. They were the first to experience the interaction between the written source (the handbook) and the material source (the church of Daming): Han Shuang, Shang Jinyu, Wang Yizhen, Ji Yu, Luo Dengke, Wang Yu, Zhang Yi'an, and Ma Qinglong.

北京—鲁汶，
2015年11月11日

Leuven-Beijing,
11 November 2015

目 录
Table of content

本书的意义和分工（代序）
MEANING AND DIVISION OF LABOR OF THE BOOK

引言——东西方建筑实践的交流
INTRODUCTION: ARCHITECTURAL EXCHANGE OF PRACTICE-BASED KNOWLEDGE

手册《传教士建造者：建议-方案》
THE HANDBOOK *THE MISSIONARY- BUILDER: ADVICE-PLANS* ·············· 6

北京大学和鲁汶大学的合作研究
RESEARCH COLLABORATION BETWEEN PEKING UNIVERSITY AND
THE UNIVERSITY OF LEUVEN ·············· 10

本书架构
THE CONCEPT OF THE BOOK ·············· 13

第一部分 手册的撰写背景
PART ONE: THE HANDBOOK IN CONTEXT

关于教堂建造和式样的书籍
ARCHITECTURAL HANDBOOKS AND PATTERN BOOKS OF CHURCHES ·············· 19

手册的作者
THE HANDBOOK'S AUTHORS ·············· 25

手册的读者群
THE HANDBOOK'S READERSHIP ·············· 44

手册的内容
THE HANDBOOK'S CONTENT ·············· 47

样式之辩：西方哥特式还是中国化
THE ISSUE OF STYLE: WESTERN GOTHIC OR SINICIZED? ·············· 85

舶来与本土：1926年法国传教士所撰中国北方教堂营造手册的翻译和研究
Building Churches in Northern China. A 1926 Handbook in Context

西方传教士眼中的中国工匠
The Perception of Chinese Workers by Western Missionaries ················· 100

技术的碰撞和转移
Encounter and Transfer of Technology ······················· 104

第二部分 手册的内容和翻译
PART TWO: THE HANDBOOK TRANSLATED

翻译者的体会
Note of the Translators ······················· 116

手册的出版信息
Published Information ······················· 118

手册的目录
Contents ······················· 120

序 言
Preface ······················· 121

第一章 选址
First Chapter: Choosing a Location ······················· 123

第二章 建筑布局和朝向的选择
Second Chapter: Choosing the Buildings' Layout ······················· 126

第三章 设计方案、工程预算和施工合同
Third Chapter: Plan — Estimate — Contract with the Contractor ············ 130

注释
Note [on Style] ······················· 133

第四章 建筑材料
Fourth Chapter: About Materials ······················· 153

第五章 砖石工程
Fifth Chapter: Masonry ······················· 169

第六章 屋顶
Sixth Chapter: The Roof ······················· 192

第七章 屋面装饰
Seventh Chapter: Roof Ornamentation ······················· 203

第八章 建筑细节的处理
Chapter Eight: Miscellaneous ······················· 210

目 录
Table of content

结 语
CONCLUSION ·· 232

第三部分 教堂建造案例分析——大名天主堂
PART THREE: THE HANDBOOK AS BUILT

天主教传教区大名
THE CATHOLIC MISSION OF DAMING ·· 285

大名法文学校
THE FRENCH COLLEGE OF DAMING ·· 292

大名天主堂
A MARIAN CATHEDRAL IN DAMING ·· 306

大名天主堂的营建
CONSTRUCTING THE CHURCH OF DAMING ······································· 319

大名天主堂的测绘
MEASURING THE CATHEDRAL OF DAMING ······································· 352

附 录
APPENDIX

附录1. 手册的法语原文
APPENDIX 1. FRENCH TEXT OF THE ORIGINAL HANDBOOK ················ 370

附录2. 建造技术地方用语
APPENDIX 2. CHINESE CONSTRUCTION TERMS MENTIONED IN THE HANDBOOK ········ 424

参考文献
BIBLIOGRAPHY ·· 435

索 引
INDEX ··· 443

舶来与本土：1926 年法国传教士所撰中国北方教堂营造手册的翻译和研究
Building Churches in Northern China. A 1926 Handbook in Context

图　录
List of Illustration

图 0.1　卜天德神父建造圣母圣心会拉花营子教堂（内蒙古），1930 年。 fig. 0.1 Father Petrus De Boeck building the Scheut mission church of Lahuayingzi (Inner Mongolia),1930. (© FVI, CHC pictures) ·················4

图 0.2　建造知识的传播：西方传教士、中国工头和中国共享建造高家营子新信徒住所（河北省），20 世纪 30 年代。 fig. 0.2 Transfer of construction knowledge: western missionary, Chinese foremen and Chinese workers building the novices' house at Gaojiayingzi (Hebei province), 1930s. (© FVI, CHC pictures) ·················5

图 0.3　手册扉页，1926 年。 fig. 0.3 The handbook's title page, 1926. (© Lyon, Bibliothèque municipale) ·················8

图 0.4　大名天主堂（河北省），法国耶稣会士在 1918—1921 年建。 fig. 0.4 Church of Our Lady at Daming (Hebei province), built by French Jesuits in 1918-1921. (© Thomas H. Hahn, 2010) ·················12

图 0.5　在理论与实践之间：建筑手册中的玫瑰花窗和大名天主堂上的玫瑰花窗。 fig. 0.5 Confronting theory and practice: rose window from the handbook and rose window of Daming's church. (© Québec, Université Laval, and THOC 2014) ·················12

图 0.6　大名天主堂测绘，北京大学考古文博学院建筑考古暑期实践班。 fig. 0.6 Measuring the church of Daming, building archaeology summer workshop of the School of Archaeology and Museology of Peking University. (© THOC 2014) ·················16

图 1.1　广州(广东省),遣使会所建法国天主教主教座堂,1863—1888 年。 fig. 1.1 Guangzhou (Guangdong province), French Catholic cathedral built by the Congregation of the Mission, 1863-1888. (© THOC 2013) ·················23

图 1.2　发表在传教士期刊上的和羹柏神父绘图，1891 年。 fig. 1.2 Father De Moerloose's drawings published in a missionary journal, 1891. (© FVI, Scheut Memorial Library) ·················24

图 1.3　一本 1900 年出版图书的封面，耶稣会东南直隶使团。 fig. 1.3 Jesuit Southeastern Zhili mission, front page of a book from 1900. (© MEP library) ·················27

目录
Table of content

图 1.4 耶稣会献县（河北省）印刷厂，手册就是于 1926 年在此印刷的。 fig. 1.4 Jesuit press of Xianxian (Heibei province), where the handbook was printed in 1926. (© ASJ France, GMC)·· 28

图 1.5 大名（河北省），耶稣会神父在法语学校执教，1920 年左右。 fig. 1.5 Daming (Heibei province), Jesuit father teaching at the French College, around 1920. (© ASJ France, GMC)·· 31

图 1.6 大名，一口 1921 年古钟上的雍居敬神父题名。fig. 1.6 Daming, Father Paul Jung's name on a bell from 1921. (© THOC 2014)·················· 32

图 1.7 沈阳（辽宁省），法国巴黎外方传教会主教座堂，梁神父建，1910 年。 fig. 1.7 Shenyang (Liaoning province), cathedral of the French Foreign Mission of Paris, built by fahter Lamasse, 1910. (© MEP archives)·············· 35

图 1.8 大名，耶稣会圣约瑟礼拜堂，修士雷振生在 1920 年左右建造。 fig. 1.8 Daming, St Joseph chapel of the Jesuits, built by brother Litzler around 1920. (© ASJ France, GMC)·· 37

图 1.9 西湾子（河北省），比利时圣母圣心会圣米厄尔及圣本笃礼拜堂，和羹柏神父建造，1930 年。fig. 1.9 Xiwanzi (Hebei province), St Michael and St Benedict chapel of the Belgian Scheut Fathers, built by father De Moerloose, 1903. (© KADOC, CICM archives)·················· 40

图 1.10 宣化（河北省），法国遣使会教堂，和羹柏神父建造，1903—1906 年。 fig. 1.10 Xuanhua (Hebei province), church of the French missionaries of the Congregation of the Mission, built by father De Moerloose, 1903-1906. (© THOC 2011)·················· 42

图 1.11 上海，法国耶稣会自助的佘山朝圣教堂，和羹柏神父设计的第一个项目，1920—1923 年。 fig. 1.11 Shanghai, Sheshan pilgrimage church promoted by the French Jesuits, first project by father De Moerloose, 1920-1923. (© KADOC, CICM)·················· 44

图 1.12 朝阳（辽宁省），圣母圣心会神父及其教区教堂，约 1935 年。 fig. 1.12 Chaoyang (Liaoning province), Scheut father and his parish church around 1935. (© FVI, BR pictures)·················· 45

图 1.13 如皋（江苏省），被罩棚覆盖的耶稣会院落，约 1920 年。 fig. 1.13 Rugao (Jiangsu province), courtyard of the Jesuit mission covered with a tent, around 1920. (© ASJ France, FCh)·················· 51

图 1.14 张官屯（河北省）及壕赖山（内蒙古），1936 年及 1930 年建造的圣母圣心会教堂。 fig. 1.14 Zhangguantun (Hebei province) and Haolaishan (Inner Mongolia), Scheut missions churches built in 1936 and 1930. (© FVI, BR picture; © THOC 2011)·················· 52

图 1.15 水东（安徽省）及扬州（江苏省），19 世纪 80 年代及 1864—1873

年建造的耶稣会教堂。 fig. 1.15 Shuidong (Anhui province) and Yangzhou (Jiangsu province), Jesuit churches built in the 1880s and 1864-1873. (© THOC 2015 and 2014) ·· 53

图 1.16 曲周（河北省），被耶稣会传教士使用的传统院落南侧，约 1910 年。 fig. 1.16 Quzhou (Hebei province), traditional courtyard house used by Jesuit missionaries, seen from the south, around 1910. (© ASJ France, GMC) ··· 54

图 1.17 仕拉乌素壕（内蒙古），为建造圣母圣心会神父住所而选砖，1938 年。 fig. 1.17 Shilawusuhao (Inner Mongolia), sorting bricks for building the Scheut father's residence, 1938. (© FVI, CHC pictures) ···························· 58

图 1.18 内蒙古，模制方形屋瓦，年代未知。 fig. 1.18 Inner Mongolia moulding square roof tiles, not dated. (© FVI, CHC pictures) ························ 58

图 1.19 海岛营子（辽宁省）及玫瑰营子（内蒙古），正在建造中的教堂，1937 年。 fig. 1.19 Haidaoyingzi (Liaoning province) and Meiguiyingzi (Inner Mongolia), churches under construction, 1937. (© FVI, CHC pictures) ············ 60

图 1.20 内蒙古，夯土施工队，年代未知。 fig. 1.20 Inner Mongolia, team of earth rammers, not dated. (© FVI, CHC picture) ··· 62

图 1.21 仕拉乌素壕（内蒙古），泥瓦匠建造圣母圣心会神父住所，1938 年。 fig. 1.21 Shilawusuhao (Inner Mongolia), masons building the Scheut father's residence, 1938. (© FVI, CHC pictures) ·· 64

图 1.22 仕拉乌素壕（内蒙古），圣母圣心会神父住所，在砌墙前竖立的门框和窗框，1938 年。 fig. 1.22 Shilawusuhao (Inner Mongolia), Scheut father's residence, door and window frames placed before the brickwork, 1938. (© FVI, CHC pictures) ··· 65

图 1.23 仕拉乌素壕（内蒙古），圣母圣心会神父住所，拱形砖门梁和窗框，1938 年。 fig. 1.23 Shilawusuhao (Inner Mongolia), Scheut father's residence, arched brick lintel and window frame, 1938. (© FVI, CHC pictures) ··· 65

图 1.24 固阳县（内蒙古），圣母圣心会礼拜堂，在上瓦前整理灰背，年代未知。 fig. 1.24 Guyangxian (Inner Mongolia), Scheut father's chapel, preparing the mud layer before placing tiles (*wa fang*), not dated. (© FVI, BR pictures) ·· 67

图 1.25 大名（河北省），铺设一所教堂的屋面，河北南部的传统仰瓦技术，2014 年。 fig. 1.25 Daming (Hebei province), making a church's roof, Southern Hebei traditional technique of concave roof tiles (*yang wa*), 2014. (© THOC 2014) ·· 67

图 1.26 内蒙古或河北北部，建造一处圣母圣心会神父住所的屋顶，地点未

目录
Table of content

知，年代未知。 fig. 1.26 Inner Mongolia or Northern Hebei, building the roof of a Scheut fathers' residence, not located, not dated. (© FVI, CHC pictures) ·········· 68

图 1.27 磨子山（内蒙古），圣母圣心会玛利亚朝圣教堂， 西式的砖墙和屋架，和羹柏神父建造，约 1910 年。 fig. 1.27 Mozishan (Inner Mongolia), Scheut fathers' Marian pilgrimage church, Western masonry and Western roof trusses, built by father De Moerloose, ca 1910. (© FVI, CHC pictures) ·········· 69

图 1.28 开封（河南省），原地区修院礼拜堂，被解体屋顶中的西式桁架，1935 年建。 fig. 1.28 Kaifeng (Henan province), chapel of the former Regional Seminary, Western trusses of the dismanatled roof, built in 1935. (© THOC 2013) ·········· 69

图 1.29 兰州（甘肃省），中式基督教样式的圣母圣心会神父住所大门，约 1932 年。 fig. 1.29 Lanzhou (Gansu province), main gate of the Scheut fathers' residence in Sino-Christian style, around 1932. (© FVI, CHC pictures) ·········· 71

图 1.30 内蒙古，圣母圣心会教堂，带有哥特窗饰和小尖塔的混合样式，1900 年以前。 fig. 1.30 Inner Mongolia, Scheut Fathers' church, hybrid style with Gothic traceries and pinnacles, before 1900. (© KADOC, CICM) ·········· 71

图 1.31 内蒙古，雕造柱头和柱础的石匠们，年代未知。 fig. 1.31 Inner Mongolia, stone carvers of capitals and bases, not dated. (© FVI, CHC pictures) ·········· 72

图 1.32 基督教石像的雕塑师，大同（山西省），1935 年。 fig. 1.32 Sculptor of stone Christian statues, Datong (Shanxi province), 1935. (© FVI, CHC pictures) ·········· 73

图 1.33 法国巴黎外方传教会长春（吉林省）教堂， 正在建造中的木柱、拱券和拱顶拱肋，约 1920 年。 fig. 1.33 Changchun (Jilin province) church of the French Foreign Missions of Paris, wooden columns, arches and lattice structure of the vaults under construction, around 1920. (© MEP Archives) ·········· 74

图 1.34 天津，圣母得胜堂（望海楼教堂），正厅中的木质拱顶，1900 年后。 fig. 1.34 Tianjin, Our Lady of the Victory church, wooden vault of the nave, after 1900. (© THOC 2011) ·········· 75

图 1.35 北京，北堂（西什库教堂）， 主教礼拜堂的拱顶，约 1900 年。 fig. 1.35 Beijing, North Cathedral Xishiku, vault of the bishops' chapel, around 1900. (© THOC. 2013) ·········· 76

图 1.36 西湾子（河北省）， 圣母圣心会修院礼拜堂， 和羹柏神父于 1902 年建造， 其家族捐赠比利时原产彩色玻璃窗。 fig. 1.36 Xiwanzi (Hebei province), chapel of the Scheut fathers' seminary, by father De Moerloose and Belgian to produced stained to glass windows donated　by his family in 1902. (© KADOC, CICM)·················· 76

图 1.37　上海，法国耶稣会圣约瑟堂，带有中文题记的哥特彩色玻璃窗。fig. 1.37 Shanghai,　St Joseph's church　of the French Jesuits, Gothic stained-glass window with Chinese inscription.　(© THOC 2011)·················· 77

图 1.38　大名（河北省），圣母圣心会神父住所，带有走廊的南立面，约 1910 年。fig. 1.38 Daming (Hebei province), Jesuit fathers' residence,　southern side with veranda, ca 1910. (© THOC 2014) ·················· 78

图 1.39　吉林（吉林省），法国巴黎外方传教会神学院，梁神父设计，约 1920 年。　fig. 1.39 Jilin (Jilin provine), seminary of the French Foreign Missions of Paris,　designed by father Lamasse, around 1920. (© MEP Archives) ·················· 82

图 1.40　上海，杨树浦教堂，和羹柏神父设计，1924 年。 fig. 1.40 Shanghai, Yangtzepoo church, designed by father De Moerloose,　1924. (© ASJ France, FCh)·················· 84

图 1.41　辽宁（辽宁省），巴黎外方传教会的法国哥特式教堂，1922 年。 fig. 1.41 Liaoyang (Liaoning province), French Gothic style church　of the Foreign Missions of Paris, 1922. (© MEP Archives) ·················· 85

图 1.42　开封（河南省），意大利圣方济各会的意大利哥特式主教座堂，1917—1919 年。　fig. 1.42 Kaifeng (Henan province), Italian Gothic style cathedral　of the Italian Franciscans, 1917-1919. (© THOC 2013) ·················· 88

图 1.43　普世一神及各因循当地传统而本土化了的教堂，中国基督教教义问答手册——《问答像解》中的图像，1928 年。 fig. 1.43 One universal God and churches inculturated in local traditions, image from　the Sino-Christian catechism book *Wendaxiajie*, 1928. (© KU Leuven, CB)·················· 89

图 1.44　中国基督教教义问答手册——《问答像解》1928 年。 fig. 1.44 Sino-Christian catechism book *Wendaxiajie*, 1928. (© KU Leuven, CB) ·················· 90

图 1.45　北京，辅仁大学，中国基督教样式的典型代表，葛利斯神父设计，1927—1931 年。 fig. 1.45 Beijing, Catholic University of Peking or Fu Jen University,　the Sino-Christian style's archetype, designed and built by father Gresnigt,　1927-1932. (© THOC 2014) ·················· 93

图 1.46　安国（河北省），中国基督教样式主教座堂，20 世纪 30 年代早期。fig. 1.46 Anguo (Hebei province), Sino-Christian style cathedral,　early 1930s. (© SAM) ·················· 95

图 1.47　西昌（四川省），法国巴黎外方传教会主教座堂之木结构，1914 年。 fig. 1.47 Xichang (Sichuan province), wooden structure of the cathedral of the French Foreign Missions of Paris, 1914. (© MEP Archives) ·················· 101

目　录
Table of content

图 1.48　内蒙古，夯筑墙体，20 世纪 30 年代。 fig. 1.48 Inner Mongolia, ramming a wall, 1930s. (© FVI, CHC pictures) ……………… 103

图 1.49　仕拉乌素壕（内蒙古），在建孤儿院的一位中国天主教 施主王亚海(音)，20 世纪 20 年代。fig. 1.49 Shilawusuhao (Inner Mongolia), Wang Yahai, a Chinese Catholic benefactor of an orphanage in construction, 1920s. (© FVI, CHC pictures) ……………… 103

图 1.50　双树子（河北省），和羹柏神父 1917 年在早期一座较小教堂旁边建造的哥特式大教堂。 fig. 1.50 Shuangshuzi (Hebei province), great Gothic church built by father De Moerloose in 1917 beside a smaller earlier church. (© KADOC, CICM) ……………… 108

图 1.51　平地泉（内蒙古），圣母圣心会哥特式教堂前合影的中国修生，20 世纪 20 年代。 fig. 1.51 Pingdiquan (Inner Mongolia), Chinese seminarians posing in front of the Scheut fathers' Gothic church, 1920s. (© KADOC, CICM) ……………… 109

图 1.52　五号（河北省），中国基督教样式的圣母圣心会教堂，1930—1933 年建。 fig. 1.52 Wuhao (Hebei province), Scheut fathers' church in Sino-Christian style, built in 1930-1933. (© ASJ France, GMC) ……………… 110

图 1.53　南泉子(吉林省)，圣母圣心会正在建造的乡村教堂，年代未知。fig. 1.53 Nantsuantse (Jilin province), Scheut fathers' village church under construction, not dated. (© FVI, CHC pictures) ……………… 113

图 1.54　舍必崖（内蒙古），圣母圣心会乡村教堂，和羹柏神父建造，1904—1905 年。 fig. 1.54 Shebiya (Inner Mongolia), Sheut fathers' village church, by father De Moerloose, 1904-1905. (© FVI, BR pictures) ……………… 114

图 2.1　手册封面，1935 年再印版。fig. 2.1 The Handbook's title page, reprint of 1935. (© Université Laval, Bibliothèque) ……………… 119

图 2.2　七份合同模板：石灰、砖、花砖、瓦、石头、木料以及 与包工头的合同，手册第 14-18 页。 fig. 2.2 Seven model contracts: lime, brick, 'flower bricks', tiles, stone, timber, and with contractor. Handbook, p.14-18. (© Université Laval, Bibliothèque) ……………… 148

图 2.3　手册图纸 1：Fig. 1～Fig. 4，模型砖，用在柱子、柱身、柱头、竖框、玫瑰花窗、檐口等处。 fig. 2.3 Handbook, plate 1: Fig. 1-Fig. 4, moulded bricks used for pillars, shafts, capitals, mullions, rose windows, cornices, etc. (© Université Laval, Bibliothèque) ……………… 235

图 2.4　手册图纸 2：Fig. 4～Fig. 5，用在竖框和玫瑰花窗处的模型砖；Fig. 6 、Fig. 8～Fig. 11，用在尖顶、檐口等地的特殊砖。 fig. 2.4 Handbook, plate 2: Fig. 4-Fig. 5, moulded bricks for mullions and rose windows, Fig. 6, Fig. 8-Fig. 11, special bricks for the spire, cornices, etc. (© Université

图 2.5　手册图纸 3： Fig. 7，用在尖塔处的砖的剖面； Fig. 12~Fig. 16，用作檐口和滴水石的砖。fig. 2.5 Handbook, plate 3: Fig. 7, section of a brick of the spire, Fig. 12-Fig. 16, bricks for cornices and dripstones. (© Université Laval, Bibliothèque) ················236

图 2.6　手册图纸 4： Fig. 17~Fig. 22，用作檐口和滴水石的砖； Fig. 23~ Fig. 36，用在小柱子上的砖；Fig. 37，用在拱门上的砖。 fig. 2.6 Handbook, plate 4: Fig. 17-Fig. 22, bricks for cornices and dripstones, Fig. 23-Fig. 36, bricks for small columns, Fig. 37, bricks for arches. (© Université Laval, Bibliothèque) ················236

图 2.7　手册图纸 5： Fig. 38~ Fig. 40，拱券和花饰窗格的细节。 fig. 2.7 Handbook, plate 5: Fig. 38~Fig. 40, profiled elements for arches and traceries. (© Université Laval, Bibliothèque) ················237

图 2.8　手册图纸 6： Fig. 41，阶式基础； Fig. 42~Fig. 46，砖墙的不同砌法。 fig. 2.8 Handbook, plate 6: Fig. 41, stepped foundation, Fig. 42-Fig. 46, brickwork, different types of bonds. (© Université Laval, Bibliothèque) ················237

图 2.9　手册图纸 7：需要用到特殊砖头的部位。 fig. 2.9 Handbook, plate 7: examples of special bricks. (© Université Laval, Bibliothèque) ················238

图 2.10　手册图纸 8： Fig. 47~ Fig. 50，砖墙的厚度： Fig. 52，有着扶壁的砖墙厚度。 fig. 2.10 Handbook, plate 8: Fig. 47-Fig. 50, thickness of brick walls, Fig. 52, brick wall with buttresses. (© Université Laval, Bibliothèque) ················239

图 2.11　手册图纸 9： Fig. 51，两砖厚的墙； Fig. 53，钟楼角上的扶壁； Fig. 55，仰瓦屋顶。 fig. 2.11 Handbook, plate 9: Fig. 51, wall of two bricks, Fig. 53, buttresses at the corner of a bell tower, Fig. 55, *yang wa* roof. (© Université Laval, Bibliothèque) ················239

图 2.12　手册图纸 10：Fig. 54,缓解负荷的辅助拱；Fig. 56,扣瓦屋顶；Fig. 57~Fig. 58,西式屋架。fig. 2.12 Handbook, plate 10: Fig. 54, relieving arch, Fig. 56, *kou wa* roof, Fig. 57-Fig. 58, Western trusses. (© Université Laval, Bibliothèque) ················240

图 2.13　手册图纸 11： Fig. 59，中式屋架； Fig. 58bis 和 Fig. 60，西式屋架的节点。 fig. 2.13 Handbook, plate 11: Fig. 59, Chinese truss, Fig. 58 bis and Fig. 60, joining of Western trusses. (© Université Laval, Bibliothèque) ················240

图 2.14　手册图纸 12： Fig. 61~ Fig. 64，西式屋架的节点； Fig. 65，用在屋脊处的特殊瓦；Fig. 66，双坡屋顶。 fig. 2.14 Handbook, plate 12: Fig. 61-Fig. 64, joining of Western trusses, Fig. 65, special tiles for the roof's ridge, Fig. 66, saddle roof. (© Université Laval, Bibliothèque) ················241

目 录
Table of content

图 2.15　手册图纸 13：Fig. 67，四坡屋顶； Fig. 68~Fig. 77，滴水石。 fig. 2.15 Handbook, plate 13: Fig. 67, 'hipped' roof, Fig. 68-Fig. 70, dripstones. (© Université Laval, Bibliothèque) ·················· 241

图 2.16　手册图纸 14：Fig. 71~ Fig. 72，滴水石和檐口； Fig. 73，窗框处的"喉咙"。 fig. 2.16 Handbook, plate 14: Fig. 71-Fig. 72, dripstones and cornices, Fig. 73, 'throat' of a window frame. (© Université Laval, Bibliothèque) ·················· 242

图 2.17　手册图纸 16：Fig. 77，钟楼尖顶； Fig. 78，钟楼顶部的十字架。 fig. 2.17 Handbook, plate 16: Fig. 77, spire of the bell tower, Fig. 78, iron cross on top of the tower. (© Université Laval, Bibliothèque) ·················· 242

图 2.18　手册图纸 15：Fig. 74，用板条做成的拱； Fig. 74bis，拱顶肋骨交接处和柱头顶部； Fig. 75~ Fig. 76，木制拱顶石； Fig. 76bis，做得好的柱头和做得差的柱头。 fig. 2.18 Handbook, plate 15:Fig. 74, arch made with planks, Fig. 74bis, junction of the vault's ribs and top of the capital, Fig. 75 -Fig. 76, wooden key stone, Fig. 76bis, good and wrong capitals. (© Université Laval, Bibliothèque) ·················· 243

图 2.19　手册图纸 17：Fig. 79，馒形式拱券砖的剖面； Fig. 80~ Fig. 81，好的拱顶做法和差的拱顶做法； Fig. 82，木楼梯的踏步； Fig. 83，旋转楼梯踏步好的做法和差的做法。 fig. 2.19 Handbook, plate 17:Fig. 79, section of an arch with torus, Fig. 80-Fig. 81, good and bad top of arch, Fig. 82, steps of a wooden stair, Fig. 83, Good and wrong steps of a cork screw staircase. (© Université Laval, Bibliothèque) ·················· 244

图 2.20　手册图纸 18：用来挂四口钟的钟架子。 fig. 2.20 Handbook, plate 18: wooden belfry for four bells. (© Université Laval, Bibliothèque) ·················· 245

图 2.21　手册原图纸 19：墙上端的檐口和扶壁，砖制玫瑰花窗和花窗中间用来装玻璃的圆形铁构件。 fig. 2.21 Handbook, plate 19: cornice on top of the wall and buttresses, brick rose window and iron circle for glazing. (© Université Laval, Bibliothèque) ·················· 246

图 2.22　手册图纸 20：窗户和门。 fig. 2.22 Handbook, plate 20: window and door. (© Université Laval, Bibliothèque) ·················· 247

图 2.23　手册图纸 21：有着砖制花饰窗格、砖制柱头和石头窗沿的哥特式窗户。 fig. 2.23 Handbook, plate 21: Gothic window with brick tracery, brick capitals, and stone threshold. (© Université Laval, Bibliothèque) ·················· 248

图 2.24　手册图纸 22：有着砖制竖框和石头窗沿的哥特式三连窗。 fig. 2.24 Handbook, plate 22: Gothic triplet window with brick mullions and stone threshold. (© Université Laval, Bibliothèque) ·················· 249

图 2.25　手册图纸 23：石头柱子、柱基和柱头。 fig. 2.25 Handbook, plate 23: stone column, basis, and capital. (© Université Laval, Bibliothèque) ·················· 250

图 2.26　手册图纸 24：侧道的壁柱，砖制柱身，石头柱基和柱子底座。　fig. 2.26 Handbook, plate 24: embedded shaft used in aisles, brick shaft, stone basis and plinth. (© Université Laval, Bibliothèque) ································251

图 2.27　手册原图纸 25：木制告解室。　fig. 2.27 Handbook, plate 25: wooden confessional. (© Université Laval, Bibliothèque) ································252

图 2.28　手册图纸 26~27：项目方案 1，古典式教堂，单一的长方形中殿和小钟楼。fig. 2.28 Handbook, plate 26-27: Classic style church's project 1, rectangular single nave with small bell tower. (© Université Laval, Bibliothèque) ································253

图 2.29　手册图纸 28~29：项目方案 2，古典式教堂，单一的长方形中殿和小钟楼。fig. 2.29 Handbook, plates 28-29: Classic style church's project 2, rectangular single nave with small bell tower. (© Université Laval, Bibliothèque) ································255

图 2.30　手册图纸 30~32：项目方案 2，古典式教堂，门口和钟楼细节。fig. 2.30 Handbook, plates 30-32: Classic style church's project 2, details of the doorway and the bell tower. (© Université Laval, Bibliothèque universitaire) ································257

图 2.31　手册图纸 33：项目方案 3，哥特式教堂，花窗的铁配件说明。fig. 2.31 Handbook, plate 33: Gothic style church's project 3, indications about iron fittings in tracery windows. (© Université Laval, Bibliothèque universitaire) ································260

图 2.32　手册图纸 34~36：项目方案 3，哥特式教堂,平面、里面、后殿细节。这个设计比较灵活能用在多处，能与大钟楼组合使用或者也能用在没有后殿的古典式教堂里。fig. 2.32 Handbook, plates 34-36: Gothic style church's project 3, ground plan and different possible elevations, detail of the sanctuary's apse. This flexible project, called 'omnibus', could be combined with a bigger tower or realized in Classic style without apse. (© Université Laval, Bibliothèque universitaire) ································261

图 2.33　手册图纸 37：项目方案 4，古典式教堂，单一的长方形中殿和方形钟楼。fig. 2.33 Handbook, plate 37: Classic style church's project 4, rectangular single nave church with square tower. (© Université Laval, Bibliothèque universitaire) ································264

图 2.34　手册原图纸 38~41：项目方案 5，哥特式教堂，单一的中殿和小耳堂，多边形后殿和钟楼。fig. 2.34 Handbook, plates 38 to 41: Gothic style church's project 5, single nave with small transept, polygonal apse, and bell tower. (© Université Laval, Bibliothèque universitaire) ································265

图 2.35　手册原图纸 42~43 和 45：项目方案 6，哥特式教堂， 单一的中殿和小耳堂,多边形的后殿和钟楼。fig. 2.35 Handbook, plates 42-43 and 45: Gothic style church's project 6, single nave with small transept, polygonal apse, and bell tower. (© Université Laval, Bibliothèque universitaire) ································269

图 2.36　手册图纸 44 和 47~49：项目方案 7，哥特式教堂，中殿和耳堂，侧道，多边形的后殿，圣器室和钟楼。 fig. 2.36 Handbook, plates 44 and 47-49: Gothic style church's project 7, nave with transept, small aisles, polygonal apse, sacristies, and bell tower. (© Université Laval, Bibliothèque universitaire)··········272

图 2.37　手册图纸 46：项目方案 7，哥特式教堂，立面装饰和边门。 fig. 2.37 Handbook, plate 46: Gothic style church's project 7, decoration of the façade and side doorway. (© Université Laval, Bibliothèque universitaire)··········276

图 2.38　手册图纸 50~54：项目方案 8，哥特式教堂，有着大后殿的中殿，耳堂，多边形的后殿，边上的礼拜堂，圣器室和钟楼。 fig. 2.38 Handbook, plate 50-54: Gothic style church's project 8, nave with large aisles, transept, polygonal apse, side chapels, sacristies, and bell tower. (© Université Laval, Bibliothèque universitaire)··········277

图 2.39　手册扉页和最后一页的小插图。 fig. 2.39 Two vignettes: one on the title page and the other on the last page.··········282

图 3.1　直隶东南部宗教代牧区地图，1929 年前。 fig. 3.1 Map of the vicariate apostolic of Southeastern Zhili, before 1929. (© ASJ France, GMC)··········286

图 3.2　早先直隶东南部宗教代牧区地图，1929—1935 年间分四个教省，大名为南部的那个教省。 fig. 3.2 Map of the former vicariate apostolic of Southeastern Zhili, divided in four ecclesiastic provinces in 1929-1935. Daming is the southern one. (© ASJ France, GMC)··········291

图 3.3　大名，从东门看法国耶稣会建造的宗教建筑群，1913 年 7 月。 fig. 3.3 Daming, building complex of the French Jesuits seen from the East Gate, picture dated July 1913. (© ASJ France, GMC)··········293

图 3.4　大名，耶稣会会士在教中国小孩弹脚踏式风琴。 fig. 3.4 Daming, Jesuit education and cultural transfer: Chinese student playing on the chapel's harmonium. (© ASJ France, GMC)··········295

图 3.5　《法中期刊》1935 年出版。 fig. 3.5 *Le Trait d'Union*, the journal of the French College's alumni, issues of 1935. (© ASJ France, GMC)··········296

图 3.6　大名，1900 年前法国耶稣会在大名建造的宗教建筑群。 fig. 3.6 Daming, building complex of the French Jesuits before 1900. (© ASJ France, GMC)··········297

图 3.7　大名，从东门看大名东街两边的天主教建筑，1920 年和 2014 年。 fig. 3.7 Daming, Catholic compounds at both sides of East Street, view from the East Gate, 1920 and 2014. (© ASJ France, GMC, and © THOC 2014)··········299

图 3.8　大名，东门附近的耶稣会建筑。 1.东城门；2.东街；3.教堂；4.学校的礼拜堂；5.以前神父的住宅； 6.法文学校所在地；7.报告厅；8.牌楼所在地。 fig. 3.8 Daming, neighborhood of the East Gate and former

Jesuits' building （© School of Archaeology and Museology, Peking University, 北京大学考古文博学院）. 1.East Gate; 2.East Street; 3.church; 4.schools' chapel; 5.former residences of the father; 6.location of the French College; 7.lecture hall; 8.location of the pailou. ······300

图 3.9　大名，从教堂钟楼看大名的学校建筑群，1920 年和 2014 年。 fig. 3.9 Daming, school complex seen from the church's tower, 1920 and 2014. (© ASJ France, GMC, and THOC 2014) ······301

图 3.10　大名，扩建前的学校礼拜堂东侧，1910 年前后。 fig. 3.10 Daming, eastern side of the school's chapel before enlargement, around 1910. (© ASJ France, GMC) ······302

图 3.11　大名，原学校礼拜堂的西侧。 fig. 3.11 Daming, western side of the former school's chapel. (© THOC 2014) ······302

图 3.12　大名，原法国耶稣会神父住宅北侧。 fig. 3.12 Daming, northern side of the former residence of the French Jesuits. (© THOC 2014) ······304

图 3.13　大名，法文学校的主立面，建于 1913 年。 fig. 3.13 Daming, main façade of the French College, built in 1913. (© ASJ France, GMC) ······304

图 3.14　圣母，"大名的守护神"，教堂钟楼和钟上的一幅中国圣画。 fig. 3.14 Our Lady of Grace, patron saint of Daming, on a chinese holy picture, on the church's tower and on the bells. (© ASJ France, GMC, and © THOC 2014) ······308

图 3.15　大名，天主堂的两块碑，左边是建造碑，右面是捐献超过 10 银元的人的名字，1921 年。 fig. 3.15 Daming, the two steles with the church's history (left) and the names of the benefactors (right), 1921. (© THOC 2014) ······311

图 3.16　大名，为募捐而做的教堂模型，1918—1919 年。 fig. 3.16 Daming, a model of the church was used for fundraising, 1918-1919. (© ASJ France, GMC) ······312

图 3.17　大名，从西北看大名天主堂完工后的照片，20 世纪 20 年代早期。fig. 3.17 Daming, church after completion, seen from northeast, early 1920s. (© ASJ France, GMC) ······314

图 3.18　吉林，巴黎外方传教会的主教座堂，梁神父建，1917—1919 年。 fig. 3.18 Jilin, cathedral of the French Foreign Missions of Paris, built by father Lamasse, 1917-1919. (© MEP archives) ······316

图 3.19　大名天主堂北侧，1920 年。fig. 3.19 Daming, church from north, 1920. (© ASJ France, GMC) ······317

图 3.20　大名天主堂西南侧，1925 年前后。 fig. 3.20 Daming, church from southwest, around 1925. (© ASJ France, GMC) ······319

图 3.21　建造中的大名天主堂，中殿的东北角，1918 年秋天。 fig. 3.21 Daming, church under construction, northeastern corner of the nave, fall 1918. (© ASJ France, GMC) ················320

图 3.22　建造中的大名天主堂，中殿北端，1918 年秋天。 fig. 3.22 Daming, church under construction, nave to the north, fall 1918. (© ASJ France, GMC) ················321

图 3.23　建造中的大名天主堂，1919 年 6 月。 fig. 3.23 Daming, church under construction, June 1919. (Les missions catholiques, © KADOC) ················322

图 3.24　大名天主堂东侧。 fig. 3.24 Daming, church from the east. (© THOC 2014) ················326

图 3.25　大名天主堂中殿南端，20 世纪 20 年代和近期。 fig. 3.25 Daming, church's nave to the south, 1920s and present. (© ASJ France, GMC, and © THOC 2014) ················330

图 3.26　大名天主堂中殿北端，20 世纪 20 年代和近期。 fig. 3.26 Daming, church's nave to the north, 1920s and present. (© ASJ France, GMC, and © THOC 2014) ················331

图 3.28　大名天主堂西南面。 fig. 3.28 Daming, church from the southwest. (© THOC 2014) ················334

图 3.29　大名天主堂中殿的砖拱。 fig. 3.29 Daming, church, brick arches of the nave. (© THOC 2014) ················335

图 3.30　大名天主堂的砖柱身和柱头。 fig. 3.30 Daming, church, brick shaft and capitals at the triforium's level. (© THOC 2014) ················335

图 3.31　大名天主堂后殿和西南角的礼拜堂。 fig. 3.31 Daming, church, apse of the sanctuary and southwestern side chapel. (© THOC 2014) ················336

图 3.32　大名天主堂中殿和耳堂的东侧窗户。 fig. 3.32 Daming, church, eastern side of the nave and transept, different types of windows. (© THOC 2014) ················337

图 3.33　大名天主堂的 13 世纪哥特式风格细节：中殿的柱头、侧道柱身的基础、拱廊、拱廊上的柱头。 fig. 3.33 Daming, church, thirteenth-century Gothic architectural sculpture: capital of nave, base of a shaft in the aisle triforium, capitals at the triforium's level. (© THOC 2014) ················338

图 3.34　大名天主堂角楼，外面红砖，里面青砖。 fig. 3.34 Daming, church, corner turret, elaborated red brickwork outside, simple blue brickwork inside. (© THOC 2014) ················339

图 3.35　大名，学校建筑上的花砖和基督十字砖、方形屋瓦（望砖）。 fig. 3.35 Daming, brick with flower and Christian cross, square roof tile (*wang zhuan*), from the school buildings. (© THOC 2014) ·· 339

图 3.36　大名天主堂塔楼的暗拱、小尖塔和塔楼尖顶。　fig. 3.36 Daming, church's tower, lower level with blind arches and upper level with pinnacles and high red brick spire. (© THOC 2014) ························· 340

图 3.37　大名天主堂钟楼的石柱头、后殿的砖柱头和石头柱子顶板。fig. 3.37 Daming, church, stone capital of the tower, brick capital and stone abacus in the apse. (© THOC 2014) ··· 341

图 3.39　大名天主堂、2014 年翻新的大名天主堂的屋顶桁架。 fig. 3.39 Daming, church, roof truss renewed in 2014. (© THOC 2014) ················ 342

图 3.40　大名天主堂后殿的哥特式木制拱顶。　fig. 3.40 Daming, church, wooden Gothic style rib vaults of the apse. (© THOC 2014) ················ 343

图 3.41　大名天主堂中殿和东侧道的哥特式木制交叉拱顶。fig. 3.41 Daming, church, wooden Gothic style rib vaults of the nave an the eastern aisle. (© THOC 2014) ··· 344

图 3.42　大名天主堂西侧道的石柱和木制交叉拱顶。　fig. 3.42 Daming, church, stone columns and wooden Gothic vaults of the western aisle. (© THOC 2014) ··· 344

图 3.43　大名天主堂中殿的横向拱，用木板、板条和石灰做成。 fig. 3.43 Daming, church, transversal arch of the nave, made of planks, lath and lime. (© THOC 2014) ··· 345

图 3.44　大名天主堂中殿的木制拱顶、拱背部分。 fig. 3.44 Daming, church, wooden vaults of the nave, extrados. (© THOC 2014) ···················· 345

图 3.45　大名天主堂中殿的窗户和 原彩色玻璃。 fig. 3.45 Daming, church, triplet window of the nave and original stained-glass. (© THOC 2014) ······ 347

图 3.46　大名天主堂后殿窗户和 花饰窗格的原玻璃遗存。 fig. 3.46 Daming, church, window of the apse and remains of original glass in the tracery. (© THOC 2014) ··· 347

图 3.47　大名天主堂后殿的原祭台、各各他山、彩色玻璃窗和横幅，20 世纪 20 年代。 fig. 3.47 Daming, church, apse with the original main altar, Calvary, stained-glass windows and banners, 1920s. (© ASJ France, GMC) ·· 348

图 3.48　大名天主堂的告解室。 fig. 3.48 Daming, church, confessional. (© THOC 2014) ··· 349

图 3.49　大名天主堂主入口处的管风琴，产自奥地利，20 世纪 20 年代。 fig. 3.49 Daming, church, Austrian organ on the gallery above the main entrance, 1920s. (© ASJ France, GMC) ··· 350

目 录
Table of content

图 3.50 大名天主堂的领洗池。fig. 3.50 Daming, church, stone baptismal font. (© THOC 2014) ······351

图 3.51 大名天主堂的木头钟架子和 在法国铸造的钟。 fig. 3.51 Daming, church, wooden belfry and bells from France in the tower. (© THOC 2014) ······351

图 3.52 大名，学者和当地工人一起讨论手册中的技术用语，2014 年 7 月 9 日。fig. 3.52 Daming, academic staff and local workers identifying together the technical words of the handbook, 9 July 2014. (© School of Archaeology and Museology, Peking University 北京大学考古文博学院） ······355

图 3.53 大名，美国宣圣会的新教徒传教士的房子，1920 年前后和 2014 年。 fig. 3.53 Daming, house of the American Protestant missionaries of the Pentecostal Church of the Nazarene under construction and present state, around 1920 and 2014. (© China Collection, Nazarene Archives, Lenexa, Kansas; © THOC 2014) ······356

图 3.54 大名，两名法国耶稣会神父和一名中国神父 在大名城墙东门上给学校照照片，20 世纪 20 年代。fig. 3.54 Daming, two French Jesuit fathers and one Chinese father taking pictures of the school from the East Gate of the city wall, 1920s. (© ASJ France, GMC) ······358

图 3.55 大名，东街，2014 年。 fig. 3.55 Daming, area of East Street, 2014. (© School of Archaeology and Museology, Peking University 北京大学考古文博学院） ······358

图 3.56 大名，大名天主堂平面，2014 年。 fig. 3.56 Daming, church, ground plan, 2014. (© School of Archaeology and Museology, Peking University 北京大学考古文博学院） ······359

图 3.57 大名天主堂仰视平面图。 fig. 3.57 Daming, church, ground plan with projection of the rib vaults. (© School of Archaeology and Museology, Peking University 北京大学考古文博学院） ······359

图 3.58 大名天主堂北立面。 fig. 3.58 Daming, church, elevation of the northern side. (© School of Archaeology and Museology, Peking University 北京大学考古文博学院） ······361

图 3.59 大名天主堂中殿的纵剖面。 fig. 3.59 Daming, church, longitudinal section of the nave. (© School of Archaeology and Museology, Peking University 北京大学考古文博学院） ······364

图 3.60 大名天主堂中殿的横剖面和假耳堂的北侧。 fig. 3.60 Daming, church, cross section of the nave and elevation of the northern side of the pseudotransept. (© School of Archaeology and Museology, Peking University 北京大学考古文博学院） ······365

~ XXXI ~

图 3.61 大名天主堂中殿和假耳堂的横剖面。fig. 3.61 Daming, church, cross section of the nave and the pseudotransep. (© School of Archaeology and Museology, Peking University 北京大学考古文博学院) ································365

图 3.62 大名天主堂中殿的横剖面。fig. 3.62 Daming, church, cross section of the nave. (© School of Archaeology and Museology, Peking University 北京大学考古文博学院) ································366

图 3.63 大名天主堂中殿、假耳堂和钟楼的横剖面。fig. 3.63 Daming, church, cross section of the nave, the pseudotransept and the tower. (© School of Archaeology and Museology, Peking University 北京大学考古文博学院) ································367

引　言
东西方建筑实践的交流
Introduction: Architectural Exchange of Practice-Based Knowledge

舶来与本土：1926年法国传教士所撰中国北方教堂营造手册的翻译和研究
Building Churches in Northern China. A 1926 Handbook in Context

20世纪早期在中国建教堂，无论对西方传教士还是对中国基督徒来说，都是巨大的挑战。教堂应该以何种样式建造？在中国当时多变的城市与乡村景观中，教堂应该具备什么样的可识别性？此外，教堂还必须建得很好，尽可能不要太贵，而且应适用于基督徒的礼拜仪式。外国传教士来到中国，他们的西方背景和他们所拥有的建筑知识，与完全不同的中国样式和结构技术发生碰撞。这样的碰撞为东西方建筑交流做出了突出贡献。

目前，关于中国清朝晚期和民国时期的基督教问题及当时受西方影响的建筑，都已经成为重要的国际研究课题。❶ 然而几乎还没有教堂建筑及其建造技术方面的研究。在20世纪之前，除了通商口岸和一些主要城市之外，西方建筑中的拱、拱顶、地基、砖砌体、石雕、木结构、塔楼、楼梯、红砖、哥特式装饰等方面的建造技术还几乎不为中国北方所知。在大多数的村庄，教堂是最早使用这些新技术和新形式的建筑，事实已经证明在教堂建造过程中有着大量中西方不同形式和技术的交叉使用。这类建筑技术知识如何由西方传教士传

Building churches in China during the early twentieth century was a great challenge for Western missionaries as well as for Chinese Christians. In what style and how identifiable should churches be in China's varied urban and rural landscapes? Moreover churches had to be well built, the least expensive possible, and appropriate to their liturgical use. Foreign missionaries came to China with their Western background and architectural references, but were confronted with Chinese styles and construction techniques, which were totally different. This encounter was a remarkable contribution to the architectural exchanges between the East and the West.

Presently, both Christianity in China and Western-influenced architecture during the late Qing dynasty and the Republic have become important international fields of research.❶ Little research, however, has been done on church architecture and building techniques. Arches, vaults, foundations, full masonry walls, stone carving, timber trusses, high towers, stairs, red bricks, Gothic ornaments, etc. were almost unknown techniques in Northern China before the twentieth century, except in the treaty ports and major cities. In most rural villages, the church was the very first building that implemented such unknown

---

❶ Nicolas Standaert (ed). Handbook of Christianity in China, Volume One: 635-1800 [M]. Leiden-Boston: Brill, 2001; R. Gary Tiedemann (ed.). Handbook of Christianity in China, Volume Two: 1800-Present [M]. Leiden-Boston: Brill, 2010.

引言——东西方建筑实践的交流
Introduction: Architectural Exchange of Practice-Based Knowledge

递,中国建造者又是如何接受的呢?

来到中国的西方传教士,有些是有建筑方面的专业知识的,但是,他们很快就遇到了实际问题(图 0.1)。他们首先要理解中国建筑的传统和技术,评估可用材料的质量,感受不同的气候条件,更好地进行建筑选址。他们也必须学习中国建筑术语及如何与中国工匠、包工头打交道。在具备了这些必不可少的知识后,传教士建造者就开始鼓励中国工匠使用西方技术。这些工头们能读懂图纸、监督工作并把新知识传递给当地工匠(图 0.2)。

西方传教士建造者和中国工匠之间的碰撞是基于彼此实践基础上的知识交流。如此碰撞产生了新的技术组合和技术"杂交"形式。从样式上来说,19世纪晚期的中国的基督教建筑多多少少是纯西方模式,到了 20 世纪 20 年代后期,其逐渐演变为本土化的中国基督教样式。❶ 这是在对中国现存教堂的实

techniques and forms. There is evidence of a wide range of hybridized forms and techniques. How was this architectural and technical knowledge transferred by Western missionaries and received by Chinese builders?

Some missionaries came to China with precise architectural knowledge, but were immediately confronted with the local reality (fig. 0.1). They first had to understand the Chinese building traditions and techniques, evaluate the quality of the available materials, experience the climate, and chose good locations for building. They had also to learn Chinese architectural terminology and how to deal with Chinese workers and contractors. After having acquired this indispensable knowledge, missionary-builders initiated skilful Chinese craftsman into Western techniques. These specialized foremen were able reading plans, supervising the works, and transmitting their new knowledge to local workers (fig. 0.2).

The encounter between the missionary-builder and the Chinese builder was thus based on exchanges of practice-based knowledge. It generated varied forms of cross-breeding and new technical combinations. From a stylistic point of view, Christian architecture in China evolved from more or less acculturated Western models in the late nineteenth

---

❶ Thomas Coomans. Indigenizing Catholic Architecture in China: From Western-Gothic to Sino-Christian Design, 1900-1940 [A] // Cindy Chu Yik-yi. Catholicism in China, 1900-Present. The Development of the Chinese

地考察和分析及对欧洲和北美现存的传教士文献的研究基础上得出的结论。

century to an indigenized Sino-Christian style from the late 1920s.^See P3❶ This stylistic evolution can be established thanks to fieldwork and analysis of the existing old churches in China, as well as research in missionary archives in Europe and North America.

图0.1 卜天德神父建造圣母圣心会拉花营子教堂（内蒙古），1930年。
fig. 0.1 Father Petrus De Boeck building the Scheut mission church of Lahuayingzi (Inner Mongolia),1930. (© FVI, CHC pictures)

———————————————（接上页 continue）

Church [C]. New York: Palgrave and Macmillan, 2014, 125-144.

引言——东西方建筑实践的交流
Introduction: Architectural Exchange of Practice-Based Knowledge

图0.2 建造知识的传播：西方传教士、中国工头和中国共享建造高家营子新信徒住所（河北省），20世纪30年代。
fig. 0.2 Transfer of construction knowledge: western missionary, Chinese foremen and Chinese workers building the novices' house at Gaojiayingzi (Hebei province), 1930s. (© FVI, CHC pictures)

然而，关于教堂建造的影像资料和文字资料都很少。因此，能够找到这本专门讲如何在中国建教堂的老手册，对我们来说是很有意义的一件事情，它值得我们去研究、翻译和出版。

However, visual and written sources about church construction are very scarce. Therefore, the discovering of an old handbook on church construction in China was such an exceptional event that it deserved to be studied, translated and published.

舶来与本土：1926年法国传教士所撰中国北方教堂营造手册的翻译和研究
Building Churches in Northern China. A 1926 Handbook in Context

手册《传教士建造者：建议-方案》
The Handbook *The Missionary-Builder: Advice-Plans*

1926年，河北省献县宗座代牧区[1]的法国耶稣会传教士出版了一本题目为《传教士建造者：建议-方案》的小册子（图0.3）。这本8开的小册子有67页文字内容和54张图。它介绍了源自北欧以砖构建筑为主的国家的西方传统哥特式建筑。这本手册用法语写成，序言写于河北省大名市，署名是"中国北方的传教士"。在手册的序言里，作者提到写此手册的"主要目的是帮助在中国的传教士建造牢固、安全、不太贵又不太丑的教堂：用极少的钱建好的教堂"。[2] 手册中除了文字，还有项目方案，提供了建造教堂的技术图纸和模型图，都是一些比较容易实现的例子。

In 1926, French Jesuit missionaries of the apostolic vicariate of Xianxian (Hebei province) published a handbook entitled *Le missionnaire constructeur, conseils-plans* (*The Missionary-Builder: Advice-Plans*) (fig. 0.3) . This small in-8° booklet contains 67 pages of text and 54 plates. It describes and promotes a type of Western traditional Gothic architecture from the Northern European brick countries. The handbook is written in French, prefaced from Daming (Hebei province), and signed by "missionaries from Northern China". In the handbook's preface, the authors explain that "*the ideal here is to help you build solidly, safely and not too expensively nor too unattractively: a pretty church for little money.* "[2] In addition to the text, in order to add example to precept, the handbook includes a pocket with technical drawings and

---

[1] 宗座代牧区（简称代牧区）是天主教会的一种教务管辖机构，设立于尚不足以达到成立教区资格的传教地区。其本质上是临时的，虽然有可能持续一个世纪甚至更久。它的最主要目的是培养足够数量的天主教徒，以能成立一个正式的教区。宗座代牧区由宗座代牧（拉丁语：Vicario Apostolico）领导，通常由一个领衔主教担任。根据天主教法典的规定，宗座代牧区直接置于教宗的管辖之下，并透过一位代表或"代牧者"（Vicar）来执行教宗管理教务的权力。教宗可任命一位邻近教区的主教兼任，或委托一位神职人员专责担任宗座代牧（https://zh.wikipedia.org/wiki/）。

[2] Le Missionnaire Constructeur. Xianxian, 1926, preface.

这本稀有手册只在最近的文献中被提到过一次，❶在我们的一篇文章里也讨论过。❷目前中国境内还没有发现。❸基于我们所掌握的资料，该手册目前在世界上只找到两本。❹

一本现存于法国里昂市（Lyon）的市图书馆，编号为 SJ H 678/42。此本可追溯到 1926 年，是手册的原始版本，但它不全，图纸缺失。它原先属于喷泉图书馆，即先前的尚蒂伊耶稣会博物馆，1999 年搬到了里昂。

另一本现存于加拿大魁北克的拉瓦尔大学图书馆的人文社会科学馆，编号为 NA 6045 M678 1935。这本是 1935 年重印本，有图纸。它曾为建筑师 Adrien Dufresne（1904—1983）收藏。这位建筑师是一位擅长设计教堂建

model plans of churches, which are considered fairly easy to realise.

This rare handbook has only been mentioned once in recent literature,❶ and discussed in one of our articles.❷ Until present we did not succeed to find any copy in China. ❺ Only two copies were traced in the world. ❻

Lyon (France), Bibliothèque Municipale, SJ H 678/42. Dated 1926, this copy is the original edition of the handbook, but it is incomplete because the plates are missing. It is coming from the former Jesuit library of Chantilly, the *Bibliothèque des Fontaines*, that has been moved to Lyon in 1999.

Québec (Canada), Bibliothèque de l'Université Laval, Bibliothèque des sciences humaines et sociales, NA 6045 M678 1935. This copy is a reprint from 1935 and preserved its illustrations. It is coming from the collection of the archi-

---

❶ Françoise Aubin. Christian Art and Architecture [A] // R. Gary Tiedemann. Handbook of Christianity in China. Volume Two: 1800-Present [C]. Leiden-Boston: Brill, 2010, 735.

❷ Thomas Coomans. A pragmatic approach to church construction in northern China at the time of Christian inculturation: The handbook Le Missionnaire Constructeur, 1926 [J]. Frontiers of Architectural Research, 2014, 3(2): 89-107.

❸ 这要感谢崔金泽，他在 2013 年 7 月在中国国家图书馆利用文津搜索引擎核实过。

❹ 需要感谢那些帮助我们找到这两本手册的人：Jeffrey Cody 和美国洛杉矶的盖蒂研究院图书馆，Luc Vints 和鲁汶大学的 KADOC 文献中心及宗教、文化和社会研究中心，Fabienne Wijnants 和鲁汶大学工学院图书馆，James Lambert 和魁北克拉瓦尔大学图书馆。

❺ With all thanks to Cui Jinze, who checked the Wenjin search engine at the National Library of China (July 2013).

❻ We would like to thank all the people who helped us to trace these copies: Jeffrey Cody and the library of The Getty Research Institute, Los Angeles; Luc Vints and KADOC Documentation and Research Centre for Religion, Culture and Society, KU Leuven; Fabienne Wijnants and Campus Library Arenberg, KU Leuven; James Lambert and the library services of Université Laval, Québec.

筑的说法语的加拿大人。❶ 这本手册能辗转到达魁北克，需要感谢法国-加拿大耶稣会，该耶稣会于 1918—1954 年期间在江苏省徐州教区很活跃。❷

tect Adrien Dufresne (1904-1983), a French-speaking Canadian who was specialised in church construction.❶ In all likelihood this copy reached Quebec thanks to French-Canadian Jesuits who were active in the diocese of Xuzhou (Jiangsu province) from 1918 to 1954.❷

图0.3 手册扉页，1926年。
fig. 0.3 The handbook's title page, 1926. (© Lyon, Bibliothèque municipale)

该手册分为 8 章：选址；建筑布局和朝向的选择；设计方案、工程预算和施工合同；建筑材料；砖石工程；屋顶

The handbook is structured in eight chapters: choice of the location; choice of the buildings' layout; plans, estimates,

---

❶ http://www.patrimoineculturel.gouv.qc.ca/rpcq/detail.do?methode=consulter&id=9892&type=pge#.VkNLTvkvehc （accessed on 1 November 2015）.

❷ Rosario Renaud. Le diocèse de Süchow (Chine), champ apostolique des Jésuites canadiens de 1918 à 1954 [M]. Montréal: Éditions Bellarmin, 1982; Jacques Langlais. Les Jésuites au Québec en Chine (1918-1955) [M]. Québec: Presses de l'Université Laval, 1979.

引言——东西方建筑实践的交流
Introduction: Architectural Exchange of Practice-Based Knowledge

（中式和西式）；屋面装饰；建筑细节的处理（涉及建筑结构的各处细节）。另外这本手册还包括插图和54张图纸。

这本手册远不仅仅是对技术知识的汇集，它还展示了西方传教士是如何理解中国建筑传统及他们如何将西方技术传递给中国工匠。它的精髓在于，它为我们提供了基于长期实践基础上的建筑知识。

除了技术方面的内容，该手册也让建筑历史学家思考了许多关于该手册写作背景的研究问题。

署名为"中国北方传教士"的作者是谁？为什么没有提他们的名字？

为什么这本手册会在1926年和1935年出版两次？这两版之间有什么不同？当时各印了多少册？

为什么这本手册出版于献县而序言写于大名（这是河北省南部的两个不同地方）？上述地区当时是否需要建造教堂？

这本手册为何用法语写成？在描述普通话和地方方言时是使用了哪种罗马化系统？这些建筑技术术语今天还能被人们理解吗？

contracts; building materials; masonry; roofs, both Chinese and Western; roof ornaments; miscellaneous, about various construction elements. The handbook also includes illustrations and 54 drawings.

The handbook, however, is more than a compilation of technical hints. It reveals a lot about how the missionaries understood Chinese building traditions as well as how they transmitted Western techniques to Chinese workers. The essence of the handbook is architectural knowledge based on practice and long experience.

Except its technical content, the handbook arouses many research questions about its context to the architectural historian:

Who are the 'missionaries from Northern China' who authored the handbook? Why are their names not mentioned?

Why was this handbook from 1926 reprinted in 1935? Are there differences between the two prints? How many copies of the handbook were printed?

Why was the handbook prefaced from Daming and printed in Xianxian, two places from the south of Hebei province? Was there a need for church buildings in that region?

Why is the handbook written in French? What kind of Romanization system is used to transcribe Mandarin and local dialect words? Are all these architectural- technical words under-

哪些人会是这本手册的潜在读者群？为什么这本手册里经常提到其实践仅限于中国北方？

这本手册里的图纸是谁画的？关于八座教堂的平面图纸是否来自真实存在的教堂建筑？

20世纪20年代和30年代在中国北方建造教堂所处的建设大背景是什么样的？是否有必要出版这样一本专业手册？

为什么这本手册推崇源自北欧以砖构建筑为主的国家的西方传统哥特式建筑？而此类建筑自19世纪90年代开始就不断受到争议，而且在第一次世界大战之后在欧洲已经完全过时。

Who was the expected French-speaking readership of the handbook? Why does the handbook often mentions that its practice is limited to Northern China?

Who made the drawings that illustrate the handbook? Are the plans of the eight model churches coming from real buildings?

What was the architectural context of church construction in Northern China in the 1920s and 1930s? Was there a need for a specialised handbook?

Why does the handbook promotes a type of Western traditional Gothic architecture from the Northern European brick countries, while in Europe this architecture was contested since the 1890s and completely outdated after World War I?

北京大学和鲁汶大学的合作研究
Research Collaboration between Peking University and The University of Leuven

对手册《传教士建造者：建议-方案》的研究，作为一个课题，促进了中国与比利时学者之间的合作。这本手册的确提供了一个契机，它把当年传教士建造者于1926年记录的实践知识和当今在中国北方开展的建筑考古的田野考察结合在一起。这样一个研究项目只

The Missionary-Builder: Advice-Plans handbook became a topic for developing research collaboration between Chinese and Belgian scholars. The handbook, indeed, gave a unique opportunity to combine the practise-based knowledge compiled by missionary-builders in 1926 with building-archaeological fieldwork

有在科学而友好的国际合作中才能得以开展。

2012年4月17日，由北京大学校长和鲁汶大学校长共同签署的合作协议促进了双方之间的合作研究。2014年1月北京大学考古文博学院和鲁汶大学工学院（考古系和雷蒙德·勒麦尔国际保护中心）签署了一项具体的合作协议。该合作协议有三个目的：①组织召开"中欧建筑考古论坛"以推动建筑考古领域的理论和方法方面的合作研究，同时也促进相关学科（如建筑结构史、欧亚建筑历史和遗产保护等）的学术合作。②促进建筑考古及上述相关学科的教学和培训。③发展建筑考古及上述相关学科的研究项目。

这项合作研究的第一阶段始于2014年7月。徐怡涛副教授和高曼士教授在大名老城组织举办了针对北京大学考古文博学院本科生的关于建筑考古的暑期培训班。在为期一周的现场培训中，北大学生测绘和分析了大名的天主教教堂，它在手册中被提到过数次（图0.4）。通过现场对建筑结构实体的学习，该培训班帮助学生掌握了那些文

in present Northern China. Such a research project could only be carried out thanks to scientific collaboration at international level.

The Memorandum of Understanding signed between the president of Peking University and the rector of the University of Leuven on 17 April 2012, stimulates bilateral research collaboration. In that general frame, a specific cooperation agreement was finalised in January 2014 between the School of Archaeology and Museology of Peking University and the Faculty of Engineering of the University of Leuven (Department of Archaeology, Raymond Lemaire International Centre for Conservation). This cooperation agreement has three aims: 1. academic cooperation and organisation of an international Sino-European Building Archaeological Forum in order to promote the theory and methods about building archaeology, as well as other related disciplines such as construction history, architectural history in Asia and Europe, and heritage conservation; 2. teaching and training; 3. research projects.

The first concrete step of this research collaboration took place in July 2014. Prof. Xu Yitao and Prof. Thomas Coomans organized a building archaeology summer workshop for undergraduate students of Peking University in the historic centre of the city of Daming. During one week the students analysed and recorded the Catholic church, which is mentioned several times in the hand-

献资料无法体现的建造和技术转换方面的知识（图0.5）。

book (fig. 0.4) . The workshop aimed learning from the material source what the archives do not reveal about construction and technological transfer (fig. 0.5) .

图0.4 大名天主堂（河北省），法国耶稣会士在1918—1921年建。
fig. 0.4 Church of Our Lady at Daming (Hebei province), built by French Jesuits in 1918-1921. (© Thomas H. Hahn, 2010)

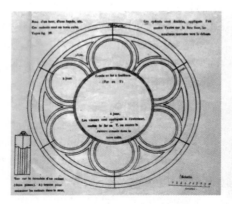 

图0.5 在理论与实践之间：建筑手册中的玫瑰花窗和大名天主堂上的玫瑰花窗。
fig. 0.5 Confronting theory and practice: rose window from the handbook and rose window of Daming's church.
(© Québec, Université Laval, and THOC 2014)

该培训班的最初成果，已于 2015 年 7 月在芝加哥召开的"建造历史国际研讨会"上做过汇报，❶ 也于 2015 年 8 月在上海召开的"建筑遗产保护国际研讨会"上做过报告。❷ 本书还包括 2014 年暑期培训班中完成的大名天主堂的测绘图纸。

这项合作研究的第二阶段是将于 2016 年 5 月在北京召开的"中欧建筑考古国际学术研讨会"。

First results of the workshop have been presented at the International Congress on Construction History in Chicago in June 2015,❶ as well as at the Architectural Heritage Preservation International in Shanghai in August 2015.❷ The present book includes the measurements of the church of Daming that have been realized during the workshop of 2014.

The next step of the research collaboration is the organization in Beijing in May 2016 of the International Forum of Sino-European Architectural Archaeology.

本书架构
The Concept of the Book

这本 1926 年出版的法语手册对研究中国近代教堂建筑具有很高的价值，于是作者决定将其翻译为英文和中文。然而，本书远不止于翻译和文本编辑，它分为以下三部分。

Studying the handbook of 1926 and being convinced of its high interest for architectural studies, resulted in the project of a Chinese translation. The present book, however, is much more than just a translation and a text edition, and therefore is built in three chapters:

---

❶ Thomas Coomans, Xu Yitao. Gothic Churches in Early 20th-Century China: Adapting Western Building Techniques to Chinese Construction Tradition [A] // Brian Bowden, Donald Friedman, Thomas Leslie, et al. Construction History Society of America (vol. 1) [C]. 2015, 523-530.

❷ Stand of School of Archaeology and Museology of Peking University, and lecture: Thomas Coomans, Western Brick Production and Construction in China: Technological Transfer, Cultural-Historical Meaning, and Conservation of a Shared Built Heritage (29 August 2015).

舶来与本土：1926 年法国传教士所撰中国北方教堂营造手册的翻译和研究
Building Churches in Northern China. A 1926 Handbook in Context

第一部分：手册撰写的背景分析

在翻译手册文本之前，理解这本手册撰写的背景非常关键。这部分内容首先简单全面地介绍了关于建筑手册和教堂样式的书籍，接着对中国的基督教历史进行了概括，主要对民国时期西方宗教在中国社会转型中的作用进行了分析。接着对手册的作者进行了考证，并对他们在当时社会背景下的动机也进行了研究。这些研究能够帮助我们理解西方传教士建造者在中国北方的天主教传教会中扮演的角色。

其后作者对手册的框架和内容进行了分析，也剖析了西方传教士眼中的中国工匠。虽然手册的章节题目没有出现此类标题，但 20 世纪 20 年代是教廷提倡基督教的中国化的时期，期间对教堂样式进行了广泛的讨论，这些都值得特别关注。

最后，作者在第一部分介绍了中西方碰撞和技术转换过程。

第二部分：手册的翻译

手册原是用法语写的，把法语原文转录给读者，必要时读者可参考原文。英文翻译和中文翻译则附加了详细

Part one: the handbook in context

Before translating the text, it is essential to understand the context in which the handbook has been written. After an introduction about architectural handbooks and pattern books of churches in general, an overview of the history of Christianity in China is sketched, with focus on the time of the Republic and the role of Western religion in the Chinese society in transformation. Identifying the authors of the handbook as well as their motivations in the precise context of the late 1920s will help understanding the role of the missionary-builders of Catholic missionary societies in Northern China.

Thereafter, the structure and content of the handbook are analyzed as well as the perception of Chinese workers by Western missionaries. Even though, it is not an explicit heading of the handbook, the issue of the style of churches in the 1920s in China, the time of Christian inculturation, deserves specific attention.

The last aspect developed is the East-West encounter and the process of transfer of technology.

Part two: the handbook translated

Because the handbook is written in French, the original text is transcribed so that the reader, if necessary, always can

的注释。

在文本翻译之后，手册的原始图纸也整体再现。

第三部分：教堂建造案例分析——大名天主堂

正如手册序言里所提到的，这本手册写于大名，手册中数次明确提到大名天主堂。法国耶稣会的传教士于1918—1921年期间建造了这座伟大的哥特式砖构教堂。2013年，大名天主堂被列为中国第七批"全国重点文物保护单位"。基于此，对这座教堂建筑进行研究并出版相关研究成果无疑是极为适时的。本书中不仅包括作者在巴黎档案馆里找到的历史资料，还包括2014年基于三维扫描技术完成的测绘图纸（图0.6）。

整个研究的基本精神基于中国和比利时学者之间的对等与合作。我们选择用中英双语出版此书，旨在其能为世界范围的读者所读，或许也能为将来更多的出版物提供一个参考。

refer to the original. The translation, both in English and in Putonghua, is enriched with explanatory footnotes.

The handbook's original illustration is integrally reproduced.

Part three: the handbook as built

As mentioned in the handbook's preface, the handbook has been written in Daming, and the authors several times explicitly refer to the church of Daming. French Jesuit missionaries built this great Gothic brick church in 1918-1921. It has been listed "national heritage site" of the People's Republic of China in May 2013 (7th list). For these reasons, it seemed opportune to make more research on this remarkable building and publish the results in this book. Our results include both historical documents from archives in Paris and the complete set of measurements of the church realised during the fieldwork of 2014 (fig. 0.6).

The spirit of the whole research is based on symmetry and collaboration between Chinese and Belgian scholars. Therefore we have chosen to make a bilingual book that could reach a worldwide audience and perhaps set up a standard for future publications.

图0.6 大名天主堂测绘,北京大学考古文博学院建筑考古暑期实践班。
fig. 0.6 Measuring the church of Daming, building archaeology summer workshop of the School of Archaeology and Museology of Peking University. (© THOC 2014)

第一部分
手册的撰写背景
Part One:
The Handbook in Context

舶来与本土：1926年法国传教士所撰中国北方教堂营造手册的翻译和研究
Building Churches in Northern China. A 1926 Handbook in Context

本部分介绍了手册《传教士建造者：建议-方案》的内容并对手册的撰写背景进行了分析。❶ 在简单介绍了关于教堂建造和式样的书籍之后，本部分介绍了手册的框架和内容组成，也对建筑材料、砖石工程、屋顶等方面进行了简单描述。该手册不仅仅是一本技术手册，它还记录了传教士如何对待中国建筑传统及如何把西方技术传递给中国工匠。

对该手册的可能作者、读者群及写作背景等方面的分析构成了本部分的主要内容。20世纪20年代，教廷改变了它的传教战略，鼓励当地传教士和牧师开展地方上的传教工作，而不是像殖民时期欧洲中心模式下的福音传教。此项"本土化"的战略旨在将基督教信仰与地方特色文化完全交融。建筑作为基督教在公共空间的最明显的实体存在，在本土化过程中遇到很大的挑战。在中国，许多传教士很抵制这项中国本土化的运动，比起中式基督样式，他们更喜欢哥特式和其他西方风格。在该手册中我们将看到这方面的论述。

The purpose of this first part is to describe, analyze and contextualize the handbook *The Missionary-Builder: Advice-Plans*.❶ After a brief introduction about handbooks and pattern books of churches, we will analyze the structure and describe the content of the handbook and its different practical aspects concerning building materials, masonry, roofs, etc. The handbook, however, is more than a compilation of technical hints-it reveals how the missionaries perceived Chinese building traditions, as well as how they transmitted Western techniques to the Chinese workers.

The identity of both its authors and its readership, as well as the context in which the handbook was produced, are essential questions. In the 1920s, the Holy See changed its missionary strategy and began promoting the principle of local churches with native bishops and priests, instead of the colonial and Eurocentric model of evangelization. This movement of 'inculturation' aimed to anchor and to integrate Christian faith within specific cultures. Architecture, the most visible expression of Christianity in the public space, was a major challenge for inculturation. In China, many missionaries resisted this Sinicizing movement, preferring Gothic and other Western styles to a new Sino-Christian style. The handbook, as we will see, participated in this debate.

---

❶ Le Missionnaire Constructeur [M]. Xianxian, 1926.

第一部分　手册的撰写背景
Part One: The Handbook in Context

关于教堂建造和式样的书籍
Architectural Handbooks and Pattern Books of Churches

建筑作为一门学科，它是对所有象征性、艺术性、技术性和实践性的知识和规则的总结和传承。在很长一段时间内，它仅限于工匠技艺的口头传承，然而在人类文明史上很早就出现了关于建筑实践的理论书籍，在19世纪和20世纪此类书籍逐渐增多。❶ 在这些书中，最古老的就是维特鲁威于公元前15年前后写成的《建筑十书》❷，这本书对西方影响非常大。在中国，要数宋朝李诫于公元1100年前后写的《营造法式》❸ 最为知名。从19世纪开始，工业革命彻底改变了西方的建造艺术和技术，出现了各类新的建筑书籍，如百科全书、分析各种建筑式样的书、技术手册及专业期刊等。这些书有些被带到了殖民地，传播了各类建筑的建造技术知识，也促进了适应于当地气候的新的建筑材料、技术、地方设计和装饰等方

Architecture as a discipline, is based on the transmission of all kinds of symbolic, aesthetic, technical and practical knowledge and rules. For a long time, it was only based on the oral transmission of craftsmanship, but the need to theorize by writing architectural treatises appeared early in the great civilizations, gradually increased, and culminated in the nineteenth and twentieth centuries.❶ Amongst the oldest treatises, the *Ten Books on Architecture* (*De architectura libri decem*) written by Vitruvius around 15 BC was the most influential in the West,❷ and Li Jie's *Treatise on Architectural Methods of State Building Standards* from around 1100 (Song Dynasty) was the most famous in China. ❸ From the nineteenth century, the industrial revolution completely transformed the art of construction in the West and new kinds of architectural publications such

---

❶ Jean-Philippe Garric, Valérie Nègre, Alice Thomine-Berrada (eds). La construction savante. Les avatars de la littérature technique [C]. Paris: Picard, 2008.

❷ 这本书目前有多种版本和译本，也有电子版。

❸ [宋] 李诫（撰）, 邹其昌（点校）. 营造法式（文渊阁《钦定四库全书》）. 北京：人民出版社，2011.

❹ There exist numerous editions and translations, including online.

❺ Li Jie. Yingzao fashi (Wenyuange edition, edited and commented by Zou Qichang) [M]. Beijing: People's Press, 2011.

面的发展。

19世纪全球人口的增长、基督教的复兴和西方的帝国主义都促进了教堂在世界范围内的扩张。工业化时期，教堂的建设崇尚理性的建造：新教堂必须要兼坚固、经济和美观为一体。

教堂建筑与宗教相关的部分体现在礼拜空间的神圣尺度和功能使用。如果古典主义时期或中世纪时期的某种特定样式能体现某特定的民族性和宗教性，该样式往往就会成为首选。例如，奥古斯都·普金(Augustus W. N. Pugin)的建筑作品和书籍在全球范围内都特别有影响力。❶ 英国的哥特复兴式成为英伦三岛和整个大英帝国的英国教会教堂的样式，❷ 而在法国殖民地和传教区（包括中国）的天主教样式则是

as encyclopedias, pattern books, handbooks and specialized journals appeared. Some of these books were spread to the colonies, diffusing technical advice for all kinds of building types and promoting new building techniques and materials, as well as home country architectural design and ornaments, more or less in accordance with the local climate.

One of the consequences of the combination of demographic growth, Christian revival and Western imperialism during the nineteenth century was the proliferation of churches across the world. At the age of industrialization, even the construction of churches was approached from a rational point of view: new churches were to be solid, economic and beautiful.

The religious aspects of church building concerned the sacred dimension of the places of worship and their liturgical use. The aesthetic dimension was in the first instance the choice of a precise style, from the classical or the medieval past, in relation to the national and religious identity of the patrons. So, for instance, the works and writings of Augustus W. N. Pugin were particularly influential worldwide.❶ English Gothic Revival became the style of the Anglican

---

❶ Timothy Brittain-Catlin, Jan De Maeyer, Martin Bressani (eds). A.W.N. Pugin's Gothic Revival: The International Style (KADOC Artes, 16) [C]. Leuven: Leuven University Press, 2016 (forthcoming).

❷ Alex Bremner. Imperial Gothic: Religious Architecture and High Anglican Culture in the British Empire, 1840-1870 [M]. City of New Haven: Yale University Press, 2013.

第一部分　手册的撰写背景
Part One: The Handbook in Context

法国哥特式（图 1.1）。❶ 甚至连比利时传教士在内蒙古建造的教堂也是比利时砖构的哥特式。❷

关于教堂建造的理论、式样和技术等方面的书能够帮助传教士、牧师、资助人和建筑师在满足需要、审美和预算的前提下做出最好的选择。法国百科全书系列《罗雷手册》中的一本就是关于教堂的建造、维护、重建和装饰的。❸ 比利时皇家学院曾将奖项授予一本由一位军事工程师写成的理性教堂建筑。❹ 英国建筑师出版了很多介绍从英国到加拿大的英国教会的各种教堂式样的书籍。❺ 新教和基督教也有其各自的手册。

Church both in the British Isles and throughout the British Empire, See P20❷ while thirteenth-century French cathedral Gothic became the style of Catholicism in French colonies and French missions, including in China❶ (fig. 1.1). Even Belgian missionaries built churches in Belgian brick Gothic in Inner Mongolia.❷

Architectural treaties, pattern books and technical handbooks for the building of churches aimed at helping parishioners, priests, patrons and architects to make the best choices, in function of their needs, taste and budget. One handbook of the French encyclopedic series *Manuels Roret* was dedicated to the construction, maintenance, restoration and decoration of churches.❸ The Belgian royal Academy created an award for a book about rational church building written by a military engineer.❹ British practicing architects published pattern books containing drawings of different church types for Anglican churches from

---

❶ Thomas Coomans. Gothique ou chinoise, missionnaire ou inculturée? Les paradoxes de l'architecture catholique française en Chine au XXe siècle [J]. Revue de l'Art, 2015,189(3) :9-19; Jean-Paul Wiest. The Building of the Cathedral of Canton: Political, Cultural and Religious Clashes [A] // Religion and Culture: Past Approaches, Present Globalisation, Futures Challenges (International Symposium on Religion and Culture, 2002: Macau) [C]. Macau, 2004, 231-252.

❷ Thomas Coomans, Luo Wei. Mimesis, Nostalgia and Ideology: The Scheut Fathers and Home-Country-Based Church Design in China [A] // History of the Catholic Church in China. From its beginning to the Scheut Fathers and 20th Century (Leuven Chinese Studies, 29) [C]. Leuven, 2015, 495-522.

❸ Jean-Philippe Schmit. Nouveau manuel complet de l'architecture des monuments religieux ou traité d'application pratique de l'archéologie chrétienne à la construction, à l'entretien, à la restauration et à la décoration des églises, à l'usage du clergé, des fabriques, des municipalités et des artistes [M]. Paris: Manuels Roret, 1859.

❹ A. Demanet. Mémoire sur l'architecture des églises [M]. Brussels, 1847.

❺ Barry Magrill. 'Commerce of Taste': Church Architecture in Canada, 1867-1914 [M]. Montreal-Kingston, 2012.

舶来与本土：1926年法国传教士所撰中国北方教堂营造手册的翻译和研究
Building Churches in Northern China. A 1926 Handbook in Context

在殖民地，主教、传教士和教众参考这些手册指导地方工匠进行教堂建造，这样就用不着花大钱请欧洲的建筑公司做设计方案了。通常这些书都会有各类制造商的广告，特别是西方教堂室内陈设品和彩色玻璃的制造商。

很多牧师也投入到关于神圣艺术和建筑的讨论中，他们中有些人写了关于教堂建造和装饰的手册。此外，许多对建筑和不同民族感兴趣的传教士还观察和记录下他们传福音地区的专业术语并撰写文章发表在各种传教士期刊上（图 1.2）。就目前我们所了解到的，《传教士建造者：建议-方案》这本手册是第一本在中国出版的关于教堂建造的专业手册。还有其他的手册，如上海耶稣会出版的《罗宾逊手册》，也介绍了很多对传教士可能有用的方方面面的信息。❶

England to Canada.See P21❺ Protestants and other Christian denominations had their own handbooks too.

In the colonies, bishops, missionaries and congregations used these books to commission local builders and craftsmen instead of ordering expensive plans from architectural firms in Europe. Often these books included advertisements of manufacturers of Western church furniture and stained glass.

Many clergymen were actively involved in the debate about sacred art and architecture and some of them wrote handbooks for the building and decoration of churches. Moreover, architecturally or ethnographically minded missionaries observed building practices, recorded the specific terminology in the regions they were evangelizing, and published articles in missionary journals (fig. 1.2). As far as we know, *The Missionary-Builder: Advice-Plans* handbook is the first specific handbook for church building to have been published in China. There were other manuals with basic hints on all possible topics useful for pioneering missionaries, such as the Robinson's manuals published by the Jesuits of Shanghai.❶

---

❶ Pol Korigan. Le manuel pratique de Robinson (2nd edition) [M]. Shanghai, 1912. This handbook is about food, medicine, and other practical things for a westerner's daily life in China.

第一部分　手册的撰写背景
Part One: The Handbook in Context

图1.1　广州（广东省），遣使会所建法国天主教主教座堂，1863—1888年。
fig. 1.1 Guangzhou (Guangdong province),
French Catholic cathedral built by the Congregation
of the Mission, 1863-1888. (© THOC 2013)

舶来与本土：1926年法国传教士所撰中国北方教堂营造手册的翻译和研究
Building Churches in Northern China. A 1926 Handbook in Context

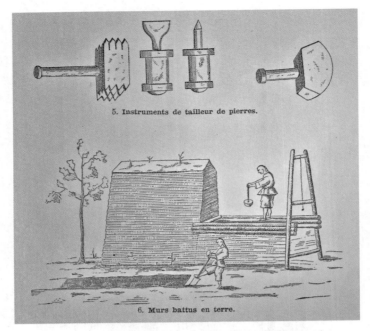

图1.2 发表在传教士期刊上的和羹柏神父绘图，1891年。
fig. 1.2 Father De Moerloose's drawings published in a missionary journal, 1891. (© FVI, Scheut Memorial Library)

手册作者在《传教士建造者：建议-方案》中只提到了两本其他手册，都是用法语写的，不是针对中国和殖民地建筑的：Étienne Barberot 写的《市政工程条例》第4版，❶ Donatien Duret 写的《宗教建筑》第1版。❷

在手册《传教士建造者：建议-方

The authors of *The Missionary-Builder: Advice-Plans* only refer to two other handbooks, both in French and not specialized in Chinese or colonial architecture: the fourth edition of the treaty of civil constructions by Étienne Barberot,❶ and the first edition of father Donatien Duret's treatise on religious architecture.❷

The reprint of 1935 adds a reference

---

❶ Étienne Barberot. Traité de constructions civiles travaux préparatoires et connaissance du sol, maçonnerie, pavages divers, accessoires de maçonnerie, béton de ciment armé, marbrerie, vitrerie, vitraux, charpente de bois, charpente métallique, couverture, menuiserie et ferrures, escaliers, monte-plats, monte-charges et ascenseurs, plomberie d'eau et sanitaire, épuration biologique, chauffage et ventilation, décoration, éclairage au gaz et à l'électricité, acoustique, matériaux de construction, résistance des matériaux, statique graphique, renseignements généraux (4th edition) [M]. Paris, 1912.

❷ Donatien Duret. Notions élémentaires d'architecture religieuse [M]. Chavagnes-en-Paillers, 1926.

第一部分　手册的撰写背景
Part One: The Handbook in Context

案》的 1935 年重印版中提到了沈锦标的书《造屋三知》。❶

to Shen Jinbiao's book *Knowledge for Building Construction*. ❷

手册的作者
The Handbook's Authors

正如手册扉页上所写，作者是"中国北方的传教士"。该手册是用法语写的，序言是在大名府写的，书是在献县出版的，这都表明了该手册与法国耶稣会有很大关联。大名（现为河北省大名县）和献县（现为河北省献县）都属于法国耶稣会教廷隶属下的天主教区：直隶东南教区，后于 1924 年更名为献县教区❸（图 1.3）。该手册的出版得到了直隶东南教区主教刘钦明和献县法国传教会神父卜鸣盛的许可，他们都隶属于法国耶稣会。❹

As mentioned on the handbook's title page, the authors were 'missionaries from Northern China'. The handbook was written in French, prefaced from *Tamingfou*, and published at the printing house of *Sien-Hsien*, what suggests strong connections with the French branch of the Society of Jesus. The cities of Daming (Daming county, Herbei province) and Xianxian (Xian county, Hebei province) belonged to a Catholic church province entrusted by the Holy See to the French Jesuits: the vicariate apostolic of *Southeastern Zhili*, renamed

---

❶ 沈锦标.造屋三知（活版）[M]. 上海:土山湾印书馆, 1902. 为什么只加了这本参考书呢？20 世纪 20 年代也出版了另外几本关于中国建筑的手册和论文。见文章: Chen Jianyu. Yingzao Fayuan（营造法原）: Two Editions by the Carpenters and the Architects in 1900s [A] // Austin Williams, Theodoros Dounas. Masterplanning the Future. Modernism: East, West & Across the World [C]. Suzhou: Transport Research Publications, 2012, 185-193.

❷ Shen Jinbiao. Zaowu Sanzhi [M] (Knowledge for Building Construction). Shanghai: T'ou-sè-wè Press, 1902. Why only this one additional reference? Several other Chinese construction handbooks and treatises were published in the 1920s. See: Chen Jianyu. Yingzao Fayuan: Two Editions by the Carpenters and the Architects in 1900s [A] // Austin Williams, TheodorosDounas. Masterplanning the Future. Modernism: East, West & Across the World [C]. Suzhou, 2012, 185-193.

❸ 关于献县和大名的教区，相关内容见第三章。 Henri-Joseph Leroy. En Chine au Tché-Ly S.E. Une histoire d'après les missionnaires [M]. Paris-Lille: Société de Saint-Augustin, Desclée De Brouwer, 1900; René Joüon. Atlas la Chine [M]. Shanghai-Zikawei: T'ou-sè-wè Press, 1930.

❹ Jean-Marie Planchet. Les missions de Chine et du Japon. 1925. Sixième année [M]. Beijing: Imprimerie des Lazaristes, 1925.

19世纪60年代，法国耶稣会在献县建了一家印刷厂，这家印刷厂后来成为一家重要的近代印刷厂（图1.4）。❷许多中国人在印刷厂工作，耶稣会兄弟负责指导和监督，在那里出版了不同领域的成千上万的书籍，有用中文写的，也有用西文写的。印刷事业不仅为传福音和教育服务，同时也传播了科学研究成果、教徒虔诚的照片等。❸

手册《传教士建造者：建议-方案》应该是集体工作的成果，所以才匿名。但是很显然的是，该手册与法国耶稣会及中国北方地区说法语的其他传教士都有关联。在手册里有数次都提到了大名主教座堂，这也是该手册里非常重要的一个建筑。因此我们来集中讨论

vicariate apostolic of Xianxian in 1924[See P25❸] (fig. 1.3). The handbook was published with the authorization (in Latin: *imprimatur*) of bishop Henri Lécroart (*Liu Qin-ming*), the vicar apostolic of Southeastern Zhili, and father Jean-Baptiste Debeauvais (*Bu Ming-sheng*), superior of the French mission of Xianxian, both French Jesuits.❶

In the 1860s, the Jesuits had installed a printing press in Xianxian that became an important modern printing office(fig. 1.4).❷ Many Chinese worked in the printing office under the supervision of Jesuit brothers, and produced thousands of books in all fields, in Chinese and Western languages. Printed matter was not only needed for evangelization and education, but also for diffusing scientific research, devotional pictures, etc.❹

The Missionary-Builder: Advice-Plans handbook thus results from collective work and is anonymously authored. It is, however, related to networks of French Jesuits and other French-speaking missionaries from Northern China. The great cathedral church of Daming is mentioned several times in the handbook

---

❶ About the vicariates apostolic of Xianxian and Daming, see: Part 3. Henri-Joseph Leroy. En Chine au Tché-Ly S.E. Une histoire d'après les missionnaires [M]. Paris-Lille, 1900; René Joüon. Atlas la Chine [M]. Shanghai, 1930.

❷ Vanves ASJ France, GMC 22/1, 57 and 22/4, 13. Also: Léon Goudallier. Une imprimerie en Chine [J]. Cosmos. Revue des sciences et de leurs applications, 1908, 57(9):524-527.

❸ 中国其他几家重要传教士印刷厂为：北京的Lazarists、上海土山湾的耶稣会印刷厂和香港（Nazareth）的法国传教士印刷厂。

❹ Other important missionary printing houses in China were the press of the Lazarists in Beijing, the press of the Jesuits in Shanghai (Tou-sè-wè), and the press of the French Mission in Hong Kong (Nazareth).

第一部分　手册的撰写背景
Part One: The Handbook in Context

参与大名天主堂建造的几位传教士，也基本能确定该手册的执笔人及三位贡献者。

and, as we will see, is the key building of the handbook. Therefore our research focused on the people involved in the construction of the church of Daming. We were able to identify the writer and three inspirers of the handbook.

图1.3　一本1900年出版图书的封面，耶稣会东南直隶使团。
fig. 1.3 Jesuit Southeastern Zhili mission, front page of a book from 1900. (© MEP library)

舶来与本土：1926年法国传教士所撰中国北方教堂营造手册的翻译和研究
Building Churches in Northern China. A 1926 Handbook in Context

图1.4 耶稣会献县（河北省）印刷厂，手册就是于1926年在此印刷的。
fig. 1.4 Jesuit press of Xianxian (Heibei province), where the handbook was printed in 1926. (© ASJ France, GMC)

神父雍居敬——手册的执笔人

Father Paul Jung, the Writer of the Handbook

基于对罗马耶稣会中央档案的研究，我们基本能确定手册执笔人的身份。每一年每个教区的主教都会向罗马耶稣会院长提交一份详细的报告。这些报告或年信都是用拉丁文写的。在1926—1927年由献县教区发来的年信里提到："新书（……）：神父雍，一卷：'传教士建造者'收集了关于怎么建造教堂、学校和住宅方面经验的信息。书的内容有66页，54幅图纸，涉及选址、气候、装饰、目前艺术家面临的最头疼的问题、材料及如何与包工头和工人打交道等方面的内容。"❶

Research in the central archives of the Jesuits in Rome revealed the identity of the handbook's author. Every year, the superior of every Jesuit province sends a detailed report to the Father General, the head of the Society of Jesus in Rome. These reports or annual letters (litterae annuae) are written in Latin. The annual letter of 1926-1927, sent by the mission of Xianxian, mentions: "New books (…): Father Jung, one volume: 'Le missionnaire constructeur' collects information based on experience about how to build churches, schools and residences. The

❶ "Novi Libri (…): Pater Jung uno in volumine: 'Le Missionnaire Constructeur' collegit praeceta quaedam experiential probate, de ratione struendi temple, scholas, domus quascumque. Textus habet 66 paginas, figuram tabulas 54; tractat de loco, de tempore, de genere ornamenti, quod nunc valde sollicitat artifices, de material, de tractatione

第一部分　手册的撰写背景
Part One: The Handbook in Context

雍神父（雍居敬）于 1863 年出生于法国，后来成为香槟省的一位传教士，1897 年被派往中国。❶ 1926 年，他是大名耶稣会和法文学校的院长（图 1.5）。❷ 雍神父不是一位建筑师，他是一名教师、管理者、传道者……，还是一位优秀的自行车手。❸ 耶稣会的神父都是有文化的人，他们传福音、传道、教学和做科研。❹ 雍神父在建造方面只有很有限的实践知识，但我们会谈到他是一位耶稣会兄弟的弟弟，而那位耶稣会兄

text has 66 pages, illustrates 54 figures; it deals with place, climate, the kind of ornament, what presently troubles artists very much, material, and the treatment of both contractors and workers." See P28❶

Paul Jung (*Yong Jujing*) was born in France in 1863, became a Jesuit of the province of Champagne, and was sent to China in 1897.❶ In 1926, he was the superior of the Jesuit community and college of Daming (fig. 1.5).❺ Thus, Father Jung was not an architect, but a teacher, an administrator, a preacher, and… an experimented bicycle rider. ❻ Jesuit fathers are intellectuals, involved with evangelization, preaching, education and scientific research. ❼ Father Jung had

────────────────────（接上页 continue）

tum mancipis tum opificium". Litterae annuae Provincia Campaniae 1926-1927, p. 425 (Rome, Archivum Romanorum Societatis Iesu — ARSI, Litterae annuae, n° 1505). The latter mentions a guide for "beginners missionary-bicycle-drives": Paul Jung. Robinson-Cycliste. Guide du missionnaire vélocipédiste débutant [M]. Xianxian, 1927.

❶ Born in Launstroff (Moselle) on 5 November 1863, entered the Jesuits in 1887, died "in China" on 23 April 1943.

❷ Jean-Marie Planchet. Les missions de Chine et du Japon. 1925. Sixième année. Beijing: Imprimerie des Lazaristes, 1925, 30. 耶稣会社区有八位神父：六位法国人（雍居敬、巴鸿勋、郎守仁、方济民、陶德民、白玉清）和两位中国人（Ming Shi-The、M. Tsoei）；还有四位修士：一位法国人（雷振声）、一位德国人（卫秉仁）和两位中国人（J. Ma 和 P. Yang）。

❸ 参考耶稣会的年度目录：Catalogus provinciae Campaniae Societas Iesu, 1987-1946, Reims: Compagnie de Jésus.

❹ 一位法国神父在 1918 年发表了一篇关于"怎么建造教堂"的文章：不是讲建筑的，而是讲怎么发展教徒的。Maurice Cannepin. Comment on bâtit une église (Esquisses jaunes) [J]. Les Missions catholiques, 1918,50 : 416-418, 429-431.

❺ Jean-Marie Planchet. Les missions de Chine et du Japon. 1925. Sixième année [M]. Beijing: Imprimerie des Lazaristes, 1925, 30. The Jesuit community was composed of eight fathers: six French (Paul Jung, Jules Bataille, Philippe Leurent, Nicolas Vagner, Charles Taranzano, Pierre Perard) and two Chinese (Ming Shi-the, M. Tsoei); four brothers: one French (Alphonse Litzler), one German (Frans Leineweber), and two Chinese (J. Ma and P. Yang).

❻ According to the annual catalogues of the Jesuits: Catalogusprovinciae Campaniae Societas Iesu, 1987-1946, Reims: Compagnie de Jésus.

❼ A French Jesuit father published in 1918 an article on "how to build a church": however, it is not about architecture, but about how to develop a congregation. Maurice Cannepin. Comment on bâtit une église (Esquisses jaunes) [J]. Les Missions catholiques, 1918, 50 :416-418, 429-431.

~ 29 ~

弟能画设计图纸和施工图纸。所以，该手册集合了从耶稣会兄弟或其他传教士建造者那里收集来的信息。这也可能就是雍神父不署名手册作者的原因。

我们对雍神父所知甚少。❶ 除了该手册，在西方传教士期刊上找到几篇署名为他的短文章，是讲大名传教区和法文学校的。❷ 这些文章写于1918—1919年间，其间大名天主堂正在建造，希望能得到西方赞助商的资金资助。在一篇由另外一位神父基于同样目的写的文章里提到雍神父负责找赞助。❸ 有意思的是，有一些文章不是雍神父自己写的但署名却是他。❹

limited practical building experience, but we will see that he was seconded by a Jesuit brother, who was able to draw plans and direct works. The handbook is therefore a compilation of information collected from Jesuit brothers or other missionary-builders. This could be the reason why father Jung did not claim the authorship of the handbook.

Little is known about father Jung. ❺ Except for the handbook, he only signed several short articles about the mission of Daming and the French School, in Western missionary journals.❷ The articles date from 1918-1919, the time the church of Daming was under construction, and solicit Western benefactors for money. In another article written with the same aim, another father of the community presents Paul Jung as the father responsible for the works.❸ It is interesting to notice that some articles, signed by father Jung, were not written by him. ❻

---

❶ 几乎找不到什么个人信息，也没有照片，只找到十封信：信102和信112都是寄到以下地址：Vanves ASJ France, GMC, coll. Desmarquest, 21; 信101,是写给教会官员的。在此要特别感谢 Barbara Baudry。

❷ Paul Jung. From the Mission-field of Taming-Fu [J]. The Far East, 1919, 6-7; 1919, 5-6 (Vanves, ASJ France, GMC, 21/1, 87-91); Paul Jung. Une grande ville chinoise qui s'ouvre à la foi [J]. Les missions catholiques, 1919, 198-200, 210-219, 223-225, 233-235, 248-249; Paul Jung. Fleurs du Céleste empire. Touchante histoire de Stanislas Wang [J]. Les missions catholiques, 1918, 317-319; Paul Jung. La joie dans la souffrance. Histoire d'une petite Chinoise malade [J]. Les missions catholiques, 1918, 534-535, 546-548. Other writings mentioned in: Robert Streit, Johannes Dindinger. Bibliotheca missionum, 13, Chinesische Missions literatur [M]. Rome-Freiburg-Vienna, 1959, 311-312.

❸ Pierre Mertens. Une grande ville chinoise qui s'ouvre à la foi. Épilogue: achèvement de la cathédrale de Taming-fou [J]. Les Missions Catholiques, 1920, 356-357, 367-369.

❹ 相关内容见雍神父于1919年7月13日写给法国传教会官员的一封信: "Vous savez sans doute, ou du moins vous avez deviné, que mon article sur Taming est tout entier du Père Mertens. Il a voulu avoir ma signature. C'est tout ce qu'il y a de moi" (Vanves ASJ France, GMC 101, Jung).

❺ Few personal data, no portrait, and about ten letters: Vanves, ASJ France, GMC 21 (letters 102 and 112) and GMC 101, letters to procurator. With special thanks to Barbara Baudry.

❻ In a lettre of Paul Jung to the Procurator of the French Mission, 13 July 1919: "Vous savez sans doute, ou du

第一部分　手册的撰写背景
Part One: The Handbook in Context

图1.5　大名（河北省），耶稣会神父在法语学校执教，1920年左右。
fig. 1.5 Daming (Heibei province), Jesuit father teaching at the French College, around 1920. (© ASJ France, GMC)

在大名天主堂现存的三口很重的铜钟上，我们找到了雍神父的名字，这实在是意外的发现（图 1.6）。基督教的钟都会有一个名字，这是非常神圣的，每口钟有一位教父和一位教母，名字用拉丁文刻在钟的外侧。这三口钟在法国安纳西铸造，之后海运到天津，接受了神父 Lécroart 的祈福，1921 年被挂在了大名天主堂里。雍神父作为学校校长是这口名为"Maria Celinia"的钟的教父。❶

Discovering Paul Jung's name on one of the three heavy bronze bells that are still hanging in the church tower of Daming was an unexpected source (fig. 1.6). Christian bells always have a name, are consecrated, and have a godfather and a godmother. It is customary to cast dedicatory inscriptions, often in Latin, on the outer side of the bells. The three bells of Daming were cast in Annecy (France), shipped to Tianjin, blessed by bishop Lécroart, and hung in the tower in 1921.

―――――――――――――――――（接上页 continue）

moins vous avez deviné, que mon article sur Taming est tout entier du Père Mertens. Il a voulu avoir ma signature. C'est tout ce qu'il y a de moi" (Vanves, ASJ France, GMC 101, Jung).

❶ "Maria Celinia"钟上刻的均为大写字母，具体内容如下: "anno domini mcmxxi benedicto papa xv regnante / ego maria celinia / in urbe taming baptizata fui / a.r.r.d.d. henrico lecroart s.j. episcopo anchialensi / ad matrem gratiae laudandam / paganosque ad lumen verae fidei vocandos: r.p. paulo jung s.j.; superiore collegii taming patrino fungente / matrina vero pia domina celinia devillers-lecroart". 另外两口钟被命名为"Maria Teresia" (1920) "'Margarita Maria" (1921), 在 1921 年 8 月 25 日的一封信里, 雍神父提到了这几口钟 (Vanves ASJ France, GMC 101, Jung).

Father Jung, as director of the college, was the godfather of the bell 'Maria Celinia'.❶

fig. 1.6 Daming, Father Paul Jung's name on a bell from 1921. (© THOC 2014)

1928 年，雍神父 65 岁，这年他离开大名，成为威县教区（河北省）主教，住在赵家庄。❷ 1931—1936 年期间，他担任天津工商学院的院长。之后，他在献县教区传教。雍神父在中国待了 46 年，于 1943 年 4 月 23 日去世，而他过世的消息却直到 1944 年才传到法国。没有人知道他具体的死亡地点和原因。

In 1928, father Jung, aged 65, left the college of Daming, became superior of the section of Weixian (Hebei province) and lived in the village of Zhaojiazhuang (Hebei province).❷ From 1931 to 1936, he was the prefect of studies at the Tianjin Commercial University. After, he served in parishes of the area of Xianxian. Father Jung lived for 46 years in China. He died 'in China' on 23 April 1943, but the news only reached France

---

❶ Full inscription on the bell 'Maria Celinia' (1921), in capitals: "anno domini mcmxxi benedicto papa xv regnante / ego maria celinia / in urbe taming baptizata fui / a.r.r.d.d. henrico lecroart s.j. episcopo anchialensi / ad matrem gratiae laudandam / paganosque ad lumen verae fidei vocandos: r.p. paulo jung s.j.; superiore collegii taming patrino fungente / matrina vero pia domina celinia devillers-lecroart". The other two bells were named 'Maria Teresia' (1920) and 'Margarita Maria' (1921). In a letter dated 25 August 1921, Paul Jung mentions the bells (Vanves, ASJ France, GMC 101, Jung).

❷ Jean-Marie Planchet. Les missions de Chine et du Japon. 1929. Huitième année [M]. Beijing, 1929, 35.

第一部分　手册的撰写背景
Part One: The Handbook in Context

in 1944. It is not known how and where he died.

梁神父——大名天主堂的设计师

有三处文献提到梁神父是大名天主堂的主要设计者，他把图纸给了雍神父。❶ 他不是手册作者，却是手册的一位主要贡献者。

梁神父不是一名耶稣会会士，他是巴黎外方传教会的一名牧师。❷ 他于1869年出生在法国，于1894年被派到中国，他在沈阳（辽宁省）和满洲的一些地方生活过。❸ 梁神父不是一名建筑师，他拥有法律学士学位，是一名杰出的社会学家。他撰写出版了几本讲述中国方言的书籍，还发明了中国方言的字母写法。❹ 作为一名年轻的传教士，他在满洲的一些村庄里建了学校、教堂、孤儿院和其他的宗教建筑。1900年爆发的义和团运动对很多教堂建筑造成

Father Henri Lamasse, the General Plans' Designer of Daming

Several sources mention that father Lamasse (*Liang*) designed the general plans of the church of Daming and gave them to father Jung.❶ He is thus not an author of the handbook, but he inspired it.

Father Henri Lamasse was not a Jesuit, but a priest of the French Foreign Missions of Paris.❷ He was born in France in 1869 and was sent to China in 1894, where he lived in Shenyang (Liaoning province) and other places in Manchuria.❸ Father Lamasse was not an architect, but held a university degree in Law and was a brilliant sinologist. He authored several reference works on Chinese dialects and invented a method of 'inter-dialectic Romanization' for or the alphabetical writing of Chinese dialects.❺ As a young missionary, he built

---

❶ Paul Jung. Une grande ville chinoise qui s'ouvre à la foi [J]. Les missions catholiques, 1919, 235; Gerbes chinoises. Les Jésuites dans la mission de Sien-Hsien depuis 1856 [M]. Lille, 1934, 65; "In memoriam. Le Frère Alphonse Litzler (1862-1948)", Chine Madagascar, Missions des Jésuites français du nord et de l'est, 1949, 13.

❷ Marcel Launay, Gérard Moussay (eds). Les Missions étrangères. Trois siècles et demi d'histoire et d'aventure en Asie [C]. Paris, 2008.

❸ 于1869年4月15日生于阿尔萨斯区(Alsace)的斯特拉斯堡市（Strasbourg），1894年成为牧师，于1952年7月19日在香港过世。 http://archives.mepasie.org/notices/notices-biographiques/lamasse (2015年11月15日查阅)。

❹ 最著名的作品是: Henri Lamasse. 新国文 Sin Kouo Wen ou nouveau manuel de langue chinoise écrite, traduit et expliqué en français et romanisé selon les principaux dialects, Hong Kong, 1920 (several later editions).

❺ His most famous work is: Henri Lamasse. 新国文 Sin Kouo Wen ou nouveau manuel de langue chinoise écrite, traduit et expliqué en français et romanisé selon les principaux dialects, Hong Kong, 1920 (several editions).

了破坏，梁神父因此有机会把他的聪明才智用到一些大型建筑的建造中，如沈阳主教座堂（图 1.7）和吉林主教座堂（建于 1917—1920 年）（图 3.18）。他还为直隶东南教区的耶稣会设计了大名天主堂和开州（河北省濮阳县）教堂。据说他也设计了满洲地区的几座小一点的教堂。除此之外，梁神父对自然科学也充满兴趣：他是一名出色的养蜂人，他在满洲种葡萄，撰写文章介绍葡萄文化，还把瑞士山羊带到满洲进行繁育，他还生产奶酪。他也是第一位在满洲拍电影的欧洲人。

1949 年，梁神父被驱逐出中国大陆，搬到了伯大尼——香港一位传教士的住宅，1952 年他在那里去世。他在中国生活了 58 年。

schools, churches, orphanages and other parish buildings in villages in Manchuria. The important destructions by the Boxers in 1900, gave him the opportunity to apply his talent to the construction of larger buildings such as the cathedral of Shenyang (Liaoning province) (fig. 1.7), and the cathedral of Jilin (Jilin province), which were built in the years between 1917 and 1920 (fig. 3.18). For the Jesuits of Southeastern Zhili, he designed the churches of Daming and Kaizhou (Puyang county, Henan province). He is said to have designed several other smaller churches in Manchuria. Furthermore, father Lamasse was also interested in natural sciences: he was a great beekeeper, acclimatized grape in Manchuria, wrote a treatise on the culture of the grape, brought goats from Switzerland to Manchuria, and manufactured cheese. He was the first European to make movies in Manchuria.

In 1949, father Lamasse was expelled from China Mainland and moved to Bethany, the house of the French Mission in Hong Kong. There he passed away in 1952, after having spent 58 years of his life in China.

第一部分　手册的撰写背景
Part One: The Handbook in Context

图1.7　沈阳（辽宁省），法国巴黎外方传教会主教座堂，梁神父建，1910年。
fig. 1.7 Shenyang (Liaoning province), cathedral of the French Foreign Mission of Paris, built by fahter Lamasse, 1910. (© MEP archives)

雷振声修士——
从技术图纸到实际施工

Brother Alphonse Litzler, Technical Drawings and Practical Experience

　　如果是梁神父做了大名天主堂的总体设计方案，雍神父对建造工程进行了监工，还需要有一个当地的建筑师来负责所有的施工图纸并且每天都在施工现场和中国包工头、工匠一起讨论和解决问题。在一篇手稿中有提到这项重要的工作后来被委派给雷振声修士，在

　　If father Lamasse designed the general plans of the church of Daming and father Jung supervised the construction works, a local architect was needed to make all the detailed plans, to visit the construction site every day, and to deal with the Chinese masters and workers.

他的讣告里也证实了这点。❶

雷振声修士于 1862 年出生在法国，他学了三年的木工后于 1885 年加入了耶稣会。❷ 雷振声修士不是一名神父，也不是一名牧师，只是一名"修士助理"，协助耶稣会房子建造材料方面的事宜。1902 年他 40 岁的时候被派往中国，在这之前他在法国里尔、布洛涅和迪尤等地的院校里做木工。义和团运动之后，中国急需一批具有手艺的传教士建造者。大名及其周边地区的耶稣会建筑大多数都是雷振声修士建造的（图1.8）。他负责设计方案并委派他的中国助手进行实施。大名政府还委派他做大名城内的污水处理系统，因为他精通各类精密水准和测量工具的使用。

在他的讣告里提到大名天主堂和开州教堂是他负责的两个主要项目，这两个项目的设计方案都是梁神父做的。雷振声修士负责所有的施工图纸、建筑材料的选择和在现场指导工匠。他

This essential coordinating role was entrusted to brother Alphonse Litzler, as mentioned in one manuscript note and confirmed in his obituary.❶

Alphonse Litzler (*Lei Zhen-sheng*), was born in France in 1862 and had studied joinery for three years before entering the Society of Jesus in 1885.❸ Alphonse Litzler was not father, thus not a priest, but was a 'brother coadjutor', helping with the material tasks of the Jesuit houses. He worked as joiner in the colleges of Lille, Boulogne and Thieu before being sent to China in 1902, at the age of 40. After the Boxer Upraising, there was an urgent need of skilled missionary-builders in China. Brother Litzler built most Jesuit buildings in Daming and its surroundings (fig. 1.8). He designed the plans and entrusted the execution to his Chinese auxiliaries. The local government of Daming commissioned him to create the city sewage system, because he was able to use precise leveling and measuring tools.

His obituary mentions that his two major works were the churches of Daming and Kaizhou, both designed by father Lamasse. Brother Litzer made all the detail and technical drawings, chose the building materials, and directed the

---

❶ Paul Bornet. Mission de Chine. Le Tche-Li S.E., 1857-1928. Mission de Sienhsien. La troisième étape de 25 ans. 1907-1932, typed manuscript of an unpublished book, p. 15 (Vanves, ASJ France, GMC 7); "In memoriam. Le Frère Alphonse Litzler (1862-1948)", Chine Madagascar, Missions des Jésuites français du nord et de l'est, 17, July 1949, p. 13-14.

❷ 于1862年5月21日生于阿尔萨斯区(Alsace)的尼费尔恩市（Niefern）。

❸ Born in Niefern (Alsace) on 21 May 1862.

也设计并建造了河间的教堂和学校，并负责献县耶稣会工坊的工作。

works. He also designed and built the church and the college of Hejian (Hokien), and directed the Jesuits' workshops of Xianxian.

图1.8 大名，耶稣会圣约瑟礼拜堂，修士雷振生在1920年左右建造。
fig. 1.8 Daming, St Joseph chapel of the Jesuits, built by brother Litzler around 1920. (© ASJ France, GMC)

1948 年的圣诞节，雷振声修士在献县过世，终年 86 岁。他在献县和大名生活了 46 年，其间，他建造和维护了当地耶稣会的多处房产。献县和大名这两个地方正是该手册的写作地和出版地。在大名，他在梁神父手下工作了十几年。因此，我们有理由相信手册里

Brother Litzler died in Xianxian on Christmas day 1948, at the age of 86. He had built and maintained Jesuit buildings during 46 years in the area of Daming and Xianxian, which is precisely the area where the handbook comes from. In Daming, he worked during more than one decade under the authority of father

大部分技术性的图纸是出自雷振声修士的手，也许他还画了教堂的模型图以便于出版。

Jung. For all these reasons, we think that brother Litzler has drawn most of the technical drawings of the handbook, and perhaps has redrawn the model plans of churches at a scale that allowed publication.

和羹柏神父——
手册的一位供稿人

手册《传教士建造者：建议-方案》中提到"一位在中国居住了多年的传教士，他具有建筑方面的知识，有着在这个国家的建造经验，他负责建造的许多教堂既具有完美的当地适应性，又具备真正的美感，这些都使得他具有这方面的发言权并且能明确清晰地表达出来"。❶ 在 1926 年，和羹柏神父是一位唯一有在中国北方建造教堂经验的传教士建造者。当年他 68 岁，他不是法国耶稣会会士，而是北京教区的一位比利时牧师。他 1885 年来到中国，建了多处哥特式教堂，还建了多处学校、传教士住宅、孤儿院等。❷

和羹柏神父于 1856 年出生在比利时根特市附近的一个地方，他在加入圣

Father Alphonse De Moerloose, the Handbook's Inspirer

The handbook refers to *"a missionary who lived a long time in China, whose architectural knowledge, experience of the country, perfect adaptation, as well as the real beauty of the numerous churches he built, authorize him to have an opinion and to formulate it."*❶ In 1926, father Alphonse De Moerloose (*He Gengbai*) was the only missionary-architect who had such experience in Northern China. This missionary, aged 68 at the time, was not a French Jesuit but a Belgian priest from the diocese of Beijing, who was in China since 1885 and had built many Gothic style churches as well as schools, residences for missionaries, orphanages etc.❷

Alphonse De Moerloose was born near Ghent (Belgium) in 1856 and had studied architecture for five years at the

---

❶ Le Missionnaire Constructeur [M]. Xianxian, 1926, 7.

❷ Joseph Van Hecken. Alphonse Frédéric De Moerloose C.I.C.M. (1858-1932) et son œuvre d'architecte en Chine [A] // Neue Zeitschrift für Missions wissens chaft / Nouvelle Revue de science missionnaire, 24(3): 161-178; Luo Wei. Transmission and Transformation of European Church Types in China: The Churches of the Scheut Missions beyond the Great Wall, 1865-1955 [D]. University of Leuven, Faculty of Engineering: Architecture, 2013, 120-195; Thomas Coomans. Sint-Lucasneogotiek in Noord-China: Alphonse De Moerloose, missionaris en architect [J]. M&L. Monumenten, landschappen en archeologie, 2013, 32: 6-33.

母圣心会之前在当地著名的根特圣鲁克学校学了五年建筑。比利时圣母圣心会也称"斯格脱神父"会，创建于1862年，计划在中国北方进行传教活动。❶ 1885 年和羹柏神父被派往甘肃，1899年就在义和团运动发生前，他搬到了河北省西湾子镇（图 1.9）。自此，他开始了在中国北方的大量建造活动，他为比利时圣母圣心会、法国遣使会、法国耶稣会和熙笃会建了很多大教堂。1910年，和羹柏神父离开了比利时圣母圣心会，成为北京教区的一名牧师，并在河北省杨家坪的熙笃会修道院附近成立了自己的工作室。他的作品有一些保存了下来，其中最重要的是于1903—1906年期间建成的河北省宣化市的主教座堂（图1.10）。❷

famous St Luke's School in Ghent before entering the Congregation of the Immaculate Heart of Mary. This Belgian missionary congregation, also called 'Scheut Fathers' was founded in 1862 with the aim of evangelizing Mongolia.❶ Father De Moerloose was sent to Gansu in 1885 and moved to the vicariate apostolic of Xiwanzi (Hebei province) in 1899, just before the Boxer Upraising (fig. 1.9). From there, he developed a considerable building activity in Northern China and built great churches for the Belgian Scheut Fathers, the French Lazarists, the Jesuits and the Trappists. In 1910 father De Moerloose left the Congregation of the Immaculate Heart of Mary, became a priest of the diocese of Beijing, and established his workshop near the Trappist abbey of Yangjiaping (Hebei Province). Only a few of his works remain, amongst which the most important is the Gothic cathedral of Xuanhua (Hebei province) built in 1903-1906(fig. 1.10).❷

---

❶ Daniël Verhelst, Nestor Pycke (eds). C.I.C.M. Missionaries, Past and Present 1862-1987 (Verbistiana, 4) [C]. Leuven, 1995.

❷ Thomas Coomans, Luo Wei. Exporting Flemish Gothic Architecture to China: Meaning and Context of the Churches of Shebiya (Inner Mongolia) and Xuanhua (Hebei) built by Missionary-Architect Alphonse De Moerloose in 1903-1906 [J]. Relicta. Heritage Research in Flanders, 2012, 9, 219-262.

舶来与本土：1926 年法国传教士所撰中国北方教堂营造手册的翻译和研究
Building Churches in Northern China. A 1926 Handbook in Context

图1.9 西湾子（河北省），比利时圣母圣心会圣米厄尔及圣本笃礼拜堂，和羹柏神父建造，1930年。

fig. 1.9 Xiwanzi (Hebei province), St Michael and St Benedict chapel of the Belgian Scheut Fathers, built by father De Moerloose, 1903. (© KADOC, CICM archives)

和羹柏神父所受的教育是基于奥古斯都·普金真实原则基础上的天主教哥特式教育。他在中国北方建的教堂完全符合普金的建筑标准，也完全符合第一次世界大战前提倡欧洲中心主义的传教士心里的理想样式。❶

The artistic education, that De Moerloose had received, was strongly Catholic, Gothic and based on the true principles of Augustus W. N. Pugin. The churches he built in China conform to Pugin's architectural canon fit perfectly in the Eurocentric missionary model prior to World War I.❶

---

❶ Thomas Coomans. Pugin Worldwide: From Les Vrais Principes and the Belgian St Luke Schools to Northern China and Inner Mongolia [A] // Timothy Brittain-Catlin, Jan De Maeyer, Martin Bressani. A.W.N. Pugin's Gothic Revival: The International Style (KADOC Artes) [C]. Leuven: Leuven University Press, 2016 (forthcoming).

第一部分　手册的撰写背景
Part One: The Handbook in Context

在 20 世纪 20 年代，和羹柏神父面对来自两个方面的批评。第一个方面的批评来自传教士自己，他们觉得哥特式教堂虽然很漂亮，但是却不适合中国北方的气候：由于大窗户和很高的木制拱顶，教堂内冬天寒冷夏天酷热。❶第二个方面的批评来自刚恒毅主教，他是 1922—1933 年期间的驻华宗座代表，他提倡教堂在中国的本土化，鼓励建造有独特中国基督教样式的建筑。❷作为一位中式建筑的倡议者，他不可能喜欢哥特式建筑。

在 1925 年的一封信里，和羹柏神父写道："我经常被人问有没有一本指导传教士实际建造工程的手册，但确实没有此类手册。主教和传教士们建议我应该写一本手册。我有计划要整理以下这些方面的内容：①一卷关于已经建成的教堂、祭台和其他教堂室内陈设品；②介绍怎么建造、怎么选建筑材料等，从项目最开始到结束。但是由于上面我提到的一些原因，我还不太敢开始

In the 1920s, however, he was criticized from two different sides. The first criticism came from the missionaries themselves, who found Gothic churches beautiful but not adapted at all to the rigorous climate of Mongolia: because of the large windows and the high wooden vaults, people were freezing in the winter and suffocating in the summer.❶ The second criticism came from archbishop Celso Costantini (*Gang Hengyi*), the apostolic delegate to China from 1922 to 1933, who implemented inculturation in China and encouraged the development of a distinctive Chinese-Christian architecture.❷ As a promoter of Chinese-inspired architecture, he could not be an adept of Gothic.

In a letter from 1925, father De Moerloose wrote: "I was often asked if there existed a practical construction handbook for missionaries, but there is nothing. Bishops and missionaries asked me to do something like that. I had the project to do this: 1° an album with existing churches, altars and other church furniture; 2° a description about how to build, with which materials, etc. from the beginning to the end of the works. But I

---

❶ Jozef Nuyts. En tournée à travers le vicariat [J]. Missions de Scheut, 1938, 218-219.
❷ Thomas Coomans. The "Sino-Christian Style": A Major Tool for Architectural Indigenisation [A] // ZhengYangwen. Sinicising Christianity [C]. Leiden-Boston: Brill, 2016 (forthcoming); Thomas Coomans. Indigenizing Catholic Architecture in China: From Western-Gothic to Sino-Christian Design, 1900-1940 [A] // Cindy Yik-yi Chu. Catholicism in China, 1900-Present. The Development of the Chinese Church [C]. New York: Palgrave and Macmillan, 2014, 125-144.

这样的工作。"❶ 和羹柏神父这里所说的原因是他感到被刚恒毅主教冒犯了，但他不想引起公开的冲突。他想写一本手册的想法被教廷的新主张给打乱了。

do not dare begin because of the reason I mentioned."❷ The reason was that De Moerloose felt offended by the criticisms from archbishop Costantini, but refused an open clash. His intention to write a handbook was deeply perturbed by the new line of the Holy See.

图1.10 宣化（河北省），法国遣使会教堂，和羹柏神父建造，1903—1906年。
fig. 1.10 Xuanhua (Hebei province), church of the French missionaries of the Congregation of the Mission, built by father De Moerloose, 1903-1906. (© THOC 2011)

---

❶ 见和羹柏神父 A. De Moerloose 于 1925 年 2 月 19 日写给 K. Van de Vyvere 的信 (Leuven, KADOC, Archives CICM, T.I.a.14.3.2).

❷ Letter of A. De Moerloose to K. Van de Vyvere, Yangjiaping, 19 February 1925 (Leuven, KADOC, Archives CICM, T.I.a.14.3.2).

第一部分　手册的撰写背景
Part One: The Handbook in Context

据此可以推测，和羹柏神父的耶稣会朋友从他那里获取了相关信息，整理并匿名发表了。和羹柏神父与天津和上海的耶稣会有密切联系，但没有找到证据证明他和雍神父有直接联系。1920—1923 年期间，就在该手册出版之前，他为上海耶稣会设计了佘山进教之佑圣母大殿（图 1.11）。一位居住在上海的西班牙神父 François-Xavier Diniz 画了施工图纸并于 1925—1935 年期间监督了建造工程的施工。

和羹柏神父在中国呆了 44 年，后来于 1929 年回到比利时。他于 1932 年 3 月 27 日在安特卫普过世。值得一提的是，19 世纪 80 年代早期他发表了一些关于中国建造技术和工艺的小文章，[1] 我们在档案中也找到一本笔记本，上面有甘肃省当地的传统农耕工具的素描，还记录了其中文名。[2] 这些都显示了他对中国建筑和技术很感兴趣。

This allows us to formulate the hypothesis that his Jesuit friends collected information from him and published it anonymously. De Moerloose had close connections with the Jesuits in China, both in Tianjin and Shanghai, but we have no proof of direct contacts with father Jung. In the years between 1920 and 1923, just before the publication of the handbook, he designed the great basilica of Our Lady of Zose (Sheshan) for the Jesuits of Shanghai (fig. 1.11). Father François-Xavier Diniz, a Portuguese Jesuit from Shanghai, designed the detail plans and supervised the construction from 1925 to 1935.

Father De Moerloose returned to Belgium in 1929, after having lived in China for 44 years. He died in Schilde (Antwerp) on 27 March 1932. It is worth mentioning that he published in the early 1890s a couple of short articles on Chinese building techniques and craftsmanship,[1] and that the archives contain a notebook of his, with sketches of traditional agricultural tools from the Gansu province and Chinese terminology.[2] These sources confirm his interest for Chinese construction and techniques.

---

[1] Alphonse De Moerloose. Construction, arts et métiers, au Kan-sou et en Chine [J]. Missions en Chine et au Congo, 1891, 34 :532-538; Alphonse De Moerloose. Arts et métiers en Chine [J]. Missions en Chine et au Congo, 1892, 37 :3-8.

[2] Françoise Aubin. Un cahier de vocabulaire technique du R.P. A. De Moerloose CICM, missionnaire de Scheut (Gansu septentrional, fin du XIXe siècle) [J]. Cahiers de linguistique. Asie orientale, 1983, 2 :103-117. Original (Leuven, KADOC, Archives CICM, Z.II.h.8).

舶来与本土：1926 年法国传教士所撰中国北方教堂营造手册的翻译和研究
Building Churches in Northern China. A 1926 Handbook in Context

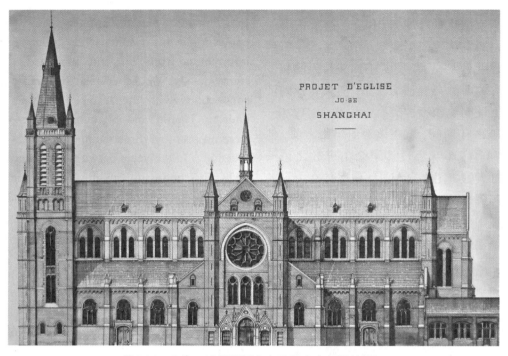

图1.11 上海，法国耶稣会自助的佘山朝圣教堂，和羹柏神父设计的第一个项目，1920—1923年。
fig. 1.11 Shanghai, Sheshan pilgrimage church promoted by the French Jesuits, first project by father De Moerloose, 1920-1923. (© KADOC, CICM)

手册的读者群
The Handbook's Readership

在《传教士建造者：建议-方案》简短的序言中说到了该手册的写作目的："该册子不是为大教堂的建造者而写的，也不是为了介绍中国艺术如何为基督教教堂做出一些适应性变化。每一个传教士都曾有义务建造教堂，即使只

The brief preface of the handbook explains who are the addressees: "*This brochure is not made for cathedral builders. This is not even an essay on the famous adaptation of Chinese art in our Christian churches. Every missionary may need to becomea builder, be it of a*

是建一所只有三间的乡村学校。有些修士，他们是谦逊的建筑师，知道怎么建房子，他们愿意分享他们的经验。……我们的目的是要为建造牢固、安全、不太贵、不太丑的教堂提供参考：用极少的钱建造一个漂亮的教堂。……最后介绍了一些具体例子，提供了一些比较容易建造的教堂的图纸和外形参考，希望能给人一些灵感和启发。当然所给的建议和说明都充分考虑了中国北方的实际条件。"

village school of only three bays. Some brothers, who aremodest architects and know about building works, want to share their experience. (...) Our aim is to help others build constructions solidly, safely, not too expensively, nor too unattractively: i. e. a beautiful church for little money. (...)Finally, adding example to precept, a pocketcontains figures and some plans of churches which are easy to execute. We hope they will be inspiring. It goes without saying that the advice and remarks concern the conditions of Northern China."

图1.12 朝阳（辽宁省），圣母圣心会神父及其教区教堂，约1935年。
fig. 1.12 Chaoyang (Liaoning province), Scheut father and his parish church around 1935. (© FVI, BR pictures)

该手册的潜在读者是那些在中国北方传教且说法语的传教士，包括满洲和内蒙古区域在内,特别是河北和山西两省内的基督教化区域的传教士（图1.12）。曾有一篇关于该手册的评论文章发表在 1927 年出版的北京天主教公告上。❶ 我们现在还无法考证当年该手册出版了多少，但可以确定的是其数量必定很有限，因为它的内容太专而且还是用法语写的。

关于 1935 年重印版我们更是知之甚少。谁还会在 1935 年的中国北方建教堂呢？自 1931 年开始满洲就被日本占领了，中国的其他地区也深陷世界经济危机、不稳定的局势和政治冲突的泥潭。1935 年的重印版，内容和页面配置都与 1926 年出版的一模一样，只有在最后一页的参考文献上加了一本传记。我们猜想 1935 年的重印版就是把 1926 年的那版重新印刷一次。印刷厂为了重印，就只是改了页数并在最后一页加了传记参考书的参考文献。❷ 该手册采用的是中国式的线装，这使得重印装订出版并不是件困难的事情。

The potential readers were thus French-speaking missionaries in Northern China, Manchuria and Inner-Mongolia especially in the Christianized areas of Hebei and Shanxi (fig. 1.12).The book was reviewed once, in the Catholic Bulletin of Peking of 1927.❶ We have no indications about how many copies of the handbook were printed, but think that the edition was limited due to the specialized topic and the French language.

The reprint from 1935 is even more of a mystery. Who was building new churches in Northern China in 1935? Manchuria was occupied by the Japanese since 1931, and the rest of the country was deeply hit by the world economic crisis, insecurity, and political tensions. The content and the layout of the handbook are identical to that of the 1926 edition. Only one bibliographical reference was added at the last page. We think that the reprint of 1935 just recycles the unsold stock of the first printing from 1926. It was usual practice for printing houses to just change the title page of a book and the last page with bibliographical references, in order to try to relaunch a publication. ❸ The handbook's Chinese-style binding with one string easily allowed for such an update.

---

❶ Aux missionnaires constructeurs [J]. Bulletin catholique de Pékin, 1927, 133-137. A first announcement was published in: Bulletin catholique de Pékin, 161, 1927, p. 109.

❷ 重印表示首印版很成功，也没有库存。另外，新的印刷日期也会让书更有吸引力。

❸ A reprint suggests that the first printing was a success and that the stocks had run out. Furthermore, a recent date makes a book more attractive.

第一部分 手册的撰写背景
Part One: The Handbook in Context

直至今天只发现了两本该手册，都是在海外发现的，这或许也说明了当年的发行是有限的，大部分保存于传教士家里的手册也没能在战争期间得以幸存。

The fact that until recently, only two copies of the book were found, both of them outside of China, suggests that the distribution was limited and that most of the copies that were kept in the missionaries' houses, did not survive the constant plundering during the wars.

手册的内容
The Handbook's Content

《传教士建造者：建议-方案》内容分8章，分别是：①选址；②建筑布局和朝向的选择；③设计方案、工程预算和施工合同；④建筑材料；⑤砖石工程；⑥屋顶（中式和西式）；⑦屋面装饰；⑧建筑细节的处理（涉及建筑结构的各处细节）。另外该手册里还包括插图和54张图纸。手册中也提到了一些中国专业术语，都是根据当年的地方说法而用罗马拼音写成的法语，其中一些是地方方言的音标拼音。本书中在相关处，给出了这些术语的汉语拼音和简体汉字。

The Missionary-Builder: Advice-Plans, is structured in eight chapters: 1. choice of the location; 2. choice of the buildings' layout; 3. plans, estimates, contracts; 4. building materials; 5. masonry; 6. roofs, both Chinese and Western; 7. roof ornaments; 8. miscellaneous, about various construction elements. 54 illustrated plates follow the text. The handbook often mentions Chinese terms, but they are written in the Romanized French form that was used in the 1920s; some are phonetic transcriptions of the local dialect. When relevant, we give a transcription of these terms both in pinyin and in simplified characters.

选址
Choice of the Location

第 1 章内容就如何为传教区和教

The first chapter deals with the

舶来与本土：1926 年法国传教士所撰中国北方教堂营造手册的翻译和研究
Building Churches in Northern China. A 1926 Handbook in Context

堂选址展开讨论。谨慎是首要原则：传教士不应该仓促地就开始建造活动，必须要保证得到新信徒的支持。比较好的做法是：先租房和发展信徒。如果买地的话，最好一开始就至少买 3 亩[1]或 0.185 公顷。不然等信徒多了，传教区发展起来后，附近居民就会抬高他们土地的价格。

该手册中提到选址时要尽可能避免以下这些情况："①不要购买村庄外的土地。它们有可能会很便宜，但那是因为没有人愿意住在那里。那边没有邻舍，有时是一边没有，经常是三边都没有邻舍。妇女和少女，尤其是刚成为教徒的妇女和少女，都会不太愿意长途跋涉去教堂祈祷。②如果该村庄有集市，教堂选址不要选在市场中心。让新教徒穿过熙熙攘攘的人群去教堂祷告，这极为不好，尤其是对妇女。集市的喧扰会影响做弥撒，尤其会影响传道。③如果街道是东西走向，选址尽可能选在一条街道的北侧。这样的话，街道北侧可以建重要建筑，包括教堂、牧师寝地和学校。主要入口、马厩、厨房可以建在南边。④千万不要选址选在低洼易发洪水的地方。如果选址靠近河边，必须要有该地夏天水位的详细信

choice of an appropriate location for establishing a mission and its church. Prudence is the first advice given: the missionary should not start building too fast and should first ensure that the converts will support the effort. It is wiser to first rent an existing building and to develop a solid congregation. When buying property, the size should be at least about 3 '*mu*' or about 0.185 hectares from the beginning. If not, the neighbors will increase the price of their land as the congregation develops.

The handbook advises against certain locations: "a) *Never buy outside the village, where land is really cheaper precisely because nobody wants to settle there. There will be no neighbors, at least on one side, often on three sides. Women and girls, especially recent converts, will not like to take long walks to go to church. b) Never buy near a market, if there is one in the village. Making one's way through milling crowds will be highly inconvenient for women going to prayer. Shouting from outside will often disturb the priests during the mass, especially when preaching. c) Preferably buy ground on the northern side of the street and settle the important buildings on the northern side of the plot: church, presbytery and schools. Build the main gate, the stables and the kitchen on the southern side. d) Avoid grounds subject to floods. Collect accurate information*

---

[1] 1 亩=666.67 平方米。

息。在河边建教区建筑的话，首先做好防洪防涝，如果这类防洪防涝工程需要基督徒自己来做的话，这种情况尤为值得注意。"接着该手册还非常明确地介绍了应该怎么夯土，怎么做好一个牢固的岸堤。

建筑布局和朝向的选择

第 2 章介绍了如何根据中国北方的气候、房子功能和好的朝向来安排院落空间内的不同建筑（图 1.13 和图 1.16）。作者建议"不要学欧洲租界，也不要学美国新教徒的做法，而应该忠实模仿当地的做法"。"最好的房子，即所谓的'北屋'或'堂屋'，它的门窗都是朝南的。可以在屋子北墙开窗，这样使得空气更流通，能让北屋在夏天更宜居。……西屋朝东：冬天自日出就会有阳光照进来，夏天自 11 点起又能避免阳光直射。东屋朝西，朝向最差，在夏天不适宜住人。所以千万不要把门窗朝西的东屋用作小教堂：或许在早上做弥撒可以，但在夏天，过了正午，就不能在这里做忏悔和救赎了。把礼拜小教堂放在朝北的南屋会好一点。在冬天如果有很多虔诚教徒来做礼拜，南屋也不至于寒冷彻骨。传教士的住房是不能放在南屋的，除非那里有一个很好的炕，尽可能把传教士的住房安排在北屋。为了保证屋子的干燥和卫生，室内地面应该

about water levels in the summertime. If the place is near a river, the only solution is banking up, especially if Christians offer to do this work. " This is followed by precise indications about how to ram earth and make a solid embankment.

Layout and Orientation of the Buildings

The second chapter defines how to arrange buildings within a rectangular compound according to orientation, functions and the Northern-Chinese climate (fig. 1.13 and fig. 1.16). The introduction advises *"not to follow the model of European concessions or of the American Protestants, but faithfully imitate the natives."* "*The best orientation is a building with doors and windows facing south, for the northern room or building and the main room or building. It is always possible to open holes in the northern back wall and get breezes during the hot summers. (...)The western house faces east: during the winter it gets light from sunrise on; during the summer it is protected from the sun from 11 am on. The eastern house, which opens to the west, has the most unfavorable orientation and is inappropriate during the summer. Never use an eastern house with doors and windows opening to the west as a small chapel: even if*

高于室外地面两英尺❶左右，或者门前踏步至少要有三步。"梁下至地面的距离不宜超过 11 英尺/3.35 米。公共教室和客厅其梁下至地面距离可至 13 或 14 英尺。

该手册也提及了那些应气候而做的相应调整及应中国风水要求按照传教所做的空间处理。对于两边都有窗户的教堂，手册里建议是南北朝向，在东西两侧安窗户，门朝南。这样做的实际原因是为了避免冬天的北风和夏天的热气：在下雨的季节，南北朝向的教堂，屋顶漏雨的可能要小。然而，传统的东西向教堂如果是小教堂又能增添美感的话也是可以考虑的：这种情况下北侧的窗户可以做得更小一号或可以减少窗户数量。因此，中国北方的教堂

celebrating an early morning mass in such a chapel would be possible, confessing in the afternoon and holding evening prayer would not. Placing the chapel in the southern house, which opens to the north, would be more convenient. During the winter, it would not be too cold if the community is fervent enough, that is to say if it is numerous. The missionary's room should not be located in the southern house, except if it is equipped with a good bedstove. The best location for his room is in the northern house. Never forget to raise the main house about two feet (at least three steps) above the level of the courtyard in order to keep it dry and healthy. " Ceilings of domestic spaces should not be higher than 11 feet / 3.35 m under the beam; ceilings of classrooms and reception halls 13 or 14 feet.

The handbook only pays attention to rational reasons related to climate, and ignores the Chinese traditional rules of space arrangement based on the flow of energy and a harmony with the environment (*fengshui*). For churches with windows on both sides, the handbook recommends a north-south orientation, with windows on the eastern and western sides, and the door facing south. Again, the reason seems to be purely practical and aims to avoid winter winds and summer heat: the roofs of north-south oriented churches leak less during the

---

❶ 1 英尺，当时相当于 33 厘米，见后文手册里"木料"部分的相关内容。

只在进深方向上有入口和窗户，朝向都是南（图 1.14，图 1.30 和图 1.53）。在中国南方的气候条件下不需要特别的朝向，教堂各侧都可以开大窗户（图 1.15）。

rainy season. Nevertheless, the traditional Christian east-west orientation of churches is acceptable for small churches with little aesthetic character: the windows at the northern side could be smaller or less numerous. So, churches in Northern China could have their main entrance and windows only at their long side, when oriented to the south (fig. 1.14). In Southern China, the climate does not require a specific orientation and large windows can be opened at all sides (fig. 1.15).

图1.13 如皋（江苏省），被罩棚覆盖的耶稣会院落，约1920年。
fig. 1.13 Rugao (Jiangsu province), courtyard of the Jesuit mission covered with a tent, around 1920. (© ASJ France, FCh)

舶来与本土：1926年法国传教士所撰中国北方教堂营造手册的翻译和研究
Building Churches in Northern China. A 1926 Handbook in Context

图1.14 张官屯（河北省）及壕赖山（内蒙古），1936年及1930年建造的圣母圣心会教堂。
fig. 1.14 Zhangguantun (Hebei province) and Haolaishan (Inner Mongolia), Scheut missions churches built in 1936 and 1930. (© FVI, BR picture; © THOC 2011)

第一部分　手册的撰写背景
Part One: The Handbook in Context

图1.15　水东（安徽省）及扬州（江苏省），
19世纪80年代及1864—1873年建造的耶稣会教堂。
fig. 1.15 Shuidong (Anhui province) and Yangzhou (Jiangsu province),
Jesuit churches built in the 1880s and 1864-1873. (© THOC 2015 and 2014)

舶来与本土：1926年法国传教士所撰中国北方教堂营造手册的翻译和研究
Building Churches in Northern China. A 1926 Handbook in Context

图1.16 曲周（河北省），被耶稣会传教士使用的传统院落南侧，约1910年。
fig. 1.16 Quzhou (Hebei province), traditional courtyard house used by Jesuit missionaries, seen from the south, around 1910. (© ASJ France, GMC)

设计方案、工程预算和施工合同

建议做一张有每栋建筑位置的总平面图。每一栋建筑，特别是教堂，应该有平面、剖面和细节处理的图纸，"这样可以避免在建造过程中发生意外和费钱的错误"。大教堂的设计方案必须得交给一位建筑师或优秀建造者来设计，之后主教可参照天主教法典给予许可。好的设计方案有助于做出好的工程预算。"做工程预算时，必须保证方方面面都考虑到，小到绑脚手架的绳子、钉椽子用的钉子，最终做预算时多出整个费用的三分之一的做法较为妥帖。"

Plans, Estimates and Contracts

An overall plan of the settlement with the location of each building is recommended. A set of plans, sections and details is required for every building, especially churches, *"in order to avoid surprises and unfortunate errors during construction."* The plans of big churches must be designed by an architect or a good practitioner and be approved by the vicar apostolic (or bishop), according to Catholic canon law. Good plans help to make a good estimate. *"When the estimate is established and nothing has been forgotten, not even a rope for scaffolding or a nail for rooftimber, you should*

第一部分 手册的撰写背景
Part One: The Handbook in Context

在介绍设计方案的部分，有一个很重要的关于样式的很长的"注释"，里面提及的关于教堂设计的观点我们之后会讨论。

需要找一位有经验、名声又好的包工头，他会雇佣工匠并对工匠的所有行为（如偷盗和恶行）负责。在天津和其他地方，有个总建筑师负责建造工程，工程是以一个总价整体外包的。这种工程总体外包的做法在其他地方并不推荐，即使是在有一个负责任的监工的情况下也不推荐。因为在那种情况下，工匠会日夜工作：建筑会很快完工，但几个月后就会裂缝漏雨。因此应该按每天计价工作，这样会花更多的时间和金钱，但效果更好。传教士应该从新信徒中挑几个信得过的、有策略的人来监工，他们其中的一位需要负责建筑材料方面的事宜。

传教士会与包工头确认工匠的人数，包括厨师、抬水的人、磨粮食的人等。所有的事情必须准确无误地写下来：用餐、往返工地还是只是回程单程、谁负责喂磨坊工的驴子、盐、烟草、犒劳、厨具、节假日、下雨天、生病、炎暑天、基督徒不工作的星期日等方面。小工和把式的比例必须要固定，理想的比

boldly increase your estimate by one third."

The chapter on plans contains an important and long 'note' about style; this crucial point about church design will be discussed later.

It would be imprudent to work without an experienced and well-respected contractor, responsible for recruiting, feeding and supervising all other behavior of the workers including theft and misconduct. In Tianjin and other places where architects are in charge of building works, people work for a flat rate. This is not recommended anywhere else, even if there is a conscientious supervisor. People working for a flat rate will work day and night: the building will be quickly erected but will crack and leak after a couple of months. Therefore it is better to work by the daily rate and spend more time and money. The missionary will chose reliable and tactful supervisors amongst the converts of his congregation, and designate one of them to be responsible for acquiring building materials.

The missionary defines with the contractor how many workers will be engaged, including the cooks, water carriers, millers, etc. Everything must be accurately written down: provisions, their transport, who will feed the donkey, salt, tobacco, feasts for workmen, kitchen utensils, great feast days, rainy days, illness, heating, Sunday leave for Christians, etc. The proportion of coolies

例是两个小工对一个把式。泥水匠和木工师傅总会有好几个徒弟，传教士在付钱时都是按人头来的，没有区别对待。这章最后提供可供参考的合同：一份建造某座教堂与包工头签的总合同（图 2.2），七份关于石灰、砖头、花砖、瓦、石头和木料的合同。

建筑材料

砖头、瓦片、木头、石灰和绳子是中国北方的主要建筑材料，几乎没有提到石头和铁。指导内容都非常实际，包括生产、质量控制、数量估测和运输的各个方面，这些内容在接下来的几个章节里都展开讨论了。

普通的中国砖头是手工做的，在窑里烧的，通常质量都不好（图 1.17）。一些重要项目需要好砖头，手册里建议在合同里明确说明所需砖的尺寸和重量，还要控制好各个流程，如要选择好的土来做砖头，要给烧砖工提供模子等，还建议建一个或几个窑并准备足够的燃料。可以通过一个铁件来检测砖的质量：如果铁件和砖头碰撞发出的声音不清脆，就说明砖不好，要退回去。砖长可作为设计教堂时的计量单位，通常一砖的长度为 8~10 寸（1 寸=3.3 厘米），按照比例就能知道砖的宽度和厚度。手册推荐用红砖，比中国青砖更经

to master workers, ideally two for one, must be fixed. Masters masons and carpenters always have several apprentices, but the missionary should pay the salaries without distinction. The chapter ends with some model contracts: a general one with a contractor for the construction of a church (fig. 2.2), and seven specific contracts concerning lime, bricks, 'flower bricks', tiles, stone and timber.

Building Materials

Bricks, tiles, timber, lime and ropes are the main building materials in Northern China; stone and iron are barely mentioned. The indications are practical and cover most aspects of production, quality control, quantity evaluation and transport, while assembling is developed in the subsequent chapters.

Common Chinese bricks are hand manufactured, fired in kilns and usually of poor quality (fig. 1.17). For quality bricks and important projects, the handbook recommends defining in the contract both precise dimensions and weight of the bricks, and to control the production process by choosing good clay, providing the shapes to the brick makers, owning one or several kilns and possessing sufficient fuel. The quality of the bricks is controlled with an iron tool: if the sound of the tool on the brick is not good, the brick will be refused. It is essential to define the length of the brick to be used as the unit of measurement or

济合算。红砖只需要烧一次，还不需要饮窑。条砖很贵，应该用在砖墙的第一层以防止上升的湿气。这类条砖也建议用在柱子顶板和砖墙部分需要受压的地方。装饰需要用特殊的花砖，这类砖需要做特殊的模子。因为这些花砖模子很容易损坏，所以必须要准备3倍于所需的量。

屋面瓦烧制时需要专门的工匠，要进行很严格的质量控制。中国屋顶用小的方形"八砖"（图1.18）。筒瓦很漂亮，但不是很好的屋瓦材料，普通的屋瓦更好。按中国泥水匠的算法，每平方英尺需要20片瓦，梁长是16英尺的一间需要大概3000片瓦。檐口和滴水处需要用特殊的瓦，烧制这些瓦需要先铸模，费用很高。

module for designing the whole church (between 8 and 10 Chinese inches, *cun*, 3.30 cm). The width and the weight are automatically proportioned to the length. The handbook recommends producing red bricks, which are much more economic than the Chinese blue bricks. Red bricks are less fired and do not need to be sprinkled with water. Hard fired bricks are precious and should be used for the first layers of the brickwork because they stop rising damp. These bricks will also be used above the abacus of columns and other parts of the brickwork where compression is high. Special bricks with profiles and ornaments require appropriate moulds and should not be carved after having been fired. Hand-moulded ornaments and capitals are fragile and should therefore be produced three times more than needed because of a high proportion of breakage.

The production of roof tiles by specialized workers is similar to bricks, but the quality control must be very strict. Little square tiles are used for making Chinese roofs (fig. 1.18). Chinese half-cylindrical tiles are beautiful but do not cover the roof well; ordinary tiles should be preferred. Chinese masons count 20 tiles per square foot; about 3000 tiles are needed for a common bay with a beam of 16 feet. Special tiles must be ordered for the bottom of the roof and for the ridges; they are moulded and more expensive.

舶来与本土：1926 年法国传教士所撰中国北方教堂营造手册的翻译和研究
Building Churches in Northern China. A 1926 Handbook in Context

图1.17 仕拉乌素壕（内蒙古），为建造圣母圣心会神父住所而选砖，1938年。
fig. 1.17 Shilawusuhao (Inner Mongolia), sorting bricks for building the Scheut father's residence, 1938. (© FVI, CHC pictures)

图1.18 内蒙古，模制方形屋瓦，年代未知。
fig. 1.18 Inner Mongolia moulding square roof tiles, not dated. (© FVI, CHC pictures)

直隶东南部的传教区，杉木料和其他木料很贵，必须经专家挑选，很早就

In the mission of Southeastern Zhili, fir and other timber are expensive, must be selected by experts, ordered long in

预定，然后从天津运过来。这个地区生长有杨树和榆树，可以用来做屋顶构件，但不能用来做家具和门窗，在使用这类木料前，建议将其浸泡在石灰水里。手册中还给出了檩条和椽子的准确尺寸。

20世纪20年代水泥还只在北京和通商口岸使用，那个时候在中国北方石灰还是做砂浆的主要材料。手册对不同种类的石灰进行介绍，包括生产过程、运输和使用方法。献县生产的石灰比较肥，需要用沙子来混合，干起来很慢。因此大教堂的墙体不要在一年之内建完，"比较理想的做法是，在第一年的秋天将墙体砌到窗子高度就停工，然后到春天再开工"（图1.19）。

在施工现场需要用到两种绳子。一种是贵的、好的、用好麻编的绳子，由专业的织绳工编织而成，另一种便宜的绳子是小工自己编的，主要用来绑脚手架。这些用过的麻绳可以重复利用：将其切碎与石灰混合，俗称麻刀灰。用来搭脚手架的杆子和木板需要花很多钱。中国北方缺乏木材和竹子，树干和枝干都是用过再重复用。

advanced, and transported from Tianjin. Poplars and elms grow in the region and can be used for roof timber but not for furniture, doors and windows. It is recommended to soak the planks and the beams in liquid lime. The handbook gives precise dimensions for the roof plates and the rafters.

Lime still was the main component of mortar in Northern China in the 1920s since cement could only be found in the treaty ports and in Beijing. The handbook gives a detailed description of the different types of quick and slaked lime, their production phases, transport and uses. Lime produced in the area of Xianxian was fat, had to be mixed with sand, and dried very slowly. Therefore, the walls of a big church could not be erected in one year: *"the brickwork would ideally be interrupted in the fall of the first year at the level of the windows and pursued in the spring of the following year."* (fig. 1.19)

Two kind of ropes were used on construction sites. Expensive, high quality ropes resisting tension were produced by rope makers, while cheap ropes were made by the coolies themselves and used for assembling scaffolding. Used ropes were recycled: finely chopped pieces were mixed with lime. The poles and planks for the scaffolding were a great expense. Because of the lack of wood and bamboo in Northern China, all branches and trunks were used and re-used.

舶来与本土：1926 年法国传教士所撰中国北方教堂营造手册的翻译和研究
Building Churches in Northern China. A 1926 Handbook in Context

图1.19　海岛营子（辽宁省）及玫瑰营子（内蒙古），正在建造中的教堂，1937年。
fig. 1.19 Haidaoyingzi (Liaoning province) and Meiguiyingzi (Inner Mongolia), churches under construction, 1937. (© FVI, CHC pictures)

第一部分　手册的撰写背景
Part One: The Handbook in Context

砖石工程

第 5 章内容比较重要，分为五个部分：地基、打夯、砖砌合砂浆、硝和防碱材料、安装门窗框。除了技术上的建议外，还有很多针对如何监督中国工匠的建议。

西式高建筑（如教堂和钟楼）的地基必须很深且要建造得很好。在地面上标出建筑的平面，之后工人会挖一个很深的地基槽。地基槽底部需要用木夯、石夯或铁夯来打夯夯实（图 1.20 和图 1.48）。对于尺度高的钟楼和教堂，在地基槽底部应该有很结实的一层，由碎砖、石头混合石灰水构造而成。在这之上，泥水匠建造一个宽度逐层递减的砌筑基础。比起砂浆，中国泥水匠更喜欢用纯石灰，但手册里坚持要求将石灰和沙子混合以做成更坚固的砂浆。好的砖砌墙体的秘诀在于用固态砂浆和湿的砌墙材料，而中国工匠更喜欢用液态砂浆和干的砌墙材料。因此很重要的一点就是要在砌墙之前把砖放在水池里让砖头饮透了，在砌墙的时候可以再用水浇一次砖，特别是在天气热的时候，在砌新砖层之前，一定要给砖浇水。

Masonry

This important chapter is divided in five parts: foundations, earth ramming, brickwork and mortar, saltpeter and insulating materials, placement of door and window frames. Beside technical and practical advice, it contains many indications for supervising Chinese workers.

Western foundations, certainly for high building such as churches and towers, must be deep and well established. After having traced the plan of the building on the ground, labourers dig the trenches to a proper depth. The bottom of the trench is rammed with wooden, stone or iron rammers (fig. 1.20 and fig. 1.48). For towers and high churches, a 'concrete' layer will be rammed on the bottom of the trenches, consisting of crushed bricks and stones mixed with lime. On this compact basis, masons will build the stepped footing of the walls. Chinese masons prefer using pure lime to mortar, but the handbook insists on mixing lime with sand, so as to obtain a more solid mortar. The secret to good masonry is having firm mortar and wet materials, but Chinese masons prefer wet mortar and dry materials. The latter will then absorb the water from the mortar, which will then dry too fast and be reduced to powder. Therefore, it is very important to soak the bricks in a basin before using them and water the bricks when laid on the mortar, especially when the weather is hot, before bricking up the next layer.

图1.20 内蒙古，夯土施工队，年代未知。
fig. 1.20 Inner Mongolia, team of earth rammers, not dated. (© FVI, CHC picture)

手册里介绍了四种当地泥水匠常做的中式墙体。质量最差的墙体是用土建成的墙：用夯土层来做打墙和土墙的地基。坯墙的寿命不会超过 20 年。有钱人会做叫作"里生外熟"的墙。外墙面可以采用表砖做法，也可以片砖到顶。因为里外双墙的沉降系数不一样，这种墙的寿命不会超过 40 年。最好的做法是"里外砖"的砖墙（图1.21）。如果没有监督的话，中国泥水匠会在内外墙面用卧砖，两墙之间胡乱用些东西填塞，他们会在每 6 或 8 块顺条砖之后放一块丁砖，用来遮住两墙中

The handbook explains the four main types of Chinese walls that the local masons are used to making. The poorest quality walls are made with earth: ram earth for preparing the foundations of a wall and the earth walls themselves. Adobe walls do not last longer than twenty years. Richer people build adobe walls with a brick outer facing. The facings could be: bricks on edge, stretchers, and in the best cases, all the way up to the top. Because of the unequal compression of both sides, these walls do not last longer than 40 years. The best brickwork is the Western full-

间的缝隙。手册里对这种做法很反感，通过图纸的方式给出了做纯砖墙的几种砌合方式：丁砖、顺丁相间、一顺一丁（图2.8和图2.10）。中国人习惯做厚墙，但对于50米高的钟楼，底部墙两砖厚、其余部分墙一砖半厚就足够了。浆是一种混合了石灰、砂子和水的液态物，泥水匠在砌下一层砖前会把浆撒到砖与砖之间的缝隙里。灌浆是一种中国本地的特有做法，效果很好，西方建筑师接受了这种好做法。

地面上升的湿气里带有硝，会侵蚀墙体。传统上，中国人用碱草（芦苇、高粱秆、各类秸秆、枯叶）和木头来对付砖碱。然而在砖墙中却不建议用这种传统做法，因为有机材料的易腐烂性会导致墙体开裂。手册里建议用一层石头、釉面砖或涂有沥青的纸板（在天津和上海可以买到）来做隔碱层。传教士可以自己做隔碱层，给草纸或席子涂上沥青，这些涂有沥青的草纸或席子可以用来做隔碱层。

brick wall (fig. 1.21). Without supervision, Chinese masons will build inner and outer facings with stretchers and fill the gap between with anything they find; every 6 to 8 layers, they will place headers. The handbook warns against this brickwork and explains with sketches the different bonds for full-brick walls: stretcher bond, header bond, cross bond, and Flemish bond (fig. 2.8 and fig. 2.10). The Chinese are used to making thick walls, but a thickness of 1.5 brick is enough for walls and 2 bricks for the lower levels of 50 m high towers. *Jiang* is a liquid mix of lime, sand and water that the masons pour into all the interstices of the brickwork before placing the next level of bricks. *Kaoantsiang* literaly 'gorging with porridge' is a specific Chinese technique that gives good results and is not rejected by Western architects.

Rising damp carries saltpeter that attacks the walls. Traditionally, Chinese use herbs (reeds, sorghum, leaves, straw) and wood to block rising damp. This is, however, not applicable in brickwork because organic materials corrupt and cause cracks. The handbook recommends using a layer of stones, glazed bricks, or bituminous cardboard ('asphaltina' or 'maltoïde') that can be found in Tianjin and Shanghai. The missionary can also make his own insulation by coating paper, or a mat, with tar.

图1.21 仕拉乌素壕（内蒙古），泥瓦匠建造圣母圣心会神父住所，1938年。
fig. 1.21 Shilawusuhao (Inner Mongolia), masons building the Scheut father's residence, 1938. (© FVI, CHC pictures)

可以在砌墙过程中或墙体完工后安装门窗框。中国人通常是在砌墙过程中安装的（图1.22）。这种做法并不推荐，因为会有一些潜在的危害。只有水平高的泥水匠能砌拱形梁（图1.23），通常中国人只知道做一根过木。建筑师在任何情况下都会选择用拱，并在图纸中标明。中国木工为了省钱会做很细的门窗框，通常不能和砖墙体很好地交接在一起。手册里强调要将

Door and window frames can be placed during the building of the wall, or after its completion. Usually the Chinese place the frames at the same time as the brickwork (fig. 1.22), which should be avoided, to prevent potential damage during the work. Only good masons can make arched lintels (fig. 1.23), whereas most Chinese only know how to place traditional wooden lintels. In any case, the architect will chose the form of the arch and trace it on plans. Chinese join-

第一部分　手册的撰写背景
Part One: The Handbook in Context

墙和窗及墙和屋顶之间的缝隙密封，这样风才进不来。

ers save money and make frames too thin, which are not well joined with the masonry. The handbook insists on the necessity of hermetic walls, windows and roofs against cold winds.

图1.22　仕拉乌素壕（内蒙古），圣母圣心会神父住所，
在砌墙前竖立的门框和窗框，1938年。
fig. 1.22 Shilawusuhao (Inner Mongolia), Scheut father's residence,
door and window frames placed before the brickwork, 1938. (© FVI, CHC pictures)

图1.23　仕拉乌素壕（内蒙古），圣母圣心会神父住所，
拱形砖门梁和窗框，1938年。
fig. 1.23 Shilawusuhao (Inner Mongolia), Scheut father's residence,
arched brick lintel and window frame, 1938. (© FVI, CHC pictures)

~ 65 ~

屋顶

第 6 章讲屋顶，介绍并比较了中国屋顶和西方屋顶的传统做法（图 1.24）。内容中没有提到用工业化生产的金属和钢筋混凝土做的屋顶，只是简单介绍了土平房的屋顶、砖墁顶的屋顶和石灰锤顶。

中国瓦房屋顶有一个很大的缺点就是太沉，因此需要一个很牢固的木构件来支撑它。瓦屋顶也有很多优点，如牢固、防雨，室内冬暖夏凉等。该手册介绍了三种中式屋顶：①用筒瓦做的屋顶，这种屋顶很漂亮但不适合用在教堂，因为会漏雨，手册里提到兖州教堂之前筒瓦做的屋顶漏雨而要重新做；②仰瓦屋顶，仰瓦之间的缝隙都不用砂浆填充，但却不会漏雨，来自广平县的工匠在做仰瓦屋顶方面最出名（图 1.25）；③扣瓦屋顶，仰瓦和扣瓦相间使用，瓦之间的缝隙用砂浆填充。这三种屋顶都需要同样的"床铺"，有两种做法：①椽子上铺八砖；②直接在檩条上放高粱秆或芦苇秆（图 1.26）。

Roofs

The chapter on roofs contrasts Chinese and Western roofing traditions (fig. 1.24). Industrial roofs in metal or concrete are not mentioned, while flat earthen houses, roofs covered with grouted bricks, and flat macadam roofs are only mentioned briefly.

The Chinese tiled roof has the great inconvenience of being heavy and therefore needs a solid timber structure. It has great qualities too: it is solid, waterproof, and preserves warmth in the winter and freshness in the summer (as opposed to roofs covered in metal sheets). Three Chinese roofs are discussed: 1. *tongwa* roofs with convex semi-cylindrical tiles is the most beautiful roof, but should not be used for churches because they leak. The author of the handbook recalls that the roof of the cathedral of Yanzhou (Shandong province) had to be replaced. 2. *yangwa* roofs with concave tiles assembled without mortar. Roofers from the region of Guangping (Hebei province) are experts in this type of roof (fig. 1.25). 3. *kuiwa* with concave and convex tiles covering each other, forming channels assembled with mortar. In the three cases, the tiles are placed on a bed of mud, which is on: a) a level of square tiles supported by rafters; b) stalks of sorghum fixed on the purlins (fig. 1.26).

第一部分　手册的撰写背景
Part One: The Handbook in Context

图1.24　固阳县（内蒙古），圣母圣心会礼拜堂，在上瓦前整理灰背，年代未知。
fig. 1.24 Guyangxian (Inner Mongolia), Scheut father's chapel, preparing the mud layer before placing tiles (*wa fang*), not dated. (© FVI, BR pictures)

图1.25　大名（河北省），铺设一所教堂的屋面，河北南部的传统仰瓦技术，2014年。
fig. 1.25 Daming (Hebei province), making a church's roof, Southern Hebei traditional technique of concave roof tiles (*yang wa*), 2014. (© THOC 2014)

舶来与本土：1926年法国传教士所撰中国北方教堂营造手册的翻译和研究
Building Churches in Northern China. A 1926 Handbook in Context

图1.26　内蒙古或河北北部，建造一处圣母圣心会神父住所的屋顶，地点未知，年代未知。
fig. 1.26 Inner Mongolia or Northern Hebei, building the roof of a Scheut fathers' residence, not located, not dated. (© FVI, CHC pictures)

中国的房架非常沉也非常贵，需要用柱子来支撑，从而形成室内空间。欧式屋架由三角形的桁架做成，可以在没有柱子的情况下覆盖10米跨度的空间（图1.27和图1.28）。建筑师必须要设计桁架，标明怎样通过不同构件做一个跨度大的空间。该手册里介绍了几种西式的桁架，并给出了组成桁架的各个构件的名字（图2.12和图2.13）。40°是最理想的屋面坡度，但35°（中国屋顶的平均坡度）坡度和26°（金属瓦顶屋面）也不错。

The Chinese timber structure is heavy and expensive and requires intermediate columns to cover large spaces. The European timber structure is composed of triangular trusses and can span about 10 m without the help of columns (fig. 1.27 and fig. 1.28). The architect must design the trusses and fix the dimension of the different pieces in relation to the span. The handbook provides several models of Western trusses, gives the names of the elements and the different assembling possibilities (fig. 2.12 and fig. 2.13). A roof pitch of 40° is the best, even though 35° (average Chinese roof) and 26° (roof covered with metal sheets) give good results too.

~ 68 ~

第一部分　手册的撰写背景
Part One: The Handbook in Context

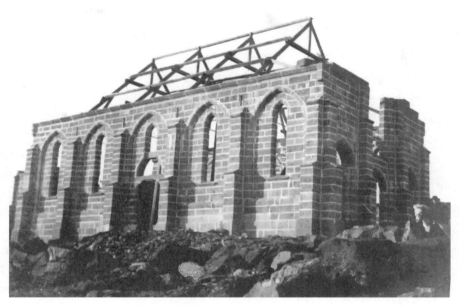

图1.27　磨子山（内蒙古），圣母圣心会玛利亚朝圣教堂，
西式的砖墙和屋架，和羹柏神父建造，约1910年。
fig. 1.27 Mozishan (Inner Mongolia), Scheut fathers' Marian pilgrimage church,
Western masonry and Western roof trusses, built by father De Moerloose,
ca 1910. (© FVI, CHC pictures)

图1.28　开封（河南省），原地区修院礼拜堂，被解体屋顶中的西式桁架，1935年建。
fig. 1.28 Kaifeng (Henan province), chapel of the former Regional Seminary,
Western trusses of the dismanatled roof, built in 1935. (© THOC 2013)

屋面装饰

第 7 章内容和写作风格与手册中介绍工艺技术的章节完全不同，对中国传统的装饰进行了讽刺批评。"用整章内容来讨论屋顶装饰……就是为了阻止你按照你周围的人所希望的那样去装饰屋顶。""每一处装饰必须要有其有用的一方面。不要为了装饰而装饰。"作者认为千篇一律的中国建筑只有通过装饰来显示区别。因为墙体不能装饰，所以所有的装饰都集中在了屋顶。作者觉得屋顶的龙头装饰一点美观性都没有，解释了中国做屋顶装饰是他们追崇的"好看文化"的一部分。屋脊部分的装饰，"福""卍"字用彩色琉璃瓦做成，被手册作者视为没有必要的奢侈之举。手册作者也批评过分使用小尖塔和垛口来装饰的做法（图 1.29 和图 1.30）。

Roof Ornaments

The chapter on roof ornaments is an un-nuanced criticism of the Chinese tradition, written in a totally different style than the technical and practical chapters. *"Devote an entire chapter to this topic (...): this is only to prevent you from decorating your roof as perhaps the people around you would like."* *"Any ornament in architecture must be motivated at the very least, it must have a useful dimension: never ornament for ornament."* The authors' point of view is that ornament is the only possible differentiation in the stereotyped Chinese architecture. Because the walls cannot be decorated, all the ornaments concentrate on the roof. They find dragonheads on top of roofs "absolutely not aesthetical" and explain contemptuously that roof ornaments belong to the Chinese 'good-looking-culture' (*haokan*). Ornaments on top of gables, characters of 'blessing' and 'ten thousand years', *fu* and *wan*, inscribed with coloured tiles on roofs, terracotta figures on ridges, etc. are regarded as unnecessary and expensive. The proliferation of pinnacles ('*nightcaps*') and battlements on churches is also condemned (fig. 1.29 and fig. 1.30).

第一部分　手册的撰写背景
Part One: The Handbook in Context

图1.29　兰州（甘肃省），中式基督教样式的圣母圣心会神父住所大门，约1932年。
fig. 1.29 Lanzhou (Gansu province), main gate of the Scheut fathers' residence in Sino-Christian style, around 1932. (© FVI, CHC pictures)

图1.30　内蒙古，圣母圣心会教堂，带有哥特窗饰和小尖塔的混合样式，1900年以前。
fig. 1.30 Inner Mongolia, Scheut Fathers' church, hybrid style with Gothic traceries and pinnacles, before 1900. (© KADOC, CICM)

舶来与本土：1926 年法国传教士所撰中国北方教堂营造手册的翻译和研究
Building Churches in Northern China. A 1926 Handbook in Context

建筑细节的处理

第 8 章，也就是最后一章，介绍了不同的建筑细节部分的处理并配有对应的图纸。我们在这里简单进行说明。

（1）滴水石是很重要的一个细节处理，应该用在窗户下面、墙体高处及墙体和屋面交接的地方。

（2）如果有可能，应在柱子底部用石头（图 1.31）。石雕是非常特别的（图 1.32）。

Miscellaneous

The last chapter contains indications about various constructive elements and refers to figures on the plates. We follow below the somewhat irrational order of the lemmas.

(1) Dripstones are indispensable elements that should be placed correctly under windows, at the top of walls and above the junction of roofs and walls.

(2) Stone, when possible, should be used for bases of columns (fig. 1.31). Stone sculpture is very exceptional (fig. 1.32).

图1.31　内蒙古，雕造柱头和柱础的石匠们，年代未知。
fig. 1.31 Inner Mongolia, stone carvers of capitals and bases, not dated. (© FVI, CHC pictures)

第一部分　手册的撰写背景
Part One: The Handbook in Context

图1.32　基督教石像的雕塑师，大同（山西省），1935年。
fig. 1.32 Sculptor of stone Christian statues, Datong (Shanxi province), 1935. (© FVI, CHC pictures)

（3）教堂的砖拱顶造价昂贵，但是可以做一个木制拱顶，不仅美观而且还更整洁。可以用木板、板条和石灰来做一个拱顶模仿品，其中心是拱顶木（图1.33~图1.35）。

（4）拱需要用特殊的砖来做，可以用简单的圆形砖做装饰。

（5）彩色玻璃窗作为教堂最漂亮的装饰，在中国做是件很困难的事情。当然可以从欧洲定制有着人物图像的彩色玻璃，需要预定小尺寸的玻璃，并要很仔细地包装（图1.36）。

(3) Vaulting churches with bricks is too expensive, but wooden vaults, even pastiche, improve both the nave's acoustics and the roof's insulation. Fake vaults are made of wooden ribs converging on a wooden key, and the compartments are made of lime on lattice work (fig. 1.33 to fig. 1.35).

(4) Arches require appropriate arch bricks, adorned with a simple round profile.

(5) Stained glass, the most beautiful ornament of a church, is extremely difficult to get in China. It is, of course, possible to order figurative glass in Europe that should be in small sections and be well packed (fig. 1.36).

舶来与本土：1926 年法国传教士所撰中国北方教堂营造手册的翻译和研究
Building Churches in Northern China. A 1926 Handbook in Context

图1.33 法国巴黎外方传教会长春（吉林省）教堂，
正在建造中的木柱、拱券和拱顶拱肋，约1920年。
fig. 1.33 Changchun (Jilin province) church of the French Foreign Missions of Paris,
wooden columns, arches and lattice structure of the vaults under construction,
around 1920. (© MEP Archives)

第一部分　手册的撰写背景
Part One: The Handbook in Context

图1.34　天津，圣母得胜堂（望海楼教堂），正厅中的木质拱顶，1900年后。
fig. 1.34 Tianjin, Our Lady of the Victory church, wooden vault of the nave, after 1900. (© THOC 2011)

灰色玻璃也能在中国找到。"即使在这种情况下，也应该放弃欧洲某些杰作的做法，而是要根据'手头所有的'，即这个国家的资源，来做好工作。当然你的作品会被烙上'中国制造'❶的标签。这又有什么关系？因为它又不是用来出口的！"

在天津可以买到铅块。在中国，有着人物图像的彩色玻璃窗只在上海土山湾的耶稣会工作室生产（图1.37）。❷

Grisaille glass can also be found in China. *"In this case you should refrain from a masterpiece made in Europe. We propose to work with the means at hand, using the country's resources. Obviously your composition will be branded 'made in China'. What is wrong with that? Since it is not an export item you are building!"*

Lead can easily be obtained in Tianjin. In China, figurative windows were only produced in the Jesuit workshop of

---

❶ 作者说这句话的背景是在20世纪20年代，而不是当今的21世纪，作者意思是说"地方当地制造"：中国制造的东西质量差不适宜于出口，只适用于中国当地。

❷ 宋浩杰. 土山湾记忆[M]. 上海：上海学林出版社，2010. 关于从比利时进口彩色玻璃窗到中国的内容见文章：Thomas Coomans. Sint-Lucasneogotiek in Noord-China: Alphonse De Moerloose, missionaris en architect [J]. M&L. Monumenten, landschappen en archeologie, 2013,32(5);6-33.

有一张图专门解释了怎么把玻璃装到一个金属窗框里（图 2.35）。

Tushanwan (T'ou-sè-wè) in Shanghai[1] (fig. 1.37).

One plate explains how to fix the glass in the window frame with metal rods (fig. 2.35).

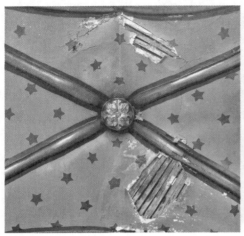

图1.35 北京，北堂（西什库教堂），主教礼拜堂的拱顶，约1900年。
fig. 1.35 Beijing, North Cathedral Xishiku, vault of the bishops' chapel, around 1900. (© THOC. 2013)

图1.36 西湾子（河北省），圣母圣心会修院礼拜堂，和羹柏神父于1902年建造，其家族捐赠比利时原产彩色玻璃窗。
fig. 1.36 Xiwanzi (Hebei province), chapel of the Scheut fathers' seminary, by father De Moerloose and Belgian to produced stained to glass windows donated by his family in 1902. (© KADOC, CICM)

（6）塔楼的尖顶为八边形，由塔楼四角的石头叠涩而形成。需用一种梯形断面的砖来做尖顶。在尖顶外侧为斜面，内侧悬挑。

(6) The spire of a bell tower has an octagonal plan and rests on squinches placed in the angles of the square tower. The bricks of the spire should have a trapezoid section: oblique on the outer side and cantilevered on the inner side of

---

[1] Song Haojie. Memory of T'ou-sè-wè [M]. Shanghai, 2010. About the import of Belgian stained-glass windows in China, see: Thomas Coomans. Sint-Lucasneogotiek in Noord-China: Alphonse De Moerloose, missionaris en architect [J]. M&L. Monumenten, landschappen en archeologie, 2013,32(5);6-33.

the spire.

图1.37 上海，法国耶稣会圣约瑟堂，带有中文题记的哥特彩色玻璃窗。
fig. 1.37 Shanghai, St Joseph's church of the French Jesuits, Gothic stained-glass window with Chinese inscription.
(© THOC 2011)

（7）勾缝有两种做法：①从下到上，就是在砌砖墙的同时勾缝；②从上到下，就是在拆脚手架的时候勾缝。建议使用第一种方法。

（8）只需要对墙体内侧粗糙部分抹泥。中国抹泥匠喜欢先抹一层泥再用石灰抹，这样可以节省石灰，但抹灰层很快就会掉落。对天花板部分的粉刷也是一样的做法，要在板条上抹上石灰。

（9）中国人喜欢在房间内部中央挖一个坑用来和泥。这种做法应该避免，因为这样做不卫生。

（10）为了防止教堂和圣器室的铺地碱化，可以用涂有沥青的黄纸铺在下面做隔离层。

(7) Grouting walls can be done from: 1° the bottom up, gradually when building up the brickwork; 2° or from the top down, when dismantling the scaffolding. The first method should be preferred.

(8) Only the inner side of the walls are roughcast and coated with lime. Chinese coaters will always prefer a first layer of mud in order to save lime, but the coating will soon fall off. The same is true for the ceilings, with lime on lattice.

(9) Chinese have the habit of mixing the mortar in a pit located in the middle of the building they are erecting. This should be avoided for obvious hygienic reasons.

(10) In order to avoid rising water and saltpeter in the brick pavements of churches and sacristies, paper coated with tar will be used as an insulating underlay.

（11）磨砖是一项既费钱又有害的工作，它把砖的那层能够防止硝的保护膜给破坏了。

（12）中国北方对于在房子南侧要不要做廊子有不同的观点，有人赞成，有人反对。廊子不应该过宽，这样在冬天光线能够进来，下午也能防日晒。在学校里，下雨天的时候廊子特别有用（图1.38）。

(11) Polishing the visible side of bricks is an expensive and harmful habit. The brick loses its protection.

(12) In Northern China, there are supporters and opponents for the construction of verandas on the south side of buildings. Verandas should not be too wide, in order to allow light to enter the building during the winter, and to protect from the sun during the summer. In schools, verandas are very useful when it rains (fig. 1.38).

图1.38 大名（河北省），圣母圣心会神父住所，带有走廊的南立面，约1910年。
fig. 1.38 Daming (Hebei province), Jesuit fathers' residence, southern side with veranda, ca 1910. (© THOC 2014)

（13）砌合墙角砖是高大建筑不可少的构造做法，特别是在有薄墙和西式屋顶的建筑上。也可以用铁构件的做法。

（14）做烟囱时一定要仔细监

(13) Clamping walls is indispensable in high buildings, especially in the case of thin walls and Western roofs. Iron anchors can be used at the level of tie-beams.

(14) The construction of stove

督：地板的横梁和任何木材都不要和烟囱烟道有接触。

（15）中国人没有做楼梯的传统。中国塔和中国多层建筑里的楼梯都不好用。建筑师必须要设计楼梯并且要画出各个细节的图纸。

（16）钟架子应该由铸钟匠来设计，因为他知道钟的重量和尺寸。

（17）昂贵的焦油总是在运输途中被船夫偷走或偷换掉。

（18）房子完工后的前几年应该确保排水沟没有被堵塞，以保证雨水能被畅通排走。

（19）可以在天津和上海买到玻璃，也可以从欧洲订购。彩色玻璃窗应该用铁丝网或磨砂玻璃来保护。

（20）家具、门窗、木架、廊子柱子和所有暴露在雨里的木材都要进行油漆。中国桐油没有亚麻籽油质量好。最好不要对教堂的木制拱顶进行油饰。

（21）木柱子的底部必须有保护措施，以防止被上升的湿气侵蚀而腐烂。石头基础、沥青纸板或涂有沥青的黄纸做在基础部分都能预防房子在20年后倒塌的风险。

chimneys must be strictly controlled: no beams of the floor and no timber may be in contact with the chimney flue.

(15) The Chinese have no tradition for making stairs. Those in towers and houses with upper floors and cellars are *absolutely impracticable*'. The architect must design the stairs and draw detailed plans of it.

(16) Wooden belfries or structures for hanging bells should be designed by the bell caster, who knows the weight and the size of the bells.

(17) Tar is often stolen by boatmen during transport.

(18) The correct flow of rainwater should be checked in the first years of the building, to be sure that the gutters work well.

(19) Glass can be found in Tianjin and Shanghai, or ordered in Europe. Stained glass windows should be protected from hail damage with galvanized iron mesh or a second clear glass pane.

(20) Furniture, doors and windows, timber, veranda posts, and all wood exposed to rain must be oiled or varnished. Chinese oil is of lesser quality than linseed oil. It is better not to paint the wooden vault of a church.

(21) The bases of wooden posts must be protected against rot, caused by raising damp. A stone base, bituminous cardboard, or paper coated with tar will prevent the house collapsing after 20 years.

舶来与本土：1926 年法国传教士所撰中国北方教堂营造手册的翻译和研究
Building Churches in Northern China. A 1926 Handbook in Context

手册中的图纸

正如该手册的副标题"建议-方案"所言，手册中的图纸是这本手册重要的组成部分。图纸部分有 54 张图纸，可以分为两类：图纸 1~24（图 2.3~图 2.26）是结构细节，图纸 26~54（图 2.28~图 2.38）是 8 座教堂的项目方案图纸。图纸 25（图 2.27）是该手册中唯一关于教堂内部陈设家具的。图纸并不难懂，结构细节部分在文中都有相关说明。关于建筑装饰的所有部分——柱体、柱头、柱基、屋脊等——都是哥特式的（图 2.7、图 2.25 和图 2.26）。项目方案中的窗户也是哥特式的，包括一个玫瑰花窗、三个石雕花格窗（图 2.21、图 2.23 和图 2.24）。有着尖拱券的教堂建筑是 19 世纪天主教传教士的世界观的一部分。

图纸 1~24 都是关于大名天主堂在柱子、柱头、石雕花格窗、砖柱体和拱顶、屋顶、尖塔、檐口和砖砌等方面的细节做法。这些图纸❶ 都是由雷振声修士所绘。所有这些方面，手册文字内容里都有说明（见本书的第二部分），也可

The Handbook's Illustration

As mentioned in the subtitle 'advice-plans', the illustrations are an important part of the book. They consist of 54 folded plates divided into two series: construction details (plates 1-24) (fig. 2.3 to fig. 2.26) and plans of eight model churches (plates 26-54) (fig. 2.28 to fig. 2.38).Between the two series, plate 25 (fig. 2.27) shows a confessional, which is the only liturgical furniture mentioned in the book. The plates, with easily understandable drawings of construction details, complement the text. All the examples of architectonic decoration —shafts, capitals, bases, profiles, ribs etc. — are Gothic (fig. 2.7, fig. 2.25 and fig. 2.26).The examples of windows are Gothic too and include a rose window, a triplet and a tracery window (fig. 2.21, fig. 2.23 and fig. 2.24).This paradigmatic association of pointed arches and church architecture was part of the nineteenth-century worldview of the Catholic missions.

Most columns, capitals, tracery window, brick shafts and vaults, roof crest, spire, cornice and brickwork details of the 24 first plates are from the church of Daming. They were thus drawn by brother Litzler. ❷ These details are described in the handbook's text (see

---

❶ Le Missionnaire Constructeur [M]. Xianxian, 1926.原标号为 1-17、19、21 和 24 的图纸。
❷ Plates 1 to 17 (fig.1-fig.53), as well as plates 19, and 21 to 24.

第一部分　手册的撰写背景
Part One: The Handbook in Context

以和大名天主堂的细节进行比较（见本书的第三部分）。

图纸中的 8 个教堂项目方案，手册内容里没有提到，每个方案都有不同的尺寸和样式。它们都是些"草案"，就是说这些项目还没有完全立项。教堂 1、教堂 2 和教堂 4 是古典主义或新古典主义样式❶，都是明显的砖构建筑（图 2.28~图 2.30，图 2.33）。这三个方案都只有中殿、圣坛后的笔直暗墙、圆拱形的窗户、壁柱和主立面上的钟楼。三张剖面图都显示了西方的屋顶结构。教堂 2 的图纸最详细：有两个外立面、一个剖面、正门的细节、钟楼的两个剖面和尖顶的细节处理。这个项目的立面和钟楼看上去与大名教区的耶稣会会士的乡村住宅里的圣约瑟小礼拜堂很像（图 1.8）。我们可以推测雷振声修士设计了这座位于大名郊区的小教堂，所以他手上有详细的方案。类似样式的教堂在离大名很远的地方也有，如建于 1922 年吉林圣约瑟神学院的小礼拜堂（图 1.39）。

Part 2 of this book) and are compared with details of the church of Daming (see Part 3 of this book).

The eight model churches are not related to the text and are examples of different sizes and styles. They are called '*avant-projets*', that is to say projects in their early stage. The churches 1, 2 and 4, are in a kind of Classical or Neoclassical style, ❷ but with visible brickwork (fig. 2.28 to fig. 2.30, and fig. 2.33). All three are single nave with a straight and blind wall behind the main altar, round arched windows, pilasters, and a small bell tower on the main façade. The three sections show Western roof structures. Church's project 2 is the most detailed: five plates show two outer elevations, one section, a detail of the main gate, two sections of the bell tower and the detail of the spire. The façade and bell tower of this project look like the St Joseph chapel of the Jesuits' country house on the outskirts of Daming (fig. 1.8). We could conjecture that brother Alphonse Litzler designed this chapel, as most of the Jesuit buildings in Daming, and therefore had detailed plans at hand. Similar style churches, however, were also built far away from Daming, such as the chapel of St Joseph seminary at Jilin, built in 1922(fig. 1.39).

---

❶ 在项目方案 3 的教堂平面上有提到"文艺复兴式"，提出要把这和高坛的形式相配，高坛用的是直墙而不是哥特式的多边形后殿。

❷ The comment written on the ground plan of the church's project 3 mentions "Renaissance style" and associates the style with the form of the choir, which should be a straight wall instead of a Gothic polygonal apse.

其他五个教堂项目方案都是哥特式的，但都有不同。教堂项目方案 3 被称为"多种样式综合体"，提供了一张哥特式的平面图和不同方向的剖面图。在平面图纸上有说明："这个教堂也可以建成文艺复兴时期的样式：不做圆形的高坛，教堂尽端就是直墙"（图2.32）。所有的哥特样式都有三边后殿，入口处上方有钟楼、尖拱券和哥特式木肋骨拱顶（图 2.31、图 2.32 和图 2.34~图 2.38）。

The five other model churches are Gothic and show variations on the same theme. The church's project 3, called 'omnibus', presents a Gothic ground plan and sections of different elevations. The comment on the ground plan mentions that "this church can also be done in Renaissance style: instead of a round choir one may end the church with a straight wall." (fig. 2.32) All the Gothic patterns have a sanctuary with a three-sided apse, a bell tower above the entrance, pointed arches and wooden Gothic rib vaults (fig. 2.31 to fig. 2.32 and fig. 2.34 to fig. 2.38).

图1.39 吉林（吉林省），法国巴黎外方传教会神学院，梁神父设计，约1920年。
fig. 1.39 Jilin (Jilin provine), seminary of the French Foreign Missions of Paris, designed by father Lamasse, around 1920. (© MEP Archives)

第一部分　手册的撰写背景
Part One: The Handbook in Context

一个教堂有中殿和后殿，但没有耳堂，另外两个是中殿加耳堂，另两个最大的教堂有中殿，且中殿两边有过道还有耳堂。立面图纸展示了高尖塔的不同外形，剖面图显示有着三中殿的教堂是"大厅教堂"，它没有天窗只有一个双坡屋顶。这些教堂的哥特式风格都很简单且是法国哥特式，我们可以在大部分从上海和江苏到满洲的法国教区里看到这种样式的教堂（图 1.40 和图 1.41）。这些图纸可能都不是和羹柏神父画的，因为他很明显是深受奥古斯都·普金的影响。

该手册的扉页和最后一页各有一幅小插图（图 2.1 和图 2.39）。这两幅小插图很有可能不是特别为手册设计的，而是属于印刷厂的。第一幅小插图画的是一座精致的会幕❶：中间是一扇画有十字架的门，左边是一个十字形的钥匙孔，上方有一个圣体光❷，两侧是哥特式小尖塔，外框是华盖，中间上方是耶稣的荆棘冠。在天主教教堂里，会幕是最神圣的地方，耶稣像就被放在那里。手册扉页用这幅小插图是想赋予教堂建造以神学意义：上帝存在于教

One has a nave and aisles but no transept, two others are single nave with a transept, and the two largest have three naves and a transept. The elevations propose different silhouettes of high spires and the sections show that the churches with three naves are 'hall churches', with no clerestory and are covered with one saddle roof. The Gothic style of these churches is simple and rather French and could be seen in most French vicariates apostolic from Shanghai and Jiangsu to Manchuria (fig. 1.40 and fig. 1.41). One could not attribute these drawings to the Gothic style of father Alphonse De Moerloose, whose style was more explicitly influenced by Augustus W. N. Pugin.

The handbook contains two vignettes, one on the title page and the other on the last page (fig. 2.1 and fig. 2.39). In all likelihood both vignettes were not specially designed for the handbook but belonged to the printing office. The first one figures an elaborate tabernacle, composed of a door with a cross in the middle and a keyhole to the left, wearing a monstrance, flanked with two Gothic pinnacles, and framed by a baldachin containing Christ's crown of thorns. In Catholic churches, the tabernacle is the most sacred place, where the

---

❶ "会幕"意为"神的居所"。 会幕记载于《旧约圣经·出埃及记》第 26 章和第 36 章当中，在士师记时代征服迦南地，直到前 10 世纪时最终按耶和华指示的样式建造了耶路撒冷圣殿，也就是所谓的第一圣殿。

❷ 圣体光，也称圣体发光、圣体皓光、圣体光座或圣体显供架，是天主教、旧天主教、圣公宗和信义宗等宗派在一些宗教仪式上面使用的一种祭具，通常为镀金银制品，正中开有一个透明的小窗，用于嵌入圣体，四周呈放射性线条，以表现出"圣体发光"的主题（https://en.wikipedia.org/wiki/Monstrance）。

堂，建造教堂就是给上帝建造住宅。另一幅小插图则混搭了中国式的立面和希腊式的山花。这幅小插图出现在最后一页倒是让人有点意外，因为手册里既批评了希腊样式又批评了中国样式。

body of Christ is conserved. Putting this vignette on the title page gives a theological meaning to church construction: God is present in the church and building churches means building houses for God. The other vignette figures a hybrid Chinese style temple facade with a kind of Greek pediment. It is rather unexpected in a handbook that, as we will see, criticizes both the Greek and the Chinese styles.

图1.40　上海，杨树浦教堂，和羹柏神父设计，1924年。
fig. 1.40 Shanghai, Yangtzepoo church, designed by father De Moerloose, 1924. (© ASJ France, FCh)

第一部分　手册的撰写背景
Part One: The Handbook in Context

图1.41　辽宁（辽宁省），巴黎外方传教会的法国哥特式教堂，1922年。
fig. 1.41 Liaoyang (Liaoning province), French Gothic style church of the Foreign Missions of Paris, 1922. (© MEP Archives)

样式之辩：西方哥特式还是中国化
The Issue of Style: Western Gothic or Sinicized?

在中国所建教堂的样式问题不只是细节而是一切问题的核心，因为建筑

The issue of the style of churches in China was anything but a detail, because

~ 85 ~

在公共空间极具易见性且它是建筑主人身份的象征。手册的序言是这样开始的:"这本册子不是为那些建造大教堂的建造者写的。这本册子也不是为了介绍中国艺术在基督教教堂上的适应性应用。"手册作者用了很长一段文字来批评这样的适应性应用,指出中国样式不适用于教堂。为了更好地理解作者的这种保守观点,有必要对20世纪20年代在中国的天主教会的社会大背景进行概述。另外我们也不能忽略那个时期中国的复杂政治和社会形势。

architecture is visible in the public space and expresses the identity of the owner. The handbook's preface opens with a statement: *"This brochure is not made for cathedral builders. This is not even an essay on the famous adaptation of Chinese art to our Christian churches."* A long passage criticizes this architectural adaptation with the evident aim of proving that Chinese style is not suitable for Catholic churches. In order to understand the handbook's conservative point of view, it is indispensable to sketch the context of the Catholic mission in China and its evolution in the 1920s. Furthermore, one should not forget the chaotic political and social situation of Republican China in the 1920s.

20世纪20年代中国教堂中国化之需要

The Need to Sinicize the Church in China in the 1920s

1920年,中国天主教传教区包含52个宗教代牧区,约200万的信徒(当时总人口约5.56亿),由1417位外籍神父和963名中国神父服务。❶ 来自西方的宗座代牧和传教修会在当地培养强烈的民族认同感,将教省如"宗教殖民地"般治理。法国身为中国天主教的保护者,深信自己有将文明带至中国的义务,自1860年后开始干预很多教案的处理。❷ 即使教廷和其他欧洲列强表

In 1920, the China Catholic mission consisted of 52 church vicariates apostolic, for a population of about 2 million Catholics (of the 556 million Chinese at that time), served by 1417 foreign priests and 963 Chinese priests.❶ The Western vicars apostolic and missionary institutes cultivated strong national identities and ruled the church provinces like 'religious colonies'. As protector of the Catholic religion in China, and convinced of its

---

❶ Robert E. Carbonneau. The Catholic Church in China 1900-1949 [A] // R. Gary Tiedemann. Handbook of Christianity in China. Volume Two: 1800 to Present [C]. Leiden-Boston, 2010, 519.

❷ 自1858年中法《天津条约》和1860年《北京条约》开始。

第一部分　手册的撰写背景
Part One: The Handbook in Context

示异议，法国仍然成功地维系保教权，并封锁了教廷尝试欲与中国建交的行动。❶ 因此，驻华宗座代表刚恒毅认为在中国的天主教传教活动，有如法国主导国家利益之下的"封建领地"。❷ 第一次世界大战后，在民国共和体制下，一股民族意识兴起，催生了后来民国政府推行的新文化运动。传教士被视为帝国主义者的代理人，这一事实对教会产生了严重的后果：基督宗教与殖民主义相勾结。

在此情况下，教廷意识到中国教堂中国化的紧迫性。目前一般著作中所使用的词汇，所谓的"本地化""本土化"或"地方化"运动，目的在于将基督信仰扎根并融入特定文化中，其实这些词汇说法在 20 世纪 20 年代时并不存在（图 1.43）。❹

duty to bring civilization to China (the '*Mission civilisatrice*'), France was involved in most missionary affairs from the 1860s.❸ Despite contestation from both the Holy See and other European nations, the French Republic succeeded in maintaining its Protectorate and blocked all attempts by the Holy See to develop direct relationships with the Chinese state.❶ Therefore, the Catholic mission in China, according to archbishop Costantini, looked like a 'feudal territory' with dominating French national interests.❷ After World War I, growing Chinese national consciousness in Republican China led to the New Culture Movement. The fact that missionaries were perceived as agents of imperialist powers resulted in a serious threat to the Church: Christianity became associated with colonialism.

In this context, the Holy See understood the urgency to Sinicize the Church in China. The movement of 'inculturation' or 'indigenization', or 'localization', all terms used in present literature, but that did not exist in the 1920s—aimed to root the Christian faith in a specific culture (fig. 1.43).❹

---

❶ Alexandre Chen Tsung-ming. Les réactions des autorités chinoises face au protectorat religieux français au cours du XIXe siècle [A] // Alexandre Chen Tsung-ming. Le Christianisme en Chine aux XIXe et XXe siècles. Évangélisation et Conflits [C]. Leuven, 2014, 125-171.

❷ Claude Soetens. L'Église catholique en Chine au XXe siècle [M]. Paris, 1997, 109-110.

❸ Since the *Sino-French treaty of Tianjin* (1858) and *the Convention of Beijing* (1860).

❹ Klaus Koschorke. Indigenization [A] // Hans Dieter Betz. Religion Pas & Present. Encyclopedia of Theology and Religion (vol. 6) [C]. Leiden-Boston, 2009, 459-460; Nicolas Standaert. Inculturation. The Gospel and Cultures [M]. Philippines, 1990.

舶来与本土：1926年法国传教士所撰中国北方教堂营造手册的翻译和研究
Building Churches in Northern China. A 1926 Handbook in Context

图1.42 开封（河南省），意大利圣方济各会的意大利哥特式主教座堂，1917—1919年。
fig. 1.42 Kaifeng (Henan province), Italian Gothic style cathedral of the Italian Franciscans, 1917-1919. (© THOC 2013)

根据教皇两封谕信的内容——1919年的"教宗文告"和1926年的"教会事务"，❶ "本地化"意味着要以本地的文化，改变一般人对天主教会太具帝国主义色彩的认知："中国神职与传教士须拥有相等的权利，中文必须是主要用语，不可批评中国人的习俗，肯定在校教育与大学之存在等。"

Based on the papal apostolic letter *Maximum illud* (1919) and the encyclical letter *Rerum ecclesiae* (1926),❶ inculturation was meant to change the too imperialistic perception of the Catholic Church by the native cultures: "*Chinese clergy and missionaries were to have equal rights, Chinese was to be the primary language, Chinese customs were not to be criticized, education in schools and universities was affirmed, etc.*❷"

---

❶ Benedict XV, 30 November 1919.Maximum illud: Apostolic Letter on the Propagation of the Faith throughout the World, Vatican; Pius XI, 28 February 1926, Rerum ecclesiae: Encyclical on Catholic Missions, Vatican. Both available on: http://www.svdcuria.org/public/mission/docs/encycl/mi-en.htm.

❷ Robert E. Carbonneau the Catholic Church in China 1900-1949 [A] // R. Gary Tiedemann. Handbook of Christianity in China. Volume Two: 1800-Present [C]. Leiden, 2010, 516-525.

第一部分　手册的撰写背景
Part One: The Handbook in Context

自 1922 年担任驻华宗座代表的刚恒毅推动传教活动的再兴，1924 年召开第一届上海教务会议，1924—1926 年开辟中国籍教士管理的新传教区，1926 年创办天主教辅仁大学及同年教宗祝圣首批中国籍主教，这些均是重要的里程碑。❶ 然而，不少代牧和年龄较长的传教士，特别是法国籍者，抵拒这一中国教会未来新方向的发展战略，并抵制驻华宗座代表刚恒毅。

Archbishop Celso Costantini, apostolic delegate to China from 1922, promoted this revitalization of the mission. Milestones were the first Catholic council of Shanghai (1924), the creation of new church provinces (1924-1926), the foundation of the Catholic University of Pekin (1926), and the consecration of the first Chinese bishops (1926).❶ Several vicars apostolic, however, and many old missionaries, especially the French ones, resisted this new vision for Catholic China and boycotted Costantini.

图1.43　普世一神及各因循当地传统而本土化了的教堂，中国基督教教义问答手册——《问答像解》中的图像，1928年。
fig. 1.43 One universal God and churches inculturated in local traditions, image from the Sino-Christian catechism book *Wendaxiajie*, 1928. (© KU Leuven, CB)

---

❶ Paolo Goi (ed.). Il Cardinale Celso Costantini e la Cina. Un protagonista nella Chiesa e nel mondi del XX secolo [C].Pordenone: Diocesi di Concordia-Pordenone, 2008; Sergio Ticozzi. Celso Costantini's Contribution to the Localization and Inculturation of the Church in China [J]. Tripod (鼎), 2008. ,28(148).

舶来与本土：1926年法国传教士所撰中国北方教堂营造手册的翻译和研究
Building Churches in Northern China. A 1926 Handbook in Context

天主教艺术与建筑的中国化

Sinicizing Catholic Art and Architecture

刚恒毅本人是位艺术家，他在去中国之前就已参与了学界有关圣教艺术的讨论活动。1922年他抵达北京，立即就开始激励传教修会，并委托中国艺术家完成基督宗教艺术作品，以推广教会建筑、艺术和礼仪家具用品的中国化（图1.32和图1.44）。

Archbishop Costantini was an artist, and was involved in the debate on sacred art long before his stay in China. As soon as he arrived in Beijing in 1922, he promoted Chinese-inspired architecture, art, and liturgical furniture by stimulating missionary institutes and commissioning Chinese artists to realize Christian art works (fig. 1.32 and fig. 1.44).

图1.44 中国基督教教义问答手册——《问答像解》，1928年。
fig. 1.44 Sino-Christian catechism book *Wendaxiajie*, 1928. (© KU Leuven, CB)

第一部分　手册的撰写背景
Part One: The Handbook in Context

刚恒毅完全排斥哥特式和其他西方建筑样式，他想推广一种得以在中国艺术和文化中生根的基督宗教艺术样式："西方式艺术在中国是一种错误的形式，将欧洲样式，如仿罗马式和哥特式艺术引进中国是一项错误。"❶ 一个特有的"中国基督宗教样式"以中国的方式，透过建筑、绘画、家具用品和其他礼仪艺术，重新表达基督宗教流传至今的传统。❷ 民国时期的中国，建筑领域比单纯所欲表达的形式和精神的问题要再复杂许多。挑选精确的建筑样式有助于推动社会现代化和寻求新身份。

在刚恒毅之前，中国教堂的建造仍旧为保守的西方式样式。❸

刚恒毅认为建筑带有表达和创造中国天主教徒身份认同的作用，而同时他面临三项挑战：第一，法籍传教士依旧推崇西式风格，并排斥所有把中国装饰用于教堂的尝试；第二，基督新教已开始在大学校园中兴建中国式校舍，并参与国民党推动在建筑领域中的中国文化复兴运动；❺ 第三，身处变动中的

Archbishop Costantini radically condemned Gothic and other Western styles and promoted a Christian style that would be rooted in Chinese art and culture: *"Western art in China is an error of style. It is an error to import European styles, Romanesque and Gothic, in China."*❶ A "Sino-Christian style" would re-express the Christian heritage in Chinese forms, for architecture, painting, furniture and other liturgical arts.❷ The field of architecture in Republican China, however, was much more complex than a mere question of forms and spirit. The choice of precise architectural styles contributed to the search for societal modernization and a new identity.

Until Costantini's action, church construction in China remained Western and conservative. ❹

Costantini faced three challenges regarding architecture as identity bearer and identity maker of the Chinese Catholics. First, French missionaries still promoted Western styles and reproved all attempt to integrate Chinese decoration in churches. Second, Protestant missions already were building university campuses in the Chinese style and were en-

---

❶ Celso Costantini. L'universalité de l'art chrétien [J]. Dossiers de la Commission synodale, 1932, 5 :410-417 (quote p. 413). Also: Celso Costantini. The Need of Developing a Sino-Christian Architecture for our Catholic Missions [J]. Bulletin of the Catholic University of Peking, 1927, 3:7-15; Celso Costantini. The Church and Chinese Culture [M]. New York, 1931.

❷ Celso Costantini. Le problème de l'art en pays de missions [J]. L'artisan liturgique, 1932, 24 :816-819.

❸ 罗马式教堂多在呼和浩特、西湾子和青岛建成。

❹ Romanesque style cathedrals were notably built in Hohhot, Xiwanzi and Qingdao.

❺ Jeffrey W. Cody. American Geometries and the Architecture of Christian Campuses in China [A] // Daniel H.

中国社会，中国天主教徒也需寻求界定他们自己的身份认同感。

驻华宗座代表刚恒毅要的是饰以中国艺术和相关家具用品并由本地神父服务的中国式教堂。他希望中国艺术家能创造中国基督宗教样式艺术，因为他深信唯有本地人方能表达一种深刻的心灵感受；而西方艺术家只能创作仿"中国风"的作品。20 世纪 20 年代初在美国接受教育的第一代中国建筑师大多为基督新教信徒，本身对教堂建筑并无太大兴趣，他们专注于为政府设计现代中国样式的官方建筑。情势紧迫下，刚恒毅于 1926 年邀请葛利斯神父来中国。在他为时不长的停留期间，葛神父设计了四幢大型建筑，分别是位于开封和香港的小修院、宣化的主徒会修院和北京天主教大学（或名为辅仁大学）（图 1.45）。这四幢具教育功能的建筑于 1927—1932 年建成。❶

gaged in the architectural Chinese Renaissance movement promoted by the Kuomintang.See P91❺ Third, Chinese Catholics sought to define their own identity in a Chinese society in transformation.

The Apostolic Delegate wanted churches in the Chinese style, with Chinese art and furniture, served by indigenous priests. He wished to entrust Chinese artists with the creation of the Sino-Christian style, because he was convinced that only natives were able to express the deep Chinese soul, while Western artists could only produce 'chinoiserie'. The first generation of Chinese architects educated in the United States in the early 1920s, most of whom were Protestant, were not interested in churches, but focused on designing modern Chinese official buildings for the Republic. Because of the urgent situation, Costantini invited father Adelbert Gresnigt (Ge Lisi) to China in 1926. During his short stay in China, father Gresnigt designed four major buildings: the regional seminaries of Kaifeng (Henan province) and Hong Kong, the seminary of the Disciples of the Lord at Xuanhua (Hebei province), the Catholic University of Peking or Fu Jen University (fig. 1.45).These four educational

---

（接上页 continue）

Bays, Ellen Widmer. China's Christian Colleges. Cross-Cultural Connections, 1900-1950 [C]. Stanford, 2009, 27-56; Jeffrey W. Cody. Striking a Harmonious Chord: Foreign Missionaries and Chinese-style Buildings, 1911-1949 [J]. Architronic, 1996, V5n3; Jeffrey W. Cody. Building in China. Henry K. Murphy's 'Adaptive Architecture' 1914-1935 [M]. Hong Kong, 2001.

❶ Thomas Coomans. La création d'un style architectural sino-chrétien. L'œuvre d'Adelbert Gresnigt, moine-artiste bénédictin en Chine (1927-1932) [J]. Revue Bénédictine, 2013, 123 :128-170.

第一部分　手册的撰写背景
Part One: The Handbook in Context

buildings were built from 1927 to 1932. See p92

图1.45　北京，辅仁大学，中国基督教样式的典型代表，葛利斯神父设计，1927—1931年。
fig. 1.45 Beijing, Catholic University of Peking or Fu Jen University, the Sino-Christian style's archetype, designed and built by father Gresnigt, 1927-1932. (© THOC 2014)

然而，"中国化的基督宗教样式"，其存在极其有限，只在绘画和教会礼仪家具用品方面较为成功。❶ 因世界经济危机之故，20 世纪 30 年代的局势不再利于大规模建造项目（图 1.46 和图 1.53）。那些于 20 世纪 20 年代带动天主教会发展的动能在此时中断了。❷ 1941 年一位天津的法籍耶稣会士 Albert Ghesquières 发表了一篇文章，论及在中国天主教传教区建筑物各种不同样式的未来。❸ 该文附和刚恒毅主张的方向，以葛利斯的作品为基础，并以更强烈明显的方式，为中国化基督宗教建筑提供了详细的建造蓝图。这篇文章因发表于战时，故太晚为建筑界所注意，而无法对教会建筑发挥影响力。

20 世纪 20 年代，刚恒毅最开始在建筑领域遭遇到的挑战是来自天主教会内部，介于保守且反对中国化的基督宗教样式的传教士与那些想把教会建筑中国化的思想自由宽容的传教士之间的对立。法籍传教士抵拒刚恒毅建筑

The Sino-Christian style, however, had a limited existence, but was more successful with paintings and liturgical furniture. ❶ Because of the world economic crisis, the 1930s were no longer favourable for ambitious building projects (fig. 1.46 and fig. 1.53). The dynamics that animated the Catholic Church in the 1920s were interrupted.❷ In 1941, father Albert Ghesquières (*Gai Sijie*), a French Jesuit from Tianjin, published an article on the future of different types of Catholic missionary architecture in China.❸ In the same spirit as Costantini, this article is based on the work of father Gresnigt and gives detailed construction plans of Sino-Christian buildings in reinforced concrete. Because of the war context, this publication came too late to make an impact on religious architecture.

In the 1920s, however, Costantini's first architectural challenge was thus part of the polarization within the Catholic mission between conservative missionaries who were against the Sino-Christian style and liberal missionaries who wanted to Sinicise church architecture. It

---

❶ Aminta Arrington. Recasting the Image: Celso Costantini and the Role of Sacred Art and Architecture in the indigenization of the Chinese Catholic Church, 1922-1933 [J]. Missiology: An International Review, 2013, 41:438-451.

❷ Claude Soetens. L'Église catholique en Chine au XXe siècle [M]. Paris, 1997, 139-150.

❸ Albert Ghesquières, Paul Muller. Comment bâtirons nous dispensaires, écoles, missions catholiques, chapelles, séminaires, communautés religieuses en Chine? [J]. Collectanea commissionis synodalis, 1941, 14 :1-81; Thomas Coomans. Une utopie missionnaire? Construire des églises, des séminaires et des écoles catholiques dans la Chine en pleine tourmente (1941) [A]//Alexandre Chen Tsung-ming. Le Christianisme en Chine aux XIXe et XXe siècles. Figures, événements et missions-œuvres (Leuven Chinese Studies, 31) [C]. Leuven, 2015, 45-79.

第一部分　手册的撰写背景
Part One: The Handbook in Context

本地化计划的态度是很明显的。我们前面已经提及佘山进教之佑圣母大殿（图1.11），这是自 1925 年以来在中国建造的最重要的教堂。❶ 1924 年在上海召开了第一届教务会议，会上刚恒毅正式启动"本地化"的新政策，但即使是1年后，上海的耶稣会士、传教士建筑师和龚柏神父还是继续推崇无中国元素的西方建筑样式。

is obvious that French missionaries resisted Costantini's project of architectural inculturation. We already have mentioned the pilgrimage basilica to Our Lady of Sheshan(fig. 1.11), the most important church built in China from 1925.❷ One year after the first Catholic council of Shanghai (1924) where Costantini officially launched the new politic of 'inculturation', the French Jesuits of Shanghai and the missionary-architect father De Moerloose still preferred a Western style building with no Chinese references.

图1.46　安国（河北省），中国基督教样式主教座堂，20世纪30年代早期。
fig. 1.46 Anguo (Hebei province), Sino-Christian style cathedral, early 1930s. (© SAM)

---

❶ 见上文关于介绍和龚柏神父的相关内容。
❷ See higher, about Alphonse De Moerloose.

舶来与本土：1926 年法国传教士所撰中国北方教堂营造手册的翻译和研究
Building Churches in Northern China. A 1926 Handbook in Context

手册中关于样式的保守观点

手册《传教士建造者：建议-方案》也参与到了这次建筑大辩论中。1926 年，雍神父、梁神父、和羹柏神父，还有雷振声修士都是属于这一代老辈传教士，这些人都已经在中国待了超过 25 年，他们都经历了义和团运动，都有着传统的传教士观点。❶ 他们不可能与刚恒毅主教发生正面冲突，因为刚恒毅主教代表了教皇的权威，但是他们可以不支持大主教刚恒毅的建筑项目，而推广他们认可的西方教堂建筑和建造技术。该手册用了很长的一个注释批评中国建筑和中国装饰，指出这两者都不适用于教堂。❷

此外，他们的观点也可以从发表于 1924 年和 1925 年的两篇文章的引述中得到印证，这两篇文章分别发表在遣使会法国传教士期刊和巴黎外方传教会会刊上。❺

The Handbook's Conservative Point of View on Style

The Missionary-Builder: Advice-Plans handbook participated in this polarized architectural debate. In 1926, fathers Jung, Lamasse and De Moerloose as well as brother Litzler belonged to the generation of the old missionaries who had been in China for more than a quarter of a century, had experienced the Boxer Upraising, and had traditional missionary views. ❸ They could not directly criticize archbishop Costantini who represented the pope's authority, but they could promote Western church architecture and construction techniques over Costantini's architectural projects. In a long footnote, the handbook criticizes Chinese architecture and ornament with the evident aim of proving that both are not suitable for churches. ❹

Moreover, their reasoning is supported by long quotes from two polemic articles published in 1924 and 1925 in the French missionary journals of the Lazarists and the Society of the Foreign Missions of Paris. ❺

---

❶ 1926 年，和羹柏神父 70 岁，已经在中国待了 42 年；梁神父 57 岁，已经在中国待了 32 年，雍神父 63 岁，已经在中国待了 29 年，雷振声修士 64 岁，已经在中国待了 24 年。

❷ 手册《传教士建造者》在长注释部分提到了教堂的平面，见手册 6~11 页。

❸ In 1926, De Moerloose, aged 70, was in China since 42 years; Lamasse, aged 57, since 32 years, Jung, aged 63, since 29 years and Litzler, aged 64, since 24 years.

❹ In a long note about the plan of churches: Le Missionnaire Constructeur [M]. Xianxian, 1926, 6-11.

❺ D.G., 1924. Quelques idées sur un art chinois [J]. Bulletin catholique de Pékin, 1924, 128-130 :127-130, 177-183,218-221; Denis-Donat Doutreligne. Causerie sur l'art chinois. Peut-on adopter indifféremment dans nos Églises certains détails d'ornementation? [J]. Bulletin de la Société des Missions-Étrangères de Paris, 1925, 42 :341-356.

这些讨论分以下步骤展开,都是建立在建筑与装饰、哥特式和中国样式彼此对立的认知基础上的。

(1)说起中国建筑,用"结构"一词比"建筑"更恰当。在当时的中国,没有建筑能够与如印度或吴哥窟那般规模宏大的文物古迹相媲美,"中国赖以立国的礼仪超越了'建筑发展'所需的核心条件:自由。自由在当时的中国是受限制的,艺术家们别无他法,只能退而追求这类二等艺术,他们比起建筑更喜欢装饰自家家园和宝塔"。

(2)"中国样式,本质上就是把一个大屋顶架在柱子上,从结构角度来看,花费极其昂贵,柱子之间的空间很难用作办公空间。这样的结构系统注定其存在必定短暂……这并不是我们基督教建筑所追求的。""如此不成熟的结构模式",外加柱子,是不能满足天主教礼拜所需要的实际空间要求的。由于拱券、穹顶和拱顶系统的运用,教堂(特别是哥特式教堂)由此具有巨大的空间,能满足天主教礼拜所需要的空间要求。"如果要把寺庙改为天主教教堂,就要在庭院上方加一个屋顶,用高墙将四周围起,只需留出开门窗的口,然后将祭台放在原先放大肚佛像的地方。"

These are the steps of the syllogistic reasoning process based on the opposition of architecture and decoration, of Gothic and Chinese styles.

(1) When speaking about Chinese buildings, the term 'construction' would be preferred to 'architecture'. In China, there is no architecture comparable to that of India or Angkor, "*because the Chinese rites that 'created this nation' have suppressed freedom, the essential condition of architectural development. Freedom was so restricted in China that the artists shifted in desperation to the secondary arts of ornamentation, preferring to adorn their homes and pagodas than creating architecture.*"

(2) "*The Chinese style, as a way of building, consists essentially of a monumental roof placed on columns, extremely expensive and very inappropriate to suit its purpose. Such a construction system is doomed to a very ephemeral existence from its inception. (...) This is not what we want for our church buildings!*" This "*eminently childish construction mode,*" with many columns, is not adapted to the open spaces that the Catholic liturgy requires. Chinese temples consist of courtyards surrounded by small halls, in which few people can gather. Thanks to arches, domes and vaulting systems, especially the Gothic one, churches developed vast spaces in accordance with the needs of Catholic worship. "*To turn a temple into a Catholic church, one need only put a roof over*

the courtyard, close the four sides with a high wall, leave the necessary openings for doors and windows, and place the altar in the space used by Buddha's throne."

（3）如果一种样式不是一种结构模式，而只是一种装饰的方式，那么很显然，中国艺术作为一种样式是可以适应性地用在欧洲建筑里的。作为装饰艺术，中国样式和教堂也能互相融合。过去天主教建筑能够对不同样式做出适应性变化，对中国样式也能一样。然而，一定要避免"迷信的设计"。快乐、长寿、蝙蝠、鱼等字样，还有中国人熟悉的其他事物，这些都能创造出他们喜欢的氛围，也会吸引他们到教堂里来。这样的装饰用在教堂是不合适的，因为这样的话，"我们信仰的呈现和教堂的设计与伟大的苦难理论几乎没有什么关联，这有损于基督教义的理想"。

(3) If a style were not a mode of construction, but just a way of decoration, it would be evident that Chinese art is a style and that it could be adapted to European constructions. As ornamental art, there is no incompatibility between the Chinese style and churches. In the past, Catholic architecture adapted itself to many different styles and could adapt to the Chinese one. Nevertheless, '*superstitious designs*' must be carefully avoided. Inscriptions of happiness and longevity, bats, fishes, and all other familiar things to a Chinese would create a sympathetic atmosphere and attract him to a church. Such ornamentation is not suitable because "*presenting to our faithful and in our churches designs so little related to the great and comforting theory of suffering, which Christ came to teach, would diminish the ideal of Christianity.*"

（4）此外，天主教教堂没有钱去建这般装饰奢侈的教堂。"因此根据目前传教区的财政状况，教堂要采用中国式的装饰样式，是比较困难的。"

(4) Furthermore, the Catholic Church had no money for building sumptuously decorated churches. "*Therefore it would be difficult to adopt the Chinese decorative style, given the current financial state of the missions.*"

（5）手册作者认为哥特式不是一种装饰样式。哥特式是一种建筑形

(5) The authors conclude that Gothic is not a style of ornament. Gothic is architecture. "*Its artistic beauty results*

第一部分　手册的撰写背景
Part One: The Handbook in Context

式。"通过对普通建筑材料的合理而经济的利用，充分实现完美的比例，这就是哥特式建筑的艺术美所在。"对哥特式建筑的崇尚与19世纪50年代和60年代维利奥雷-勒-杜的法国理性理论相呼应。❶

很显然，批评中国建筑结构的"不成熟"和其存在的不长久性，还有中国装饰的昂贵，这些都是拒绝将天主教建筑中国本土化的一种含蓄表示。"哥特式"一词虽仅局限于文字描述，但8个教堂草案里的外形和设计都是西方哥特式的。把大名天主堂作为参考案例就明确表达了作者在样式选择上的倾向。

面对教廷的官方政策，手册作者有抵抗情绪，这就能更好地理解为什么手册的作者最后选择匿名方式。一方面，耶稣会在传教士事务上需要服从教宗，因此不能批评教皇的本土化政策，而应该和大主教刚恒毅合作，匿名方式是处理这类敏感问题的一种"耶稣会方式"。另一方面，他们也不能提和羹柏神父，因为他已经与刚恒毅产生冲突，刚恒毅由于和羹柏神父反对其意愿而对他进行压制制裁。至少，可以确定的是，刘钦明主教，他也是献县的耶稣会的名誉主教，他认识和羹柏神父和雍

only from the rational use of common building materials, used economically and in exactly sufficient proportions to fulfill the goal." This plea for the Gothic echoes the French rationalist theory of Viollet-le-Duc from the 1850s and 1860s.❶

It is obvious that this condemnation of Chinese architecture as being 'childish' and not for the long term, and Chinese decoration as too expensive, was an implicit refusal of any attempt to Sinicize Catholic architecture. Even if the use of the word Gothic is limited in the text, the figures and the plans of the eight model churches are Gothic and Western. Reference to the church of Daming is a concrete expression of the authors' stylistic preference.

One understands better why the authors of such resistance to the official policy of the Holy See preferred remaining anonymous. On the one hand, Jesuits are bound by their vow of obedience to the pope and therefore should not have criticized the papal inculturation policy and should have collaborated with archbishop Costantini. Anonymity fits in with the 'Jesuit manner' of dealing with delicate questions. On the other hand, they could not mention father Alphonse De Moerloose, who had already clashed with Costantini, and expose him to sanctions against his will. At least, bishop

---

❶ Eugène-Emmanuel Viollet-le-Duc (1814-1879), author of the famous Dictionnaire raisonné de l'architecture française du XIe au XVIe siècle, 10 vol., Paris, 1854-1868.

神父，也了解整个过程，他也许授权了该手册的写作和出版。

Henri Lécroart, the Jesuit vicar apostolic of Xianxian, who knew both Alphonse De Moerloose and Paul Jung, covered the whole operation and perhaps commissioned the book.

西方传教士眼中的中国工匠
The Perception of Chinese Workers by Western Missionaries

在需要用到传统工艺的时候，手册作者对有经验的中国工匠称赞有加。"关于中国木构，可以完全相信当地好工匠的经验"（图 1.47）。❶ 在建造西方建筑时，中国工匠在有人监督的情况下可以做得很好，而通常他们只做到"差不多"。❷ 要对筑岸堤的农民、烧砖的工人（"这些人责任心时好时坏，特别善于做各种承诺……"）进行监督，对夯地基、砌筑砖墙、做勾缝、做屋顶等工序进行监督。❸ 当然，像世界其他地方一样，监督的一个主要原因就是要保证工匠是按照合同要求来做的，这样可以避免工程缺陷、工资方面的不满和讨论等。因此，该手册建议要签订很详细的合同，明确规定责任事项和工资待遇，并参照使用由某位建筑师设计的

The authors of the handbook praise experienced Chinese workers when using traditional techniques. "*For Chinese timber, one can rely on the experience of good local workers.*"❶ (fig. 1.47) When working on a Western construction, the Chinese are still capable of doing good work on condition that they be closely monitored. Often the work is 'more or less' well done (*chabuduo*).❷ One should keep a close watch on peasants making embankments, brick makers ("*people with an elastic conscience, especially generous in promising…*"), coolies ramming foundations, masons, joiners, roofers, etc.❸ Of course, like everywhere in the world, the main reason for supervision was to ensure that the works were done according to contract, to prevent

---

❶ Le Missionnaire Constructeur [M]. Xianxian, 1926, 46.
❷ Le Missionnaire Constructeur [M]. Xianxian, 1926, 19, 29, 41, 42, 60.
❸ Le Missionnaire Constructeur [M]. Xianxian, 1926, 2, 12, 21, 29, 33, 41.

第一部分　手册的撰写背景
Part One: The Handbook in Context

整套方案。❶

defects, dissatisfaction and discussions about payments. The handbook advocates contracts that define responsibilities and salaries, and plans designed by an architect. ❶

图1.47　西昌（四川省），法国巴黎外方传教会主教座堂之木结构，1914年。
fig. 1.47 Xichang (Sichuan province), wooden structure of the cathedral of the French Foreign Missions of Paris, 1914. (© MEP Archives)

工资和其他花费必须要严格且诚实地按照合同来。为了避免误解和罢工，传教士代表总是与包工头签一个合同，让他来负责管理所有的工匠。在工程不是以整体定价而是按日薪计算的情况下，工匠们总是慢慢悠悠地干活。"要能够容忍他们称之为'磨工夫'的行为，有时候为了显示洋气，他

Salaries and other payments must be fixed contractually and honored fairly. In order to avoid misunderstanding and strikes, the missionary delegates the management of the workers to the contractor and signs a contract with him. Because the work is not done for a flat rate but for a daily rate, workers will tend to take their time. "*Stoically take your*

---

❶ Le Missionnaire Constructeur [M]. Xianxian, 1926, 5, 11, 16, 63.

们会说'磨洋工'（浪费欧洲人的时间）……千万不要生气，不要为此太生气或经常生气……而是要去找一个好的监工。" ❶

传教士可能会从工程一开始就发怒，从夯地基开始（图 1.48）。"建议不要从一开始就紧张兮兮。对所有人而言，打夯就像是赶集。有好吃的，每天有馍馍（白馒头），时不时犒劳（有酒有肉），但会有人几乎不工作，他们只打号子……不要无限度地容忍懒人，最好不要在现场说某人懒或和个人说什么。就跟包工头说某人有些累了，这个懒人就会自己改正或他会被以其他什么理由除名，这样做，可以保住所有人的脸面。"❷

选择一个称职的监工很关键。传教士可以从他教区的基督徒里选出一两个来，要非常了解他们的性格（图 1.49）。"监工在处理关系时会把捏分寸，既不会粗暴对待工人，也不会无来由让他们丢面子，同时也会坚持自己的要求。如果不谨慎处理好与工人的关系，就可能会有人故意搞破坏。" ❸ 要

share of what they call 'mouokoung-fou' (tempus tenere); when they want to speak elegantly, they say 'mouo yang kung' (wasting time to the Europeans). (...) Do not be angry, too demanding, too hard nor too impatient, often so (...) but start by looking for good inspectors." ❶

The missionary would, indeed, become angry from the beginning of the works, when the ramming of the foundations start (fig. 1.48). "*It is necessary not to give in to anxiety. For everyone, ramming is a fair. The food is good, 'mouo-mouo' (white bread) every day and 'k'aolao' (feasts with meat and brandy) from time to time, but people work very little, despite or because of the shouting (...) This is not a reason to tolerate laziness, but it is better not to say anything on the spot, or personally. Indicate the lazy worker to the contractor, and you will see: the lazy worker will correct himself or will be eliminated for another reason, and so all faces will be saved.*" ❷

The choice of good inspectors is crucial. The missionary will chose one or two Christians of his parish and will be very attentive to their character (fig. 1.49). "*It is important that inspectors deal tactfully with their people, do not rush the workers and do not make them unnecessarily lose face, but remain firm in their demands. Without these precau-

---

❶ Le Missionnaire Constructeur [M]. Xianxian, 1926, 12.
❷ Le Missionnaire Constructeur [M]. Xianxian, 1926, 30.
❸ Le Missionnaire Constructeur [M]. Xianxian, 1926, 12.

第一部分　手册的撰写背景
Part One: The Handbook in Context

给监工经济报酬，不要仅仅给他们精神鼓励，如比划一个十字。他们会对疏忽和偷盗进行监督，而这些都是经常发生的。❶

tions: mind sabotage!" ^(See p102)❸ Inspectors will be rewarded with money, not with pious objects such as a crucifix. They will also focus on negligence and theft, which are common occurrences. ❶

图1.48　内蒙古，夯筑墙体，
20世纪30年代。
fig. 1.48 Inner Mongolia, ramming a wall, 1930s. (© FVI, CHC pictures)

图1.49　仕拉乌素壕（内蒙古），
在建孤儿院的一位中国天主教
施主王亚海（音），20世纪20年代。
fig. 1.49 Shilawusuhao (Inner Mongolia), Wang Ya-hai, a Chinese Catholic benefactor of an orphanage in construction, 1920s.
(© FVI, CHC pictures)

如果工匠没有或几乎没有西方建筑建造方面的经验，应该对他们进行监督并让有经验的工匠进行指导。好学、有技能和有耐心的工匠会很快融入学习过程中并且学得很快，与那些固执、

Workers with no, or little, experience in Western building techniques should be supervised closely and coached by experienced workers. Curious, skilled and patient workers would learn and progress rapidly, contrary to

---

❶ Le Missionnaire Constructeur [M]. Xianxian, 1926, 13, 64.

懒惰和无知的家伙正好相反。中国工匠受其建筑传统和文化背景的影响，他们很难改变他们的习惯。"他总是本能地反对：这是新的。"❶ 举个例子，他不理解为什么西方建造者不喜欢青砖，他会想："他们是'笨洋人'，只需要红砖。"❷ 对中国人来说，一栋房子必须要好看，这比坚固更为重要。❸

手册里也强调了维护建筑的必要性，而维护又不是中国人的传统："就中国人的习性，维护是令人厌倦的，无论是衣服、机器或房子的维护，他们都不喜欢。如果事关公共建筑的维护，你会碰到的困难可能是物质上解决不了的。看看那些衙门、塔、皇家陵墓、道路、桥梁、城墙。但愿我们那穷酸的教堂不要变成那样！"❹

stubborn, lazy and negligent ones. The building traditions and the cultural background of the Chinese worker would not incline him to change his habits. "*He will instinctively make opposition: it is new.* "❶ For example, he would not understand why Western builders dislike blue bricks and would think: "*One must be 'yangren' (foreigner = fool) for requiring red bricks.* "❷ For a Chinese, a building must be good-looking (*haokan*) rather than solid.❸

The handbook's conclusion stresses the need to maintain buildings, which is not part of the Chinese tradition: "*Unfortunately nothing seems more repugnant to the Chinese temperament than maintenance, whether it be a dress, a machine or a building (...) Ancient public buildings, city walls, bridges, roads etc. are ruinous; may churches not become so.* "❹

技术的碰撞和转移
Encounter and Transfer of Technology

基督教传教区每天都上演着西方传教士和当地老百姓因彼此间文化不

The Christian mission is based on an encounter between different cultures,

---

❶ Le Missionnaire Constructeur [M]. Xianxian, 1926, 35.
❷ Le Missionnaire Constructeur [M]. Xianxian, 1926, 21.
❸ Le Missionnaire Constructeur [M]. Xianxian, 1926, 23, 31, 41, 51, 61.
❹ Le Missionnaire Constructeur [M]. Xianxian, 1926, 66.

第一部分　手册的撰写背景

Part One: The Handbook in Context

同而产生的碰撞。在中国，这类碰撞是很艰难的，19世纪40年代至20世纪40年代期间经常发生冲突。❶ 不管怎样，这类碰撞远不仅仅是以上帝、宗教和西方利益为名义的传福音，它也带来其他领域，如日常生活、知识、艺术和科学技术等方面的互动和交流。手册《传教士建造者：建议-方案》的主要成就在于它集合了建筑和结构领域因碰撞而产生的经验，从混凝土建筑、建造技术到处理与中国工匠的社会和专业关系，再到建筑样式的讨论等。在本书的第三部分，对该手册进行了深入分析，将"理论实践"与已在中国建成教堂的设计方案、建造工程的文献照片及大名天主堂进行比较。对中文和河北方言中的建筑术语也进行了研究。❷

在基督教提倡中国化的背景下，手册《传教士建造者：建议-方案》必须从两个方面切入：样式和结构。说及样式，手册阐述了来自"老一辈"传教士

between foreign missionaries and local people. In China, this encounter was difficult and lead to recurrent clashes from the 1840s to the 1940s.❶ Nevertheless, the encounter was much more than evangelization in the name of a god, a religion, and Western interests. It generated interactions and exchanges in all possible fields of daily life, knowledge, art, science and technology. The main achievement of *The Missionary-Builder: Advice-Plans* handbook is precisely to have compiled the experience from encounters in the field of architecture and construction, from concrete architectural and technical aspects, to social and professional relationships with Chinese workers, and statements about architectural styles. In the second and third parts of this book, the handbook will be analyzed more in depth by comparing its 'theorized practice' and its plans, with construction details of churches built in China, archival pictures of building works, and with the cathedral of Daming. Architectural terminology in Chinese and the Hebei dialect has been studied as well. ❸

In the context of Christian inculturation, *The Missionary-Builder: Advice-Plans* handbook must be considered from two architectural points of views:

---

❶ R. Gary Tiedemann (ed.). Handbook of Christianity in China. Volume Two: 1800 to Present [C]. Leiden-Boston: Brill, 2010.

❷ 相关内容见附录2。

❸ See Appendix 2.

的抵制情绪，如和羹柏神父就反对大主教刚恒毅和教廷提倡的中国式基督教样式。手册鼓励使用西方建筑，特别是哥特式。中国基督徒对建筑样式又是什么样的态度呢？要回答这个问题，仅仅参考西方的文献资料是不太容易得到答案的。

（1）保守派的传教士确信中国基督徒只需要西方建筑，和羹柏神父在1924年的一封信里这样写道："基督徒不需要中国式教堂。他们认为这种样式适用于宝塔。如今中国人也按照欧洲方法来建造，当然建得都很差，但其理想是像欧洲人那样建造。北京、保定、张家口的建筑和地方上的学校都是欧洲式的……宗座代表（刚恒毅）特别反对哥特式，同所有意大利人一样，他崇尚拉特朗圣约翰大殿❶的样式。我所碰见的那些爱思考的传教士和主教都对建造中国式教堂深信不疑，他们都没有建造实践方面的经验。"❷雍神父在1919年大名天主堂建造期间给他的上司写过一封信，他在信中批评了中国样式并提出："一座真正的哥特式教堂，只要朝向好，在这个国家同样会冬暖夏凉。我

style and construction. About style, the handbook expresses the resistance of the 'old school' and missionaries such as father De Moerloose against the Sino-Christian style promoted by archbishop Costantini and the Holy See. The handbook pleads for Western architecture, especially Gothic. What was the attitude of Chinese Christians towards architectural styles? Answering this question is not easy, from only Western sources.

(1) Conservative missionaries were convinced that Chinese Christians only wanted Western architecture, as father De Moerloose wrote in a letter dated 1924: *"The Christians do not want Chinese churches. They say that this style is good for the pagodas. Today's Chinese also build in the European way, of course with a bad result, but the idea is to build like Europeans. In Peking, Paotingfu* (Baoding), *Kalgan* (Zhangjiakou) *buildings and even local schools are European. (...) The Apostolic Delegate (Costantini)is particularly against Gothic; like all the Italians, his model is St. John Lateran. All the right-thinking missionaries and bishops I have seen are convinced that there is no question of building Chinese churches, all the more they are not practical.* "❷ In a letter to his superior, written during the construction

---

❶ 拉特朗圣约翰大殿是天主教罗马总教区的主教座堂，罗马总主教（教宗）的正式驻地（教座），也是罗马四座特级宗座圣殿（拉特朗圣约翰大殿、圣伯多禄大殿、圣母大殿和城外圣保禄大殿）中最古老、排名第一的一座，享有全世界天主教会母堂的称号。

❷ Letter of Alphonse De Moerloose to father Karel Van de Vyvere, Yangjiaping, 28 August 1924 [original in Dutch] (Leuven, KADOC, CICM Archives, T.I.a.14.3.2).

们的信徒都非常喜欢，因为它不是塔。"❶（图 1.50 和图 1.51）

（2）葛利斯神父在 1928 年的一篇文章里表达了反对的观点："中国天主教徒也许的确是出于敬重，对西方牧师有偏爱，隐藏了他们对事物的一些真实想法，甚至他们会假装更喜欢外国建筑；这样的情况也许是因为他们那与生俱来的谦恭，也许也是因为他们担心会对他们虔诚崇拜的那份感情造成伤害。因此，传教士如果仅仅从表面理解这些，是很不明智的。"❸（图 1.52）

works of Daming in 1919, father Jung criticizes the Chinese style: "*A real Gothic church, in this country, is cool in summer and warm in winter, if they are oriented as it must be. Our Christians love them very much, precisely because they are not pagodars.* "❷ (fig. 1.50 and fig. 1.51)

(2) The opposite point of view is exposed by father Gresnigt in an article from 1928: "Chinese Catholics may, indeed, out of deference for the evident predilections of the foreign priest, conceal their real feelings on the subject, and may even go so far as to feign a preference for foreign architecture; but in such cases they are speaking out of innate courtesy and from a fear to wound the sensibilities of one for whom they have sincere respect. It would, therefore, be exceedingly unwise for the missionary to take expressions of this kind literally." ❸ (fig. 1.52)

---

❶ 相关内容见雍神父于 1919 年 7 月 13 日写给教会官员的法文信(Vanves, ASJ Fance, GMC 101, letters to procurator, Jung)。

❷ Letter of Paul Jung to the Procurator, Daming, 13 July 1919 [original in French] (Vanves, ASJ Fance, GMC 101, letters to procurator, Jung).

❸ Adelbert Gresnigt. Chinese architecture [J]. Bulletin of the Catholic University of Peking, 1928, 4:41.

舶来与本土：1926年法国传教士所撰中国北方教堂营造手册的翻译和研究
Building Churches in Northern China. A 1926 Handbook in Context

图1.50　双树子（河北省），和羹柏神父1917年在早期一座较小教堂旁边建造的哥特式大教堂。
fig. 1.50 Shuangshuzi (Hebei province), great Gothic church built by father De Moerloose in 1917 beside a smaller earlier church. (© KADOC, CICM)

有意思的是，传教士建筑师的认识与中国基督徒的想法完全相反。是不是所有的中国人，不管是不是基督徒，都有着同样的观点呢？这样的问题就不在这里深入讨论了。然而值得一提的是，目前中国基督徒明显更喜欢西方样式，对后现代的哥特式风格更是偏爱。❶

那是不是就说明宗座代表刚恒毅

It is interesting that both missionary-architects had totally opposite perceptions of what Chinese Christians thought. Did all the Chinese people, however, Christian or not, have the same opinion? This rhetorical question will not be developed further here. It is, however, worth noticing that in today's China, Christians build churches with explicit references to Western styles and a predilection for postmodern Gothic forms. ❶

Did Costantini's inculturation the-

---

❶ Thomas Coomans. Gothique ou chinoise, missionnaire ou inculturée? Les paradoxes de l'architecture catholique française en Chine au XXe siècle [J]. Revue de l'Art, 2015, 189 :9-19; Thomas Coomans. Die Kunstlandschaft der Gotik in China: eine Enzyklopädie von importierten, hybridisierten und postmodernen Zitaten [A] // Heiko Brandl, Andreas Ranft, Andreas Waschbüsch. Architektur als Zitat. Formen, Motive und Strategien der Vergegenwärtigung [C]. Regensburg, 2014, 133-161.

第一部分　手册的撰写背景
Part One: The Handbook in Context

当年提倡的中国化理论彻底失败了呢？21世纪中国的基督教特性是不是还是需要借由其外国根源来体现呢？

ory and efforts completely fail? Does the identity of Christianity, in twenty-first-century China, still need to express its foreign roots?

图1.51　平地泉（内蒙古），圣母圣心会哥特式教堂前合影的中国修生，20世纪20年代。
fig. 1.51 Pingdiquan (Inner Mongolia), Chinese seminarians posing in front of the Scheut fathers' Gothic church, 1920s. (© KADOC, CICM)

另外，也是该手册的主要目的，就是介绍建造实践的经验，这些经验都是由雷振声修士与和羹柏神父在长期实践基础上总结出来的。尽管该手册自始至终是一种比较性的西方思维，但是对中国建造技术和建筑材料也给予了关注。其中有一些都是结合具体的环境来讨论的。因此，我们可以说该手册在一

The other aspect, and also the main purpose of the handbook, was construction practice, in all likelihood based on the long experience of brother Litzler and father De Moerloose. Even if comparative Western thinking is patent from the beginning to the end of the handbook, a great deal of attention is paid to Chinese building techniques and materials. Some

~ 109 ~

定程度上体现了文化的互相渗透和互相影响——对特定文化环境（即中国文化背景）的适应性能力。在选址和建筑朝向选择方面，都是结合中国地理条件或风水❶来考虑的。

of them are recommended in precise circumstances. Therefore we could conclude that the handbook reveals a certain acculturation —the ability to adapt to a given cultural milieu—the Chinese context. This is particularly true in the choice of locations and in the orientation of buildings, according to geomancy or fengshui. ❷

图1.52　五号（河北省），中国基督教样式的圣母圣心会教堂，1930—1933年建。
fig. 1.52 Wuhao (Hebei province), Scheut fathers' church in Sino-Christian style, built in 1930-1933. (© ASJ France, GMC)

---

❶ 风水在该手册里只出现过一次，是在讲中国屋面装饰的时候提到的。Le Missionnaire Constructeur [M]. Xianxian, 1926, 51.

❷ Fengshui is only mentioned once, in the chapter about Chinese roofs' decoration. Le Missionnaire Constructeur [M]. Xianxian, 1926, 51.

第一部分　手册的撰写背景
Part One: The Handbook in Context

除了选址和朝向，作者似乎没太关注中国北方地区的特殊气候，他们仿佛觉得有着大窗户和大空间的哥特式教堂适用于世界所有地方。但有些传教士觉得哥特式教堂很麻烦："（和羹柏神父）没有完全考虑我们区域的特殊气候条件。这些有着薄墙的高教堂，窗户很多，木式天花板很薄，外形很好看，但那些在零下30摄氏度到零下35摄氏度或特别炎热的天气还必须去教堂的牧师和信徒们真是可怜。在里面完全会被冻僵或被闷热死；在众信徒礼拜时伸出胳膊给他们分圣餐时，简直就是折磨。牧师和基督徒对此都深有体会；然而如此漂亮又不切实际的样式却很难消失……但愿上帝保佑我们，不要让那些保守的艺术家们因为他们的艺术准则而把这些没用的苦难介绍到蒙古传教区来。"❶

从现代性影响的角度看，手册《传教士建造者：建议-方案》是很保守的。一方面，西方作者拒绝本土化和任何向中国式基督教样式的风格演变。另一方面，他们也没有提及任何现代技

Except for the location and orientation, the authors do not seem to have paid great attention to the specific climate of Northern China, as if Gothic churches with great windows and vast spaces were appropriate to all places in the world. Some missionaries, however, found Gothic churches totally inappropriate: "[De Moerloose] *did not pay enough attention to the particularities of our region and its excessive climate. These high chapels, with thin walls, many windows, and thin wooden ceilings, have pleasant forms, but the priest and the churchgoers are to be pitied when they are obliged to gather to pray, by minus 30-35°C or in oppressive heat. You are either freezing or suffocating; staying with spread arms during the canon of the mass is torture, as is serving communion to crowds. Priests and Christians have all experienced this problem; however the vogue of this so beautiful, but so unpractical, style was slow to disappear. (…) May God protect us from conservative artists who, because of their art principles, introduced useless suffering in our missions of Mongolia.* "❶

From the point of view of the impact of modernity, *The Missionary-Builder: Advice-Plans* handbook is rather conservative. On the one hand, the Western authors refused inculturation and any stylistic evolution to a Sino-Christian

---

❶ Josef Nuyts. En tournée à travers le Vicariat [J]. Missions de Scheut, 1938, 218-219.

舶来与本土：1926 年法国传教士所撰中国北方教堂营造手册的翻译和研究
Building Churches in Northern China. A 1926 Handbook in Context

术，如钢筋混凝土、金属的使用，石头雕刻和色彩的使用等。❶ 在 20 世纪 20 年代中期，除了一些大城市和通商口岸，中国人对当时西方新的构造技术知之甚少。在很有限的经济预算下，地方教会几乎不可能建什么教区教堂。有一点很难相信，经常出入北京、天津和上海的和羹柏神父，他与西方建筑公司一直有联系，他也经常阅览建筑期刊，但他竟然不知道钢筋混凝土和钢结构技术，来自大名-献县区域的雍神父和雷振声修士也是，他们也与天津耶稣会有密切联系，他们知道天津的西方现代建筑，但他们都没有在手册提及任何新建筑材料和构造技术，这点很难让人理解。

该手册介绍了一种来自北欧以砖构建筑为主的国家西方传统哥特式建筑。在欧洲，这类建筑自 19 世纪 90 年代起就备受争议，在第一次世界大战后完全被摒弃了。而在同时期的中国农村，该手册里提及所需的建造技术也在

style. On the other hand, they say nothing about modern techniques of reinforced concrete, use of metal in general, stone carving, and the use of colors. ❷ In the mid-1920s, recent technological advancements in the West were indeed largely unknown in China, except in the great cities and the treaty ports. They were impossible to afford by local congregations building a parish church with a limited budget. It is, however, hard to believe that father De Moerloose, who was often in Beijing, Tianjin and Shanghai, who had contacts with Western construction firms and who read architectural journals, was not aware of the technical possibilities of reinforced concrete and iron structures. The same is true for father Jung and brother Litzler from the Daming-Xianxian area, who had close contacts with the Jesuits of Tianjin and knew this city and its modern Western architecture.

This allows us to conclude with a paradox. The handbook describes and illustrates a type of Western traditional Gothic architecture from the Northern European brick countries. In Europe, this architecture had been contested since the 1890s and was completely outdated after

---

❶ 钢筋混凝土只在该手册里提到过三次，但实际上指的不是钢筋混凝土而是"中国式混凝土"——土和石灰或砖屑和石灰的混物。压型钢板做的屋顶在该手册里提到过三次，但没有讲具体细节。关于"中国式混凝土"的介绍，可见文章： Ian Svagr. Le béton chinois [J]. La Cité. Architecture, Urbanisme, 1923, 4(5):91-92.

❷ Concrete is only mentioned three times in the handbook (p. 29, 30-31), but it is rather 'Chinese concrete' —a mix of earth and lime, or crushed bricks and lime— then reinforced concrete. Galvanized metallic roof sheets are mentioned three times (p. 35, 48 and 49) but are not considered in detail. About 'Chinese concrete': Ian Svagr. Le béton chinois [J]. La Cité. Architecture, Urbanisme, 1923, 4(5): 91-92.

当地难以实现。在中国北方的许多村庄里，建造一座哥特式教堂是完全不同的建筑体系的表达，与地方传统相悖，它被视为"外国的"或"舶来的"（图 1.53 和图 1.54）。

World War I. At the same time in rural China, the techniques described in the handbook remained difficult to realize. In many villages of Northern China, building a Gothic church was the expression of a completely different architecture, which could be qualified as 'foreign' or 'imported' in contrast to the local tradition (fig. 1.53 and fig. 1.54).

图1.53　南泉子（吉林省），圣母圣心会正在建造的乡村教堂，年代未知。
fig. 1.53 Nantsuantse (Jilin province), Scheut fathers' village church under construction, not dated. (© FVI, CHC pictures)

传教士需要人来建造他们的教堂，他们需要那些在西方建造技术方面受过培训的中国工匠。因此，他们可以被视为实现建筑技术转移的代理，但不是那些20世纪20年代在通商口岸提倡装饰风艺术和现代性的一群西方建筑师和工程师的代理。❶

Missionaries needed people for building their churches, and trained Chinese workers in Western traditional building techniques. They should therefore be considered as agents of technological and architectural transfer, but not as agents of modernity like Western ar-

---

❶ Laura Victoir, Victor Zatsepine (eds). Harbin to Hanoi: The Colonial Built Environment in Asia, 1840 to 1940 [C]. Hong Kong: Hong Kong University Press, 2013; Thomas Coomans. China Papers: The architecture archives of

chitects and engineers who developed Art Deco and Modernism in the 1920s in the treaty ports.^{See p113}❶

图1.54 舍必崖（内蒙古），圣母圣心会乡村教堂，和羹柏神父建造，1904—1905年。
fig. 1.54 Shebiya (Inner Mongolia), Sheut fathers' village church, by father De Moerloose, 1904-1905. (© FVI, BR pictures)

——————————————（接上页 continue）

of the building company Crédit Foncier d'Extrême-Orient (1907-1959) [J]. ABE Journal. European architecture beyond Europe, 2014, 5:689, http://dev.abejournal.eu/index.php?id=689.

第二部分
手册的内容和翻译
Part Two:
The Handbook Translated

舶来与本土：1926 年法国传教士所撰中国北方教堂营造手册的翻译和研究
Building Churches in Northern China. A 1926 Handbook in Context

翻译者的体会
Note of the Translators

这部分内容是将该手册内容完整地翻译成英文和中文。手册是用法语写的，法语原文可见附录 1。手册的翻译工作并不容易，主要原因如下。

最主要的一个原因是雍神父的写作风格，他的写作接近口语化描述，而不是学术的规范书面表达。作为法文学校的一名教师和教堂的一名传道者，他演讲时会用幻灯机，边放映图片边演讲，在当时他是一个很受欢迎的演讲者，他很善于口头表达。然而，这样的表达很难翻译，特别是当他要对一些技术操作层面进行书面描述的时候。本书的中文和英文翻译尽可能地接近雍神父的原有写作风格，因此很多时候不是规范的学术用语。

第二个主要原因是特定的建筑词汇。全世界不同语言使用的建筑词汇是非常不同的，无论是学术文献还是建造工程的地方用语也都不同。《圣经》里巴别塔❶的故事讲述的就是建造工程

In the following pages, the handbook is translated into English and into simplified Chinese. A transcription of the handbook's French original text can be found in Appendix 1. For several reasons, translating the handbook was not an easy task.

The main reason is father Jung's style, which often looks more to oral speaking than written academic French. As a teacher at the French College, a preacher in the church, and a popular lecturer using a magic lantern, father Jung was used to oral communication. Such a style is difficult to translate, especially when the author tries to give a literary dimension to technical descriptions. The English and Chinese translations respect as much as possible father Jung's style, what explains that they not always are pure academic language.

The second main reason, as would be expected, is the specific architectural vocabulary. In all the languages in the world, the use of architectural vocabulary varies tremendously from academic

---

❶ 巴别塔（the Tower of Babel），也译作巴贝尔塔、巴比伦塔，或意译为通天塔，巴别在希伯来语中有"变

第二部分　手册的内容和翻译
Part Two: The Handbook Translated

中因为工匠使用的语言不同而造成的混乱，最后通天的巴别塔当然没有建成。翻译该手册的两名人员的专业背景都是建筑历史和建筑遗产保护，他们花了很多时间来讨论技术用语的准确翻译。❶ 手册中有用法语表达的技术用语，也有雍神父按照法国语音写的中国技术用语，但他不是按汉语拼音写的，这些技术用语很多都是当年工匠的方言，具体可见附录 2。在英文翻译中，这些中文技术用语及地方的名字都用汉语拼音来表达。

该手册的翻译工作分两步进行，首先是 Thomas Coomans 把法语手册直译为英文，再由吴美萍将英文翻译成中文，前者的母语是法语，后者的母语是中文；接着由 Whitney de Courcel 对英文翻译进行校正，她的母语是英文，而且她熟练掌握法文，因此她能完全参照法语原文对英文翻译进行润色。

Thomas Coomans 和吴美萍也就历史、文化、宗教方面的术语进行了讨论，并对一些术语进行了注释（脚

treaties to local dialects on the building works. The myth of the Tower of Babel is based on the tongues' confusion on construction works. The two translators of the handbook, both architectural historians and conservationists, spent many hours discussing on the accurate translation of technical terms. ❷ This had to take in account both the French technical terms and the Chinese technical terms written by father Jung in a Romanized form that is not pinyin but French phonetical. These terms are listed in Appendix 2. In the English translation, these terms as well as the names of places in China are mentioned in pinyin.

The translators of the handbook worked in two steps. First Thomas Coomans made a literal translation of the handbook from his mother tongue French to English. This English text was translated by Wu Meiping to Chinese her mother tongue. Thereafter, Whitney de Courcel, a native English speaker, revised the English translation.

Wu Meiping and Thomas Coomans also discussed about which historical, cultural or religious terms required an

──────────────（接上页 continue）

乱"之意。据《圣经·创世记》第 11 章记载，当时人类联合起来兴建希望能通往天堂的高塔。为了阻止人类的计划，上帝让人类说不同的语言，使人类相互之间不能沟通，计划因此失败，人类自此各散东西。

❶ 主要用的专业字典有：Melchior de Vogue, Jean Neufville. Glossaire des termes techniques (Introduction à la Nuit des Temps, 1) (4th revised edition, La Pierre-qui-Vire), 1989; Guo Qinghua.中国建筑英汉双解词典 / A Visual Dictionary of Chinese Architecture. Mulgrave: Image Publishing, 2007.

❷ Several specialized dictionaries were used: Melchior de Vogue, Jean N eufville. Glossaire des termes techniques (Introduction à la Nuit des Temps, 1) (4th revised edition) [M]. La Pierre-qui-Vire, 1989; Guo Qinghua. Zhongguojianzhu Ying Han shuangjiecidian / A Visual English-Chinese Dictionary of Chinese Architecture [M]. Mulgrave: Image Publishing, 2007.

注）。有时候对一方很容易的词对另一方来说却不是，这样的讨论本身就是一项很有成效且能互相促进的文化互动。基于此，英文翻译和中文翻译的脚注不完全相同。

该手册中的原始图纸都会在第二部分呈现（图2.1~图2.39）。对手册内容的翻译会参考引用原始图纸及第三部分的一些图纸，这样的交叉引用也有助于更好地理解手册中的经验、理论和实际的教堂案例，特别是对大名天主堂的认识和理解会进一步加深。

explanatory footnote. Often what was obvious for the one was not for the other. This resulted in a fruitful and stimulating cultural interchange. For this reason, the explanatory footnotes are not the same in the English and the Chinese translation.

The complete original illustration of the handbook is reproduced in this chapter (fig. 2.1 to fig. 2.39). References to these figures are mentioned in the text as well as references to illustrations from chapter 3 in order to facilitate the crossed links between the handbook's theory and the practice, especially from the church of Daming.

手册的出版信息
Published Information

题目：传教士建造者：建议-方案
作者：中国北方的传教士
印刷地：献县，1926年（1935年重印）（图0.3和图2.1）
印刷授权：❶
卜鸣盛神父，耶稣会会士，直隶东南部传教区区长。
刘钦明神父，耶稣会会士，波摩莱

The Missionary-Builder: Advice- Plans
by missionaries from Northern China.
Xianxian Press. 1926 [reprint 1935] (fig. 0.3 and fig. 2.1).

License to print:❷
Father Jean Debeauvais, Jesuit, regular superior of the mission (of Southeastern Zhili).
Father Henri Lécroart, Jesuit, titular bishop

---

❶ 天主教教会声明有教会某成员写成的关于宗教或教义问题的书籍文稿均可以印刷出版，但要经过审查确定其中并没有违背天主教的言论，最后的授权方当地主教，主教批准则可出版。

❷ Declarations by Catholic authorities that writings on questions of religion or morals by a member of the Church may be printed. This means that censors charged with examining the writings granted a declaration of no objection. Final approval is given by the bishop of the place of publication.

(Anchialus)的领衔主教，献县宗座代牧。　　of Anchialus, vicar apostolic of Xianxian.

第二部分　手册的内容和翻译
Part Two: The Handbook Translated

图2.1　手册封面，1935年再印版。
fig. 2.1 The Handbook's title page, reprint of 1935. (© Université Laval, Bibliothèque)

舶来与本土：1926 年法国传教士所撰中国北方教堂营造手册的翻译和研究
Building Churches in Northern China. A 1926 Handbook in Context

手册的目录
Contents

前言/5	Preface
第 1 章/选址/6	First Chapter: Choosing a Location
第 2 章/建筑布局和朝向的选择/8	Second Chapter: Choosing the Buildings' Layout
第 3 章/设计方案、工程预算和施工合同/10	Third Chapter: Plan, Estimate, Contract with the Contractor
设计方案/10	Plan
工程预算/10	Estimate
施工合同/14	Agreement with the Contractor
第 4 章/建筑材料/16	Fourth Chapter: About Materials
砖/16	Bricks
瓦/18	Tiles
木料/18	Timber
石灰/19	Lime
麻绳/20	Rope
脚手架（杆子）/21	Scaffolding Poles
第 5 章/砖石工程/22	Fifth Chapter: Masonry
地基/22	The Foundations
打夯/22	Ramming
砖石工程/23	Masonry
砂浆/23	Mortar
墙/23	Walls
中式墙/24	Chinese Walls
砖的砌合/25	Brick Bonds
墙的厚度/25	Thickness of the Walls
隔碱层/26	Saltpetre
装窗框和门框/28	Installing Door and Window Frames
第 6 章/屋顶/30	Sixth Chapter: The Roof
中式屋顶/30	The Chinese Roof
欧式屋架/31	The European Roof Frame
屋面坡度/33	Pitch of the Roof
第 7 章/屋面装饰/34	Seventh Chapter: The Roof Ornamentation
第 8 章/建筑细节的处理/37	Eighth Chapter: Miscellaneous
滴水石/30	Dripstones
拱顶/30	Vaults

第二部分　手册的内容和翻译
Part Two: The Handbook Translated

砖拱/30	Brick Arches
彩色玻璃窗/30	Stained Glass Windows
钟楼尖顶/30	The Spire of a Bell Tower
勾缝/30	Grouting of Masonry
抹泥/30	Roughcast or Coating
和泥/30	Mix Earthmortar
保护铺地/30	Protecting the Flooring
磨砖/30	Polish the Bricks
廊子/30	Veranda
墙角砖砌合/30	Clamping
灶囱/30	Fireplaces
楼梯/30	Staircase
钟架子/30	Belfry
焦油/30	Tar
排雨水/30	Flow of Rainwater
玻璃/30	Glass
油漆/30	To oil
土坯墙的梁下柱/30	Columns under the Beams
结语/40	Conclusion

序　言
Preface

这本册子不是为那些建造大教堂的建造者写的。这本册子也不是为了介绍中国艺术在基督教教堂上的适应性应用。

每一个传教士都有可能面临建造教堂的任务，即使只是建一所只有三间的乡村学校（图 1.24 和图 1.26）。这些谦逊的建造者，有些是我们的兄弟，有些甚至是建筑师，他们都曾做过这类建筑工程，他们希望能分享他们的经验。

This brochure is not made for cathedral builders. This is not even an essay on the famous adaptation of Chinese art to our Christian churches.

Every missionary may find themselves obliged to build something, if only a village school of three bays (*jian*) (fig. 1.24 and fig. 1.26). It is with these modest builders, that we, brothers, even more modest architects, but already confronted with many building projects in China, would like to share their experience.

建造教堂最重要的当然是要牢固。如果能兼顾功能使用，如果一个小教堂在坚固的同时也能够美观就更好了。这本册子就是想为建造牢固、安全、不太贵、不太丑的教堂提供一些参考：用极少的钱建造一个漂亮的教堂，就像阿巴贡的宴会❶一样。

刚到中国的传教士，会面临如同建巴别塔一样的挑战：语言不通的困惑。因此，我们认为，在给出准确建议的同时，非常有必要对一些专业术语进行解释说明。

最后也附加了一些具体例子，主要是那些比较容易建造的教堂案例。我们提供了一些图纸和方案，希望能给人们一些灵感和启发。

这本册子里面所给出的建议和说明，适用于中国北方大多数地方。

愿圣约瑟❸保佑此书。

大名，1926 年 3 月 19 日

The first priority, of course, is to build solidly. If the pleasant can be combined with the useful, if a little church can be attractive while remaining firmly set on its foundations, all the better. So, the ideal here is to help you build solidly, safely and not too expensively nor too unattractively: a pretty church for little money, like Harpagon's feasts. ❷

The missionary newly arrived in China, will find himself struggling with the consequences of the construction of the Tower of Babel: the confusion created by different languages. As well as the technical advice given to builders, we thought it appropriate to add, as we go along, the translation of the professional terminology.

Finally, to add example to precept, we include a pocket containing figures and some plans of churches, which are fairly easy to realize. One can at least find inspiration from them, if nothing is better.

It goes without saying that the following advice and remarks, apply mostly to the conditions of Northern China.

May St Joseph bless these pages. ❹

Daming, 19 March 1926

---

❶ 阿巴贡是法国作家莫里哀(1622—1673)的著名戏剧《吝啬鬼》中的角色，他是吝啬鬼的典型。

❷ Harpagon is a character figuring 'greed' in L'Avare, a famous play by Molière (1622-1673).

❸ 圣约瑟是《新约圣经》记载中耶稣的养父、圣母玛利亚的丈夫，他生活在纳匝勒，以木匠手艺维生。天使曾经多次在梦中给他指示，他每次都无条件地立即遵行。据《圣经》描述，他被称为义人，怀有体谅别人的心并深爱着自己的儿子。

❹ St Joseph, the worldly father of Christ, is the patron of the carpenters, the workers, and the families.

第二部分　手册的内容和翻译
Part Two: The Handbook Translated

第一章　选址
First Chapter: Choosing a Location

千万不要着急忙慌地在一个新基督教国家开始建设。❶ 即使在很大一部分人已经接受洗礼的情况下，基督教义也可能会分崩离析，尤其是在全家人还没有全部受洗、妇女还是非信徒或这些新信徒几乎都是穷人的情况下。当然基督教社区也可能扩展得很快，原先建好的教堂很快就容不下更多的教徒。

比较理想的做法是，时机好的时候以抵押担保，这样就有足够的回旋空间，即使情况有变，也不会因为失败而导致金钱上的损失。

建造教堂对新教徒是巨大的考验。他们有可能会拒绝捐赠，也有可能每个人捐赠的数额不一样，这多多少少取决于他们的信仰。由此，会有人嫉妒，也会有人丢面子。如果让他们负责购买建筑材料，或者他们只是中间商，这种时候他们往往会面临良心的考验。那些买回来的建造教堂的材料可能会被挪为私用，有人可能只是暂时借

Do not begin building too fast in the new 'christianities'.❶ The 'christianity' could still crumble, even with a considerable number of baptized, especially if they are not entire families, or if the women remain pagan, or if these neophytes are almost all poor. Inversely, the Christian community could develop rapidly and the premises would soon become too cramped.

The ideal, when there is a good opportunity, is to take out a mortgage. You have sufficient room, pending new converts, and the money is not lost in case of failure.

Furthermore, undertaking a building is often too daunting a task for brand new Christians. They will refuse to contribute, or will contribute in a very unequal way, depending on the strength of their faith. Hence creating jealousy, or loss of face. But above all they will risk weighing down their conscience, if they are asked to buy building materials, or simply serve as an intermediary for such purchases. Materials purchased by the

---

❶ A 'christianity', in French a 'chrétienneté', is both a Christian community and a Christian settlement, which could be a quarter in a town or in a village, as well as an entire Christian village.

的，也有人是故意忘记还回来，或者更简单地只是说被偷了。新基督徒会说天父所有即为其子民所有，这也是他们对"诸圣相通功"❶教义的理解。

关于买地，如果可能的话，最好一开始就至少买两三亩。要不然，等将来教区要扩张的时候，附近居民就会抬高土地的价格。

选址时要尽可能避免以下情况。

（1）不要购买村庄外的土地。它们有可能会很便宜，但那是因为没有人愿意住在那里。那边没有邻舍，有时是一边没有，经常是三边都没有邻舍。妇女和少女，尤其是刚成为教徒的妇女和少女，都不太愿意长途跋涉去教堂。

（2）如果该村庄有集市，教堂选址不要选在市场中心。让新教徒穿过熙熙攘攘的人群去教堂祷告，这极为不好，尤其是对妇女。集市的喧扰会影响

mission or the Christian community will be used for private purposes; they will say that the materials are 'borrowed'; but they will forget to return them; or they will simply be stolen. New Christians will willingly say that what belongs to the father belongs to the children as well, commenting in their way on the dogma of the Communion of Saints. ❷

When buying land, if possible, it is better to buy at least two or three acres (*mu*) at the start. Otherwise, later, when the community will have grown, the neighbours will make you 'favourable' prices, at a much higher price…

As much as possible, the following locations should be avoided:

(1) Do not buy outside the village. The properties are, of course, much cheaper precisely because nobody wants to settle there. There will be no neighbours, at least on one side, often on three sides. Women and girls, especially when new Christians, will not gladly make such a long journey for going to church.

(2) Do not settle in the middle of a market (*tsi*), if it is a village with a market. The disadvantage would be the same, and even more serious for women,

---

❶ "诸圣相通功"是指所有教会成员，包括所有生者死者、殉道圣人及宣信者自己，总之是一切领过洗的信徒的相通。对"诸圣相通功"较实物性的解释是将"诸圣"理解为"诸圣物"，包括信仰及一般圣事或特祝圣过的圣体的分享(http://baike.baidu.com/view/243152.htm)。

❷ The Communion of Saints is the spiritual union of all the members of the Christian Church, both the dead in heaven and the living on earth.

做弥撒❶，尤其会影响讲道。

（3）如果街道是东西走向，选址尽可能选在一条街道的北侧。这样的话，街道北侧可以建重要建筑：教堂、本堂神父住宅和学校。主要入口、马厩、厨房可以建在南边，相关内容我们会在后面"南屋"部分进一步讨论。

（4）最后，最重要的一点，千万不要选址选在低洼易发洪水的地方。如果选址靠近河流，必须要有夏天水位的详细信息。这种情况下，还是需要填土筑堤，这可以交给基督徒们做。要不然，所有相关的费用，包括劳工费用，都要提前做好预算并纳入总的开支预算里。

通常，这些填土筑堤工程的做法如下：第一层土 20 厘米厚，夯实，浇上大量水；然后做第二层同样厚的土层，夯实，浇上大量水；一层一层，以此类推。

如果这些岸堤承重很大，必须不停

if they were obliged to cross these teeming crowds for going to prayer. Shouts and noise from outside will often disturb the priest during Mass and especially during the sermon.

(3) If possible, preferably buy a location on the northern side of the street (if it is a street that runs from east to west). Choose a good place to the north, for the main buildings: the church, the rectory, the school. The main gate, the stables, and sometimes the kitchen, can be built to the south, on *nanou*, as we will discuss further.

(4) Finally, more than anything else, you must avoid land that is too low and thus subject to flooding. You must thoroughly inquire about the water level in summer, especially if the location is near a river. If that is the case, the possibility remains to bank up, especially if the Christians offer to do the job. Otherwise the price, sometimes considerable, of this labour, should be estimated in advance, and should be taken into account in the forecast of expenditure.

This is how these embankments should be executed: they will be made of layers of about 0.20m, carefully rammed; then they will be heavily watered; then a second layer of the same thickness will be made, and rammed, and watered, and so on.

You must then load these embank-

---

❶ 弥撒是圣教会最重要、举行得最多的礼仪。其目的是成圣礼圣血、祭献天主，向天主表示钦崇、感恩、祈求和赎罪(http://www.catholic.org.tw/m99121kh/Catholic.html)。

地浇上大量的水，接着重新夯实，通常需要用具有很宽轴距的夯土工具。

如果要在这种低洼地建教堂，把这些填土筑堤工程完全交给农民做而没有任何监工，这种做法是极其不明智的。

ments, water them copiously, then ram them again, and then use a very wide footing for any building.

It would be supremely unwise to let the peasants make such embankments on their own, especially when it comes to the construction of a church.

第二章　建筑布局和朝向的选择
Second Chapter: Choosing the Buildings' Layout

在中国的这个区域，建筑有个好朝向是非常重要的，这往往与建筑用途相关。请不要学欧洲租界，也不要学美国新教徒的做法，我们只需要忠实模仿当地的做法（图 1.13）。

北方有俗语说"有钱不盖东南房，冬不暖来夏不凉"❶。有钱人是不会盖东屋和南屋的，冬天不暖和，夏天不凉快。

此俗语所言不假。最好的房子，即所谓的"北屋"或"堂屋"，它的门窗都是朝南的。如果想让朝向更完美，就需要向东南偏斜 10 度。这样的话，冬天屋里很早就会有阳光照进来，夏天下

In this part of China, it is of utmost importance to give a good orientation to the buildings depending on their use. Do not use the European concessions, nor the Protestants from America, as models, but faithfully imitate the locals. (fig. 1.13)

"*You qianbugai dongnan fang, dong bunuan, xia buliang.*" "A rich man does not build a house to the east, nor to the south; in winter, it is not hot; in summer, not fresh."

So wills the Chinese proverb, and rightly so. The best orientation is considered that of a northern house or main house (*beiwu* or *tangwu*) whose doors and windows face south. And if you really want to aim at perfection, you

---

❶ 这里按照直译是"有钱不盖东南房，冬不暖来夏不凉"，而查到的说法是"有钱不住东南房，冬不暖来夏不凉"。译者认为可能是当时撰写此册子的人搞错了。

午屋里很早就不会有烈日照进去了。如果朝向偏斜西南方，情况就正好相反。

如果想让北屋在夏天更宜居，可以在屋子北墙开窗，使得空气更流通。北墙开窗确实能使空气更流通，但或许也会让屋子在冬天不宜居，如果窗户造得很严实，或者是双层窗户，就不会有这个问题了。此外，也可以开与人等高的小窗户，夏天可以打开而冬天可以关闭并封严。

北屋其后便属西屋好，西屋朝东。冬天自日出就会有阳光照进来，夏天自 11 点起又能避免阳光直射。

相比而言东屋朝向最差，往往朝向西。东屋在夏天不适宜住人。千万别把东屋用作小礼拜堂（它的门窗只能朝西开）。或许在早上做弥撒可以，但在夏天，过了正午，就不能在这里告解❶和救赎❷了。

should turn about 10 degrees to the southeast. So, you will have the winter sun very early in the morning and in summer you will be spared its direct rays earlier in the afternoon. The direction towards the southwest would naturally create the opposite effect.

If you want to make these houses (*beiwu*) more pleasant in summer, you can place windows in the northern wall in order to catch the breeze. But this breeze will be unpleasant in the winter, except if the windows are very well made, or are double windows. And even then! One could also be satisfied with simple air inlets or very small windows at breast height; they could be open in the summer, and closed and caulked in the winter.

After the northern house comes the western house (*xiwu*), whose orientation faces east. In the winter it is sunny from sunrise. In the summer, from eleven o'clock on, it is out of the direct rays of the sun.

The eastern house (*dongwu*) has the worst orientation because it is turned to the west. These houses are virtually uninhabitable in the summer. Therefore it must never be chosen for small chapels (doors and windows on one side, that is

---

❶ 告解是天主教的一种宗教仪式，信徒在神职人员面前忏悔自己的罪过，以求得天主宽恕，并得到神职人员的信仰辅导和生活道理（http://baike.baidu.com/subview/405100/11377596.htm）。

❷ 救赎理论一般都讲及与造物主的关系，以及如何与造物主重新合一。这些理论大致可以分为普世救赎论及善功救赎论两种。

普世救赎论主要认为：救赎是适用于所有人的、无条件的。救赎的因由并非因为后天的努力，而是因为造

把礼拜堂放在朝北的南屋会好一点。如果教徒都很虔诚，也就是说会有很多教徒来做礼拜，这样南屋在冬天也不至于寒冷彻骨。传教士的住房是不能放在南屋的，除非那里有一个很好的炕，尽可能把传教士的住房安排在北屋。

为了保证屋子的干燥和卫生，室内地面应该高于室外地面 2 英尺❸左右，或者门前踏步至少要有三步。

相反，屋顶不易过高，梁下至地面的距离不宜超过 11 英尺。有俗语说"高处不要做高，低处要做高"❹。当然这只是说私人房屋。一些公共教室、大的客厅可以更高：可至 13 或 14 英尺。

to say only opening to the west). One could use these *dongwu* in the morning, for Mass. But in the summer, after midday, you could not use it for confession,❶ or to celebrate salvation.❷

It would be better to put the chapel in the southern house (*nanwu*) whose orientation faces north. In the winter, it will not be too cold, if the 'christianity' is. . . fervent, that is to say if there are many people! The missionary's house, however, should not to be put in the *nanwu*, unless it has a good stove. He should live in the northern house if possible.

To keep this house dry and sound, do not forget to raise it about two feet above ground level, or at least three steps at the front door.

But do not raise the ceiling too high; eleven feet under the beam (*liang*) is a height that should not be exceeded. "*Not high from the top, but high from the bottom*" says the Chinese proverb. Of course, this concerns only the private rooms. The common classrooms and the large reception halls (*keting*) should be higher: 13 or 14 feet under the *liang*.

（接上页 continue）

物主的恩宠或慈悲。因此，大凡所有符合救赎条件的人，都可以得到救赎。善功救赎论认为救赎是要通过后天的不断努力和补救才可以得到(https://en.wikipedia.org/wiki/Salvation)。

❶ In the Roman Catholic church, confession is the acknowledgement of one's sins to a priest, who forgives in the name of God. The sacrament of penance (presently also called sacrament of 'reconciliation') is one of the seven sacraments of the Roman Catholic Church. Usually the confession requires a specific piece of furniture: the confessional. See plate 25 of the handbook. (fig. 2.25 and fig. 3.34).

❷ Salvation, in French 'le salut', is a prayer performed by the congregation in the parish church.

❸ 手册原注文为"1 英尺，相当于 30.33 厘米"。

❹ 译者没有查到这句俗语，是按照法语原文"Pas élevé par le haut, mais haut par le bas"翻译而成的。至于是否有此俗语及其正确说法目前还尚未能考证。

对于两侧有窗户的大教堂,最好的朝向是南北向,这样的话窗户往东西向开,门可以放在北边或放在南边更好。冬天里,全天都会有阳光照进教堂来,夏天里,两边窗户使空气畅通。如果教堂结构的类型和尺寸都允许的话,可以通过巧妙安装窗帘使得教堂室内更舒适。

经观察,南北朝向的教堂在雨季漏雨的可能性要比东西朝向的教堂小。

对于那些窗朝南北、门朝东西的大教堂,夏天可能比较好,但在冬天,北风刮起来后,教堂里就会很冷。我们知道冬季是传教活动和教堂节日庆祝频繁的季节。

那些东西朝向的教堂也有它们的优点,为了改善这类教堂冬天很冷的情况,有人提出了以下两个改进方法,这两个方法仅适用于样式简单的小尺寸结构。

(1)在北墙,每隔一个窗户即以凹墙代替窗户而保留窗户外形形制,那上面可以挂具有启发性和教育性的宗教绘画作品。

For large churches, with windows on both sides, the best orientation seems to be the north-south direction, that is to say with windows opening to the east and the west, and the door or doors, placed to the north or better still, to the south. In the winter, you will have the sun all day; in the summer, thanks to the windows on both sides, you will enjoy a refreshing breeze. When the type and size of construction allows for it, do not forget an intelligent use of curtains to modify sunlight or breezes.

It was also generally observed that north-south oriented churches leak less during the heavy rainy season, than east-west oriented churches.

For these large churches, the orientation with windows opening to the north and the south, with doors to the east or west, can be advantageous in the summer, but in the winter, the church will become icy cold, as soon the wind blows from the north. Keep in mind that winter is the season of missions and of many church festivals.

For the east-west oriented churches, which have their good points, some people propose the following two arrangements, which would only be acceptable for small size constructions with less style:

(1) either, on the northern side, replace every other window with a recessed wall, that imitates the form of the window, and on which one could hang large instructive and edifying posters;

（2）或者在北墙保留与南墙同样数量的窗户，只是缩小北墙窗户的尺寸，可比南墙窗户小 2 英尺，这样的话，南北墙的窗拱上沿在同一高度上，而同时北墙窗户的下沿离地面会较南墙窗户的下沿高出 2 英尺。

(2) or, by keeping the same number of windows on the northern side as on the southern wall, but making them smaller, for example by 2 feet, at the northern side than at the southern. Their upper arch will be at the same height on both sides, but the windowsills will be higher, for example by 2 feet, from the floor, at the north side.

第三章　设计方案、工程预算和施工合同
Third Chapter: Plan — Estimate — Contract with the Contractor

设计方案

所有建筑，即使再简单，也需要有一个设计方案。很显然，如果要建一个大教堂，就必须找一个建筑师，不管他有没有正式的文凭证书，都必须是专业的建筑师。参考《天主教法典》❶ 第1162条，这些设计方案必须上交并获得宗座代牧的同意。很多有柱子的教堂，其建成往往只凭神父脑子里的设计方案。很可惜我们不太注意到这些。

Plan

For all construction, no matter how modest, make, or have made, a plan. It is obvious that for a large church, you should go and see an architect, with a formal degree or not, along as he is a known professional. The plans must be approved by the apostolic vicar. See the canon law, Canon 1162.❷ More than one church with columns was built while the plan existed only in the mind of the fa-

---

❶ 这里指 1917 颁布的《天主教法典》，由教宗本笃十五世正式颁布。该法典共有 2414 条教规，分五卷：总则、人法、物法、诉讼法、罪与罚。卷下为编，编下为题，题下为条，条下为项。这是基督教历史上的第一部法典。这部法典不仅适用于天主教会，对于东方教会的法律发展也产生了颇大的影响（https://zh.wikipedia.org/wiki/）。

❷ The canon law (in Latin ius canonicum), is the body of laws governing the Roman Catholic Church. The articles of the canon law are called 'canons'.

设计方案不能只是线条所绘的平面，还应包括立面、框架结构和细节部分，如钟楼、大门、窗户、拱顶等。

This plan should not only be a ground plan, consisting of a drawing with single lines. Elevations should be added, as well as plans of the timber frame, and detailed plans of the tower, the portal, the windows, the vaults, etc.

在这些设计方案里，对墙的厚度、墙砖砌法（以指导那些缺乏经验的泥水匠）等都要有说明，这样可以避免施工过程中不必要的麻烦和错误。如果是建一个建筑群的话，有一张总体平面图是很好的，可以对每一栋建筑的位置加以确定和说明。对建筑群所处的地理位置及与所在地区的关系也应该加以说明，即使只是一个简单的介绍说明也很好。

On these plans, one should mention the thickness of the walls, the bound for the less experienced masons, etc. to avoid surprises, or unpleasant mistakes during construction. It is good to make a general plan for the entire property, designating the location of each building. The topographical position of the property in relation to the rest of the place should be roughly indicated as well.

工程预算

Estimate

如果有很详细的设计方案，整个工程的开销预算就不难做了。做工程预算时，必须保证方方面面都考虑到，小到绑脚手架的绳子、钉椽子用的钉子，以及提供给工人使用的工具、给工人的奖金等，做最终预算时要多算出整个费用的三分之一，这种做法较为妥帖。如果做工程预算时粗枝大叶，就会落得与福音书中某人同样的遭遇❶，因为错误估

If the plans are well made, it will be relatively easy to establish an estimate of expenditure. When the estimate is established, and when you are sure to have forgotten nothing, even ropes for scaffolding, nails for the rafters, tools and bonuses for the workers, then you should increase the estimate by a third. Without this precaution, you will risk the fate of the man in the Gospel who could not fin-

---

❶ 圣经里的一个寓言故事，见"路加福音"14 节 28~30 句："你们哪一个要盖一座楼，不先坐下来做一下预算看能不能盖成呢？等做好地基，又不能建成，恐怕看见的人都要笑话他，说：'这个人开了工，却不能完工。'"

算工程费用而导致建筑不能完工。❶

遵循谨慎之原则，谨遵教会法之细则，就不会过分依赖于那些"可能"的捐赠。另外，建教堂时，有必要细读《天主教法典》中关于"圣所"的细则❷。

ish his building because he had miscalculated the costs. ❶

A rule of prudence and even of canon law will convince you not to rely on 'possible' alms. Moreover, when constructing a church, it will be useful to read again the excerpts from the code of canon law about sacred places (*locis sacris*). ❸

---

❶ One of the parables in Christ's teaching. Gospel of St Luke 14, 28-30: "For which one of you, when he wants to build a tower, does not first sit down and calculate the costs to see if he has enough to complete it? Otherwise, when he has laid a foundation and is not able to finish, all who observe it begin to ridicule him: 'This man began to build and was not able to finish'."

❷ 《天主教法典》第三编"圣所与节期"第 1205 条至第 1243 条是对"圣所"的说明，并分第一章教堂、第二章圣堂及私人小圣堂、第三章朝圣地、第四章祭台、第五章墓地进行说明。参见：http://www.vatican.va/chinese/cic_zh.htm。

❸ Roman Catholic Code of Canon Law: http://www.vatican.va/archive/cdc/index.htm Book IV, chapter III, title I, chapter I, canons 1214-1220: Churches. Can. 1214: "By the term church is understood a sacred building designated for divine worship to which the faithful have the right of entry for the exercise, especially the public exercise, of divine worship."; Can. 1215 §1-3: "No church is to be built without the express written consent of the diocesan bishop. The diocesan bishop is not to give consent unless, after having heard the presbyteral council and the rectors of the neighbouring churches, he judges that the new church can serve the good of souls and that the means necessary for building the church and for divine worship will not be lacking. Although religious institutes have received from the diocesan bishop consent to establish a new house in the diocese or the city, they must also obtain his permission before building a church in a certain and determined place."; Can. 1216: "In the building and repair of churches, the principles and norms of the liturgy and of sacred art are to be observed, after the advice of experts has been taken into account."; Can. 1217 §1-2: "After construction has been completed properly, a new church is to be dedicated or at least blessed as soon as possible; the laws of the sacred liturgy are to be observed. Churches, especially cathedrals and parish churches, are to be dedicated by the solemn rite."; Can. 1218: "Each church is to have its own title which cannot be changed after the church has been dedicated."; Can. 1219: "In a church that has legitimately been dedicated or blessed, all acts of divine worship can be performed, without prejudice to parochial rights."; Can. 1220 §1-2: "All those responsible are to take care that in churches such cleanliness and beauty are preserved as befit a house of God and that whatever is inappropriate to the holiness of the place is excluded. Ordinary care for preservation and fitting means of security are to be used to protect sacred and precious goods."; Can. 1221, §1-2: "Entry to a church is to be free and gratuitous during the time of sacred celebrations. If a church cannot be used in any way for divine worship and there is no possibility of repairing it, the diocesan bishop can relegate it to profane but not sordid use. Where other grave causes suggest thata church no longer be used for divine worship, the diocesan bishop, after having heard the presbyteral council, can relegate it to profane but not sordid use, with the consent of those who legitimately claim rights for themselves in the church and provided that the good of souls suffers no detriment thereby."

第二部分　手册的内容和翻译
Part Two: The Handbook Translated

注释
Note [on Style]

对教堂的中国样式做一些讨论是必要的。请允许我在这里引用1925年在香港出版的第42期《外方传教士公报》❶中一篇关于建筑的文章中的结论。按目前的中国艺术学的现状，"建筑"一词有点夸大，"结构"一词更为准确。据我们在中国所看到的情况，那些所谓的高级工程师，他们都不懂"结构"之外的东西。现在，还很难预测什么（胡克神父❷会不会在甘肃省境内沙漠覆盖的遗址内有什么新发现）。

It would be appropriate to deal here with the question of the Chinese style for our churches. Allow us to quote the conclusion of an article edited in the *Bulletin des Missions Étrangères*❸ published in Hong Kong, No. 42, June 1925, where it concerns architecture. In the present state of the science of Chinese art, the word 'architecture' may be pompous. The word 'construction' would be more appropriate. What we see currently in China, said a senior engineer, does not exceed the limits of 'construction'... At present, it is hardly possible to augur anything (of what might be discovered in the many ruins buried in sand in the deserts of Kansu and in the countries covered by Father Huc❹).

---

❶ 法国传教士协会有一份期刊《巴黎外方传教会》，由香港拿撒勒出版社出版。有关香港的巴黎外方传教士方面的内容可参考2007年由香港大学出版社出版的乐艾伦撰写的《伯大尼与纳匝肋：英国殖民地上的法国遗珍》一书。

❷ Évariste Huc（1813—1860），法国天主教神父，遣使会的传教士，著有《西藏、蒙古和中国：1844、1845和1846在西藏、蒙古和中国旅程回忆录》（Souvenirs d'un voyage dans la Tartarie, le Thibet, et la Chine pendant les années 1844, 1845 et 1846）（1850年出版）和《中华帝国》（L'Empire Chinois）（1854年出版）。

❸ Journal of the French missionary society Missions Étrangères de Paris (MEP) published on the Press of Nazareth in Hong Kong. See: Alain Le Pichon. Bethanie and Nazareth: French Secrets from a British Colony [M]. Hong Kong, 2007.

❹ Évariste Huc (1813-1860), French Catholic priest, missionary of the Congregation of the Mission, author of influential books on Tibet, Mongolia and China: Souvenirs d'un voyage dans la Tartarie, le Thibet, et la Chine pendant les années 1844, 1845 et 1846 [M]. Paris, 1850; and L'Empire Chinois, Paris, 1854.

另外，在中国找不到像印度或吴哥窟那般规模宏大的文物古迹，因为中国建筑所用的材料不是石头，也因为中国赖以立国的礼仪一直与对龙和虎的崇拜相关，那里缺少"建筑发展"所需的核心条件：自由。Baudrillart 主教❶说过"自由是艺术的空气"；但自由在中国是受限制的，艺术家们别无他法，只能退而追求装饰这类二等艺术，比起建筑他们更喜欢装饰自家家园和宝塔。

说起木头、石头、丝绸、瓷器等上面的装饰，有一篇文章值得一读，就是《闲话中国艺术》❸，作者在书中阐述了自己的观点。

"我知道有一个小教堂独具特色：外观和立面都有着逼真的中国印记。我进去过一次，为装饰感到惊奇，我想中国装饰艺术的使用是关键。我无法描述当时的感觉，一开始感到很吃惊，接着感到不舒服，当我看见里面只

Moreover, you will never find grandiose monuments, comparable to those in India or Angkor, because the materials used were not stone, and because the Chinese rites that 'created this nation,' together with the Dragon and Tiger superstitions, took away the essential condition of architectural development: freedom. 'Freedom is the atmosphere of art,' said Bishop Baudrillart❷; but freedom has been so regulated in China, that artists fall back in desperation on the secondary arts of ornamentation, preferring to decorate their homes and pagodas than make them artistic constructions. (Loc. cit. , p. 343).

As for the decoration on wood, stone, silk, porcelain, etc. , here is the opinion of the author of *Talk on Chinese art,*" talk that is well worth being fully read. ❹

"*I know a little church with a unique character: the outside, the facade, has a definite Chinese charm. I went in once, curious to admire the interior decoration that, I thought, would be the key, the model to follow, for Chinese decorative arts. I could not repress a feeling of as-*

---

❶ Alfred-Henri-Marie Baudrillart（1859—1942），法国天主教神父，法国天主教学院的校长，1935 年任罗马天主教红衣主教。

❷ Alfred-Henri-Marie Baudrillart (1859-1942), French Catholic priest and historian, rector of the Catholic Institute of Paris, auxiliary bishop of Paris (1921), cardinal of the Roman Catholic Church (1935).

❸ D. Doutreligne 神父在 1925 年《巴黎外方传教士》期刊发表文章《闲话中国艺术》（Causeries sur l'art chinois），见 341-357 页。

❹ Father Denis-Donat Doutreligne (1881-1929), of the Foreign Mission of Paris, missionary in Guizhou province, author of the article: Causerie sur l'art chinois. Peut-on adopter indifféremment dans nos Églises certains détails d'ornementation? [J]. Bulletin de la Société des Missions-Étrangères de Paris, 1925,42 : 341-356.

有对快乐和长寿的祈盼，夹杂着最异想天开的一些字谜和谐音文字，如上面所说的（蝙蝠、鱼、钓鱼、中国'寿'（即长命）等）。进入这个圣所的中国人被他所熟悉的装饰包围，他可能会感到温暖，他所呼吸的周边空气会让他对该古迹想表达的祈盼深有同感。他在那里感到很轻松，他喜欢那里并被吸引。但是，这是否有损基督教教给我们信徒的理想——圣殿的装饰与耶稣教给我们的伟大的苦难理论❶几乎跟这些没有关系。在地球上快乐生活，如此低级趣味，不就像令人失望透顶的低级道德吗？"相关内容见：

（1）1924 年 4 月、5 月和 6 月期间出版的《北京天主教公报》中的三篇长文，见 127，177 和 218 页。

（2）1925 年 6 月期间的《巴黎外方传教士协会公报》第 42 期第 431 页。

这是一个在中国居住多年的传教

第二部分　手册的内容和翻译
Part Two: The Handbook Translated

tonishment at first, then uneasiness, when seeing inside only wishes of happiness and long life, mixed with the most fanciful inscriptions, riddles and puns mentioned above (bat, fish, fishing, Chinese character shou long life, etc.). The Chinese person who enters this sanctuary is undoubtedly surrounded by designs that are familiar to him; perhaps he feels warm; the atmosphere creates a favourable impression of what this monument represents. He will feel at ease coming here; familiar things attract, too. But does presenting such designs to our faithful and in our temples not diminish the ideal of Christianity? These designs are so little in relation to the great and consoling theory of suffering, ❷ which Christ came to teach. Happiness here on earth, to live life, to simply enjoy oneself, isn't that proning depressing and too undemanding morals?" (Loc. cit. , p. 555) See:

(1) *Bulletin catholique de Pékin,* three long articles published in the issues of April, May and June 1924, p. 127, 177 & 218.

(2) *Bulletin de la Société des Missions Étrangères de Paris,* N° 42, p. 431, issue of June1925.

This is the opinion of a missionary

---

❶ 基督教认为人类所受的苦难在宗教中起着很重要的作用。耶稣为救众生而被钉死在十字架上。因此，苦难是一种积极的体验，它可以帮助人类实现更高的生活目标，借由苦难人类也能得以与上帝相遇。

❷ Christianity believes that human suffering plays an important role in religion. Jesus suffered and was killed for the lives of other people. Therefore, suffering can be a positive experience in the case of achieving a higher meaning of life and encountering God.

舶来与本土：1926年法国传教士所撰中国北方教堂营造手册的翻译和研究
Building Churches in Northern China. A 1926 Handbook in Context

士的观点。❶ 他具有建筑方面的知识，他在这个国家有建造经验，他在这里负责建造的许多教堂具有完美的当地适应性又具备真正的美感，这些都使得他具有这方面的发言权并且能明确清晰地表达出来。

"中国样式，本质上就是把一个大屋顶架在柱子上，从结构角度来看，花费极其昂贵，柱子之间的空间很难用作办公空间。这样的结构系统注定其不会长期存在。佛教徒视宝塔之倒塌如生命之不定和脆弱。所以，在日本，连续几千年，著名的伊势神宫每20年就要重建一次。❸ 但这并不是我们基督教建筑所追求的。"

"此外，如此昂贵又不稳定的系统是不太可能满足天主教礼拜所需要的空间要求的。木料和木柱的长度，决定了它只能建很小尺寸的建筑，如亭子、宝塔大小的房间等。"

"再看希腊建筑，它的比例如此完美，线条如此纯粹，但即使如此也无法由之产生天主教建筑。希腊帕特农神

who spent many years in China.❷ His architectural knowledge, his experience of this country, the perfect adaptation as well as the real beauty of the many churches he built here, give him the right to have an opinion and to formulate it:

"*The Chinese style, essentially a monumental roof placed on columns, as a mode of construction, is extremely expensive and very ill-adapted to fill its office. Such a construction system is doomed in advance to be very ephemeral. Buddhists are free to see, in the periodic collapse of their pagodas, a symbol of the instability and the fragility of human things. So, in Japan, for thousands of years, the famous pagoda of Ise is regularly rebuilt every twenty years.❸ But this is not what we want for our religious buildings!*"

"*Furthermore this extremely expensive and unstable system is completely unsuitable for the practical conditions required by Catholic worship. Based on the usual length of the wooden beams and posts, it is only suitable for buildings of very small dimensions: kiosks, rooms the size of a pagoda, etc.*"

"*For a similar reason, Greek architecture, so admirable in its proportions and purity of lines, has proved absolutely

---

❶ 根据作者目前的有限考证，作者认为这位传教士建筑师可能是和羹柏神父，见第一部分相关内容。
❷ In all likelihood, father Alphonse De Moerloose. See Part 1, p. 38-43.
❸ Ise Grand Shrine, Japan. See: Jean-Sébastien Cluzel, Nishida Masatsugu (eds). Le sanctuaire d'Ise. Récit de la 62e reconstruction [C]. Brussels, 2015.

第二部分 手册的内容和翻译
Part Two: The Handbook Translated

庙❶如此一个小型的奇迹，它的尺度却很难增大，原因很简单，石门楣、架于柱子之间的整块巨石，这些形成了古希腊结构的主要特征，也是帕特农神庙实际能到达的最大长度。"

"可能有人会以巴黎的马德莱娜教堂❸为例而进行反驳。然而，它只是希腊风格的一个低劣模仿品，它的梁不再是整块巨石，而是通过石头的连接并通过内部铁架达到了令人满意的长度。罗马拱券的大跨度，很显然也是不为希腊人所知的。"

"这个例子几乎没有被模仿，证明它并不可效仿。天主教建筑中被大量模仿和重建的是罗马的巴西利卡，一个有着异教徒起源的建筑形式。什么原因呢？因为这种形式的建筑有一个大空间，也许是天意，在充满异教徒的罗马发现的这个教堂，❺极其符合天主教礼拜仪式的需要：一大群虔诚的信徒在一个大空间内聚会，他们都能看见位于大

incapable of generating Catholic buildings. The Parthenon is a tiny marvel that cannot grow, ❷ for the simple reason that the stone lintels, monoliths placed from one column to the next, and characterizing the construction of ancient Greece, reach at the Parthenon their maximum practical length."

"One could object with the example of the Madeleine church in Paris. ❹ Agreed, but it is only a poor forgery of Greek style in which lintels are no longer monolithic, but made of sufficient length by adding stones together and consolidating them with iron frames. In this church, Romanesque [round] arches are used for large spans, which were absolutely unknown to the Greeks."

"The fact that such an example has rarely been imitated, proves that it is not an example to follow. What has been imitated and reproduced a huge number of times, in the construction of Catholic churches, is the Roman basilica, a building of pagan origin. Why? Because this mode of building a huge room, which, through a provision of Providence,❺ the Church found the model of in

---

❶ 帕特农神庙（Parthenon）是古希腊雅典娜女神的神庙，位于兴建于公元前 5 世纪的雅典卫城。它是现存至今最重要的古典希腊时代建筑物，一般被认为是多立克柱式发展的顶端；雕像装饰是古希腊艺术的顶点，此外还被尊为古希腊与雅典民主制度的象征，是举世闻名的文化遗产之一。

❷ The Parthenon in Athens, temple built from 447 to 438 BC, the masterpiece of Ancient Greek architecture and the absolute reference of Classical and Neo-Classical architecture.

❸ 马德莱娜教堂（the Madeleine church）为新古典主义风格。

❹ The Madeleine church in Paris, built in Neo-Classical style in 1807-1842, looks like a Greek temple.

❺ Christians consider the Providence or 'Divine Providence' as God's intervention in the world, both to maintain the natural order of the universe and to help the faithful Christians.

厅一头的高坛和祭台。"

异教徒的宗教仪式，至少在中国，很少需要聚集在一起（一年一两次），他们没有秩序，没有管理，没有神圣感，没有时间观念，就像在集市，不在塔里面，而在一个塔前离得很远的大庭院里。这些庭院，两侧环绕又长又窄的边房，由位于中心线的桥、门廊、牌坊等隔开，中轴线两边立着刻有文字的巨大的石碑，所有这些形成了美感，塔的美感。庭院四周的房子围成一圈，才使得这个空间有价值。如果要把这里改为天主教教堂，就要在庭院上方加一个屋顶（图 1.13），用高墙将四周围起，只需留出开门窗的口，然后将祭台放在原先放大肚佛的地方，撤去木隔板和纸隔板，然后放上圣礼栏杆❶。

把中国宝塔改建为天主教堂期间，那些"赶集"的异教徒们，在远处

pagan Rome, was wonderfully suited to the needs of Catholic worship: gathering in a vast space a large number of the faithful, all having a good view of the choir and of the altar, that is placed at one end of the hall."

Pagan worship, at least in China, gathers crowds on very rare occasions (once or twice a year), without order, special dress, dignity, or schedule, *like at the fair*, and not inside the pagoda, but in the vast courtyards preceding the pagoda as far as the eye can see. These courtyards-flanked by long and narrow side buildings, cut by bridges, triumphal arches, monumental gateways, and lined, along the middle axis, by huge steles with scholarly inscriptions- form the beauty of… the pagoda. Here, it is the frame that makes the value of the picture. To turn this into a Catholic church, one need only to place a roof over the courtyard (fig. 1.14), close the four sides by a high wall, with the necessary openings for doors and windows, and replace the fat Buddha with the altar, after having removed the wooden and paper partitions where the communion rail will be. ❷

Until the Chinese pagoda is transformed into a Catholic church, the pagan

---

❶ 圣坛栏杆将普通信徒坐的地方和只有祭司和神父才能进入的圣坛空间进行隔离。圣坛栏杆不仅起到空间隔离的作用，那里也是信徒领取圣餐的地方。

❷ Communion rails or altar rails mark a separation between the nave, where the people sit, and the sanctuary with the altar, where only the priest and the servants can enter. The communion rail not only delimitates the space, but it is also the place where the people come for receiving communion.

观音菩萨❶或大肚佛的法眼下❷，还是照常在庭院嬉闹、喧哗、买卖各种小物件。在宝塔里面，一些比丘和小和尚跪在佛像前，边敲着磬❸边用梵文念着经。也有一些很特殊的信徒，特别是那些虔诚的妇女，她们缓缓挪步前行烧香，烧完香后，偷瞄一眼比丘，❹礼拜仪式就算结束了：别样的礼拜，别样的风格！

回头再看巴西利卡和由它转变而成的供上帝的圣所。一开始，教堂正厅的宽度受屋顶木梁长度的限制，这是事实。经过不断摸索，最后发现哥特式拱顶（交叉拱）的做法，完美地解决了空间跨度的问题，并可以把教堂的宽度（包括长度）无限延长，尤其是教堂中殿的跨度。这样的结构系统尤其适用于天主教教堂，它以最大限度的经济性、强度和耐久性使得各类教堂建筑得以

crowd that comes 'to the fair' continues to bustle in the courtyards, shouting, purchasing and consuming its modest provisions, under the watchful and distant eye of Guanyin Pusa❶ or of the pot-bellied Fuwo.❷ Inside the pagoda, some *bhikkhu* and young monks kneel before the idol, chanting their prayers in Sanskrit, and accompanying themselves on small iron bells (the *qing*). In small groups, the faithful, especially women, come to prostrate themselves and light some incense sticks. A glance to the *bhikkhu*,❹ and the pilgrimage is over: different worship, different practices!

Let us come back to the Roman 'basilica' and its transformation into a temple of the true God. At first, it is true, the width of the main nave was limited by the length of the wooden beams used for the ceiling. The trials and errors, and finally the discovery of the Gothic vault (rib vault), which was the perfect solution to the problem, has extended virtually indefinitely the length of the church, especially that of the main nave. Thus

---

❶ Guanyin Pusa, the Bodhisattva of Compassion or Goddess of Mercy.

❷ The statue of the Buddha with a big belly and a smiling face is always located in the first hall of a temple. This Buddha can bear all the worst problems of the people, his smile can face everything and everybody, showing patience and tolerance, humor and happiness.

❸ 佛教寺院中所使用的一种乐器和法器。有铜制的引磬，是寺院中所使用的打击乐器，又称手磬。形似酒盅，直径约 7 厘米，形状与仰钵形坐磬相同，置于一根木柄上端，木柄长约 35 厘米，用细长铜棍敲击。也有用铁铸造而成的磬，少有铜制，口大底圆，外形似锅而深，底呈半球形。规格大小不一，常以寺院的规模和等级而定，小者如碗，大者似缸。在一般寺院中，以磬高 20～25 厘米、磬口直径 30～40 厘米者居多。演奏时，铁磬置于佛案一侧，需放置在软性圆垫或布圈之上方可稳定。磬口朝上，奏者右手执木槌敲击磬壁而发音。小者音色明亮，大者音色深厚，有庄严肃穆之感，是僧人或世俗礼佛必击的乐器。在佛经的法器乐谱中，磬有专用的敲击符号。

❹ Buddhist monk or 'bonze'.

建成，从卑微的乡村小教堂到宏伟的主教座堂。

我们宗教建筑中唯一能与罗马巴西利卡抗衡的样式是由君士坦丁堡的圣索菲亚大教堂❶派生的中央圆顶加希腊十字❷的做法。这种做法有它的美丽之处，但已被证明很不实用。

至于从其他样式借鉴而建的教堂，我觉得全世界范围内能找到1%或2%就不错了。中国样式，只能勉强算为建筑，它总是有着不成熟的结构，如果这种中国样式也能给教堂建造以借鉴经验的话，这在我看来是极不可能的。

但是我需要澄清一点，我所说的中国样式是指中国的建造方法。如果中国样式不是一种建造方法，而只是中国艺术中的某种装饰，那么这个问题就完全不一样了。这种中国艺术风的装饰的存在是不容否认的，它为适应欧洲建筑而

was invented the form of construction so eminently suited to Catholic worship, that allows for building churches, with maximum economy, strength and durability, from the humblest village church to our large cathedrals.

The only style in our religious buildings that creates some competition to the Roman basilica, is the style of the Greek cross with a central dome, derived from Hagia Sophia of Constantinople. ❸ This style has its beauty, but has proved to be much less practical.

As for the churches borrowed from other styles, as far as their construction is concerned, I do not think they represent more than one or two percent, worldwide. To hope that the Chinese style, barely existing as architecture, and in any case eminently childish as a mode of construction, will one day, for China, contradict our experience, seems, to me, pure illusion.

However, I want to clarify that in everything I have just said, 'Chinese style' means the Chinese way of building. If 'Chinese style' would mean, not a way to build, but a kind of ornamentation proper to Chinese art, the question would then be totally different. The existence of

---

❶ 圣索菲亚大教堂是拜占庭式建筑最佳的现存范例，其保持着最大教堂的地位达1000年之久。这座教堂内一共使用了107根柱子，最大的圆柱高19~20米，直径约1.5米，以花岗岩所制，重逾70吨。与主要使用大理石的希腊建筑及主要使用混凝土的罗马建筑不同的是，圣索菲亚的主要建筑材料为砖块。教堂内部的空间广阔，教堂正厅之上覆盖着一个最大直径达31.24米、高55.6米的中央圆顶。如何在立方体的建筑上放置圆形穹顶，一直是古代建筑学的上的难题。圣索菲亚大教堂则完美解决了这个问题。

❷ 希腊十字是由两根等长边在中间相交而成的十字图形。

❸ Christian cathedral dedicated to the Holy Wisdom, built by Emperor Justinian in 532-537, in present Istanbul.

第二部分　手册的内容和翻译
Part Two: The Handbook Translated

做些适应性调整也不是没有可能的，它到底也是有着逻辑和艺术传统的，甚至也是有宗教蕴意的。

全世界的天主教建筑，它们的平面和结构，很多都是学罗马巴西利卡的。一般的话，教堂的建造都会根据所在国家的自然条件和艺术环境做一些适应性调整，首先是结构（这个方面中国不能算入），其次是装饰（中国艺术可以发挥其作用）。需要指出的是，哥特式建筑师们在装饰柱头时，总是参考当地的叶状装饰，而从来不参考外来的叶状装饰（如希腊的毛茛叶）。

在中国，没有比模仿更值得推荐和称道的，当然这种做法也有前提，就是在模仿的同时应该避免使用那些迷信的装饰模式——这往往又不是很容易的事情。

建筑的装饰部分（壁画、木工、石雕、琉璃瓦等）可以模仿中国艺术，但前提是这些装饰部分必须是真实存在的，要建于例如一栋包含纯装饰预算的豪华建筑中，而不是用石灰石膏做的很寒酸的仿制品。

this Chinese art of ornamentation is undeniable; its adaptation to the European way of building does not seem impossible, and finally it is in the logic of things and art traditions, even religious.

There are, indeed, few Catholic churches in the whole world whose plan and construction method are not derived from that of the Roman basilica. There are few churches, too, that do not show, first in their construction (but here the Chinese art is too poor to count), and second in their ornamentation (here the Chinese art can have a role to play), some adaptation to the natural conditions and to the artistic atmosphere of the countries where they are built. It was noted, for example, that the Gothic architects, in decorating their capitals, always referred to the models that their own flora provided (for example the Greek acanthus leaf) and never to a foreign flora.

So nothing is more praiseworthy and of better taste than imitating, in China, this same way of doing things, on the condition, of course, that at the same time any borrowing of superstitious ornamental designs be avoided, which will not always be easy.

However, to imitate Chinese art in the decorative part of the building (painting, woodwork, carved stones, glazed wall tiles, etc.) supposes that such a decorative part exists, as in a luxury building, with a budget for pure ornamentation, which is not just miserable

事实上，我们现在不鼓励建奢华的建筑和豪华的教堂（我们不要忘记最近教廷的告诫）。你或许会说："那些已经在中国建成的主教座堂和大教堂，它们难道不宏伟吗？"我会说："不，这只是一种错觉，源于欧洲艺术，特别是哥特式风格。"哥特式风格所体现的艺术和美主要体现在：普通材料的合理利用，正确措施的恰当实施并以最经济的方法满足实际需要。很少或几乎没有（因为"极少相当于没有"❶）装饰，以沈阳的主教座堂❷为例（图 1.7），除了彩色玻璃之外，它的装饰部分没花什么钱。

因此，如果中国艺术只能从装饰上介入我们的教堂，我觉得这纯粹就是一

plaster or lime imitations.

But we are not presently building luxury buildings and sumptuous churches (and the Holy See recently advised us not to forget this). You could say: "*And such cathedral, such large church, already built in China, are they not magnificent?*" I would reply: "*No, this is an illusion that stems from the very nature of European art, in particular from the Gothic style.*" The latter produces an impression of art and beauty by just the intelligent use of the most ordinary materials, in just the right amount, to achieve, in the most economical way, the purpose of the building. Little or nothing (because "*very little counts for nothing*"❸) is sacrificed to ornamentation, for example in the Cathedral of Shenyang. ❹ (fig. 1.7) We could say that ornamentation has no part in the expenses, except for the stained glass.

So, if Chinese art can be introduced into our churches from a decorative point

---

❶ 拉丁成语"parum pro nihilo reputatur"用来说明道德问题，即一个人若做了一件不重要的事情，就可以视之为不存在。整个讨论围绕"不重要"展开（如小罪就会被认为是无罪）。手册作者在手册中用这个来说明装饰问题，我们可以理解为：很有限的装饰就是没有装饰，很少或几乎没钱被用来装饰教堂。

❷ 又名南关天主教堂，是全国重点文物保护单位。最早的教堂在 1900 年的义和团运动中被毁，现存教堂为 1912 年法国主教利用《辛丑条约》中的庚子赔款在原址上重建的。见第一部分内容和图 1.7。

❸ The Latin sentence "parum pro nihilo reputatur / very little counts for nothing" is used in morals with the meaning that if one does something of minor importance, it could be considered as not existing. The whole discussion is than about what is "minor importance" (for example when a minor sin would be considered as not being a sin)… The author of the handbook refers to this sentence in his reasoning about ornamentation in churches. One can understand it as: very limited decoration is like no decoration; or as: very little or no money has been spent to decorate a church.

❹ In 1900, the Boxers demolished the Sacred-Heart cathedral of Shenyang (Liaoning province) and killed the French bishop in his cathedral. A new cathedral was built with indemnity money from the Chinese state. Father Lamasse was involved in the works. See Part 1.

项奢侈的开销，根据我们传教会的现有的财务状况不应该考虑使用这类装饰。在我们的宗教建筑中适当引入这类奢华的装饰，这将是我们的继任者所要想的问题。

施工合同

雇用自己人施工，或者雇用那些需要救济的信徒施工，是很不明智的做法。这些所谓的善举会给施工带来许多小麻烦。那些信徒往往不会服从包工头的指挥，他们会认为他们是你的自己人，包工头也会嫉妒你的干涉，他不敢拿本来工人工资中属于他的那一份份例钱。

应该让包工头（领头的）自己去雇用所有他需要的人，包括厨师等，他需要对他雇用来的人的所有行为负责，如施工、偷盗、品行恶劣等。

可以给整个施工工程定个总价（"定价"）❶，就像天津和其他大城市一样的做法，每个工程有一个总负责的建筑师。考虑到这个国家的风俗习

of view, that is to say as a purely non-essential expenditure, then I do not think I am wrong in thinking that this expense is not to be considered possible, given the current state of affairs and finance of our missions. It will be for our successors to think about, when it becomes appropriate to introduce a little extravagance in our religious buildings.

Agreement with the Contractor

It would be unwise to recruit people for the foreseen work yourself, or to give work as a labourer to one or another poor Christian you want to help. During the course of the work, this short-sighted charity could become the source of many small problems. Your 'protégé' will not obey the contractor; he will consider himself as your man; or the contractor will be jealous of your interference; he will not dare take his normal percentage on the salary of this worker, etc.

The *lingtoude* (entrepreneur) should have the task of hiring all his people, including the cooks, and he should take the responsibility for their actions from all points of view: work, theft, misconduct, etc.

The construction can be done 'at entreprise' = work for a set price as is the practice in Tianjin and elsewhere where the work is led by a responsible architect. Apart from these particular places, given

---

❶ 通常有两种做法，一种是每天付每个工人固定工资，另一种是给工程一个总报价。

舶来与本土：1926年法国传教士所撰中国北方教堂营造手册的翻译和研究
Building Churches in Northern China. A 1926 Handbook in Context

惯，在其他地方不建议采取这种形式的合同，即使有一个很好的监工也未必可行，就看工人有没有责任心，施工中有没有无故离开工地的行为。如果工程以一个总价整体外包，往往会很快完工，工人日出之前就开工，他们只在中午休息片刻，即使是炎热的七月份也是如此，天黑后他们才收工。工程以一个总价整体外包可以节省费用，费用往往会比按天计算工钱的要低。但是无此虑必有他忧：在直隶平原及其周边地区，一到七八月份就有雨季，这个时候，等建筑完工后过不了多久你要担心的事情就来了——墙倒塌歪闪、屋顶漏雨或房子其他部位坍塌，一旦碰到这些问题，你就要重新计算之前省的那些钱是否合算了。❶

所以宁可多花一点钱，每天付每个人固定工资（"日薪"），这样比较好。当地工匠称之为"磨工夫"，有时会说得更洋气，称之为"磨洋工"，对这种做法要学会忍耐。就让他们拿着空筐慢踏踏地走吧。如果看见两个工人一起搬六块砖头还要在那寻找平衡点的情况，千万不要生气，不要为此太生气或经常生

the customs of the country, this form of contract is not advisable, even if one has very good supervisors (*jiangong*), who are conscientious and involved in the work, and who would not leave the site throughout the whole construction period. Done for a flat rate, the building will rise very quickly since the workers will be at work before sunrise. They will just rest a few moments at noon, even in July, and they will only descend from the scaffolding when it is dark. You will feel you are saving money and your expenses will be lower than if they work 'at a day' rate. You will worry some; but when the building is completed, wait for one of these typically heavy rains that fall on the plains of Zhili, in July and August; you would not have to wait long to be deeply worried. As to the money you thought you saved, you will re-calculate everything when looking at the walls that crack and warp, and at the roofs and the rest that will crumble…

It is better to work by day and spend a little more money. Resist to what they call *mo gongfu* (*tenere tempus*); when they want to speak elegantly, they say *mo yanggong* (take the time to foreigners). Let them wear empty baskets and walk slowly in procession. Do not get too angry, neither too much nor too often, when you see two men expertly combining the

---

❶ 手册作者在手册中提到工程以一个总价整体外包，工人会更努力更快地完成工程，但不能保证工程的质量。整体外包费用是低，但工程质量有问题，后期的维修工程会产生大量的花费。因此还是按天计算工钱的做法比较好。

第二部分 手册的内容和翻译
Part Two: The Handbook Translated

气。为了尽量避免对这些"体育活动"进行大声斥责的情况发生，需要找一两个好的监工。监工不一定是基督徒，但必须是能信得过的聪明家伙。这样的人不会太多，但感谢上帝，在我们的信徒里还是能找到的。施工中如果能建立一定的等级会更好，比如，在有几个监工的情况，你就需要有人来管好他们各自不同的脾气。

这些监工中的一位也会是采购员。比起其他方面，采购的事情更需要传教士的掌控。监工从每个工人身上拿到的份例钱，他们称之为佣钱，我们是不易查知的，因为我们并不是经纪人。他们就像那些为圣教堂工作的教理讲授员❶一样，是按月领工资的。要承诺他们会有一个好的经济回报，而不仅仅是精神鼓励，比如比划一个漂亮的十字。如果有一个好的监工，即使是一个大工程，20 银圆❷也不会浪费。监工在处理关系时也会把捏分寸，既不会粗暴对待工人，也不会无来由让他们丢面子，同时也会坚持自己的要求。如果不谨慎处理与工人的关系就会有人故意破坏，比如，地基的某个角不夯实啦，或者在工程尾期屋顶铺瓦时不留神啦，这

balance of their six bricks' weight. In order to avoid as much possible too striking abuse in this kind of 'sport', first find one or two good supervisors. I do not mean one or two Christians, but two clever guys on which you can rely. They do not abound, I know; but thank God, they are not impossible to find among our believers. If there are several supervisors, you should establish a certain hierarchy and have an eye to the different temperaments (*piqi*).

One of these supervisors should simultaneously be the buyer. Here the personal control of the missionary is required more than everywhere else. The bonuses, *yongqian* as they call it, cannot be perceived by our people, who are not brokers, *jingjiren*. Like the other catechists, ❸ who also work for the Holy Church, they will be paid every month. Promise them a good financial reward, not just a beautiful crucifix. For a big construction, even twenty piasters will not be lost, if the supervisors have well done their job. Supervisors should deal tactfully with their people, they should not rush the workers, do not unnecessarily make them lose face, but at the same time should be firm in their demands. Without such precautions, mind sabotage. It is so easy to forget to ram a cor-

---

❶ 教理讲授员是那些负责教理问答的人，主要对受洗前的小孩或大人进行教理教育。教理讲授员是值得信赖的基督徒，他们的信仰是经过考验验证的，他们因传授教理工作领取工资。

❷ 银圆是用银铸造的一种流通货币。

❸ Catechists are Christians who teach catechism to children or adult converts before their baptism or communion. Catechists are reliable Christians, whose faith is improved.

~ 145 ~

些对工人们都是小菜一碟。

如果包工头是一个知名人士，如果大家证实他是一个好人，而且是一个诚实的人，那就更好了。如果他有名，应该就会有好的推荐人，其他传教士建筑者提供的信息是不能忽视的。

包工头会赚工人的钱：只要包工头做得不过分，这也情有可原，毕竟他也履行了他的责任。他拥有别人没有的技能，也比最好的工人懂得多。当工程完成后他买了三亩地，别人会指责他贪污，你也不用太在意，这是常人的嫉妒心理，而不是说包工头真的贪污。

一个工程总共需要雇多少人，有多少把式❶多少小工，还有厨师、磨坊工、抬水的人数，这些都要和包工头协商好。另外，以下这些方面也要规定清楚：用餐、往返工地还是只是回程单程、谁负责喂磨坊工的驴子、盐、烟草、犒劳费、厨具、节假日、下雨天、生病、炎暑天、基督徒不工作的星期日等方面。所有这些方面都要写下来，所谓：与诚实的人谈生意，要有合同；与无赖谈生意，谈都不要谈。

ner of the foundations, and especially to have a 'distraction', at the very end of the construction, when laying the tiles of the roof.

It is preferable that the contractor would be a known man who has proven to be a good worker and a honest man. If he is known, at least he should have good references. The information provided by other missionary-builders will not be disregarded.

This contractor will gain on his workers: this will only be justice if the gain is not excessive, since his responsibility is engaged. He is expected to show that he knows better than the best workers and have skills that others do not have. Therefore do not be too impressed by the charges [gossip] against him when, at the end of your construction, he will purchase three '*mou*' ground.

Agree with the contractor about the total number of workers to be employed and the proportion of laborers (*xiaogong*) compared to the masons (*bashi*), as well as the number of cooks, millers, water carriers. Be also precise about: provisions, round journey or return only, who will feed the mill's donkey, salt, tobacco, better meals as a reward (*kaolao*), kitchen utensils, main festival days, rainy days, illness, heating, not working days for Christians on Sunday, the pagans, etc. All these thing should be written, following the principle: "*with honest peo-*

---

❶ 把式都有一门吃饭的好手艺，他手下会有好几个小工。

第二部分　手册的内容和翻译
Part Two: The Handbook Translated

ple, writing; with rascals, no business."

通常每个泥瓦匠有两个小工。把式、瓦工头,特别是做细木的工匠头,他们通常有一些徒弟。这些徒弟从他们师父(这个地区也称老师或师傅)那拿不到多少钱,但是你付钱时并没有区分师傅和徒弟,都是按每人每天固定工资给的。

Usually there are two unskilled workers by one mason. The *bashi*, master masons, and especially master carpenters, almost always have with them a few apprentices, *tudi*. They do not receive much from their master (*laoshi* or *shifu*, according to the region); but you pay the salary of the master without distinction both to the master and the apprentice.

让包工头和监工知道他们有经济回报是好事。如果钱用到位,就物有所值:20 银圆对于一个重要的房子来说实在不算太多。

It is good policy to point to a monetary reward to the contractor as well as to the supervisors. If it was merited, it will be well invested money: 20 piasters will not be too much for a building of a certain importance.

包工头有可能会把利润和可能的亏损计算错。那些工人,如果他们是吃东家的饭,就会吃的很多。另外,还可能发生各种情况,比如:工人有可能会偷面包和粮食、亲戚朋友的来访、粮食价格的突然上涨等情况,谁能料到呢?不要轻易为这位亲爱的先生的呻吟动容,但是如果你确实发现他的工程亏损但又不是他的过错造成的,按照白纸黑字来履行合同会很残忍,那也要按合同约定来履行,这个不幸的人可以选择逃走或不支付他的工人的工钱吧,这样的事在大公司屡见不鲜。

It may be that the contractor miscalculates its profits and possible losses. The workers, when they 'eat the boss' (according to their expression *chi dongjia*), have a great appetite. Moreover there may be theft of bread, grain, visits from relatives and acquaintances, a surge in the price of grain; what else do I know. Do not be too easily impressed by the groans of this dear man. If, however, there is evidence that he is working at a loss and that it is not his fault, it would be cruel to stick to the letter of the contract, and give this unfortunate man the alternative to flee or not to pay his workers, as is happens in some big companies.

(手册里给出了六份合同样板,包括订购石灰、砖、花砖、瓦、石头的合同,还有第七份与包工头签合同的样板。)(图 2.2)

[Here the handbook gives six model contracts in Chinese, for ordering lime, brick, 'flower bricks', tiles, stone, and timber, as well as a seventh general contract with a contractor] (fig. 2.2).

(1) 石灰

立包約人　　今包到大名

天主堂塊灰　斤言明　秤每　斤大洋　元要塊不
要麵並言明送至　　堂內至本年　月　日送齊越期拆
半價除原價外至於脚力捐用堂內一概不管以後若有違約或有
他種情形保人情愿担負完全責任空口無憑立包約為証

保人
中人
民國　年　月　日
　　　　　　　　　　　　　當日付洋　元
　　　　　　　　立

(2) 磚

立包約人　　今包到大名
天主堂磚　萬係長　寸寬　寸厚　寸言明每仟大
洋　　元其中若有不符尺寸頂水腿磚紅磚等一概不要無亮
音者退換每百個許代封頭五個並言明（後節與石灰同）

图2.2　七份合同模板：石灰、砖、花砖、瓦、石头、木料以及与包工头的合同，手册第14-18页。
fig. 2.2 Seven model contracts: lime, brick, 'flower bricks', tiles, stone, timber, and with contractor. Handbook, p.14-18. (© Université Laval, Biblio-

(3) 花磚

天主堂花磚二宗計小柱礅　個十字磚　個小柱礅每個京錢　文

十字磚每個　文並言明（後節與石灰條同）

立包約人　今包到大名

(4) 瓦

天主堂瓦　宗計行瓦　片勾簷　片滴水　價錢尺寸列後行

瓦四角齊整勾簷滴水五角全有並言明送至堂內至本年　月　日

送齊越期拆半價除原價外至於腳力捐用堂內一概不管以後若有違約或有他種

情形者保人情愿担負完全責任空口無憑立包約作証

立包約人　今到包大名

計開

　　行瓦長　寸寬　寸厚　寸　每仟京錢　文

　　勾簷滴水同上　　每個京錢　文

保人

中人

民國　年　月　日　立

(5) 石头

立包約人　今包到大名

天主堂青石料　宗計石柱子　根石柱頭　個石柱礅　花條石　丈尺

寸價錢列後並言明送至　堂内至本年　月　日送若越期拆半價除原

價外至于脚力捐用等堂内一概不管送到后石工照样子做活作工石工系核日工

合算言明每日每人京錢　文堂内除每人日薪及來時有盤費並柴火外其除

事項壹概不管無犒賞以後若有達約或有他種情形者保人情願擔負完全責任恐

口無憑立字作證

計開

　　石柱身高　尺圓徑　尺寸每個價洋　元

　　石柱礅高　尺寬　尺厚　尺每個價洋　元

　　石柱項高　尺寬　尺厚　尺每個價洋　元

　　條石寬　尺厚　寸每尺價　元

　　當日付洋　元

　　　　保人

　　　　中人

民國　年　月　日　立

图 2.2(c)
fig. 2.2(c)

(6) 木料

立包約人　今包到大名

天主堂白楊木料　宗計大梁　根檁　根叉條　根寸板　丈二寸板　丈尺寸價錢列後紅楊虫眼不要並言明送至堂內至本年　月　日送暨除原價外至於脚力挪用等堂內一概不管以後若有達約或有他種情形者保人情願担負完全責任恐口無憑立包約為証

　計間

大梁長　丈大頭　尺小頭　尺每根價洋　元

檁長　仗大頭　尺小頭　尺每根價洋　元

叉梁長　丈大頭　尺小頭　尺每根價洋　元

椽子長　尺活　寸見方每根價洋　元

寸板長七尺厚一寸每寬一尺價洋　元

二寸板長七尺厚二寸每寬一尺價洋　元

木板量時量一大面量一小面

當日付洋　　元

　　　　保人

　　　中人

　　　　　　立

民國　年　月　日

图 2.2(d)
fig. 2.2(d)

(7) 立包約人　　今包修大名

天主堂路南大堂一所言明按日工做活每有一大工即有二小工錢不分大工小工

每人每日京錢　　文平日堂內只管柴火住處吸烟喂驢堂內管草不管料主日

奉教者罷工四大瞻禮全體罷工日或病人養病日堂內每人只給飯費

文下雨時亦是管飯無工錢惟下雨時矩促者則按鐘點扣算每十天一犒賞四大點

禮犒賞上樑調脊犒賞犒賞時每人四兩肉四兩粉菜斤半饜饜二兩酒青菜足用油

鹽足用其餘不管來時管盤費一上泥工木工一概在內至大堂完告成為止恐口不

願立字作証

民國　年　月　日　　立

　　　　　　　　　　中人

图 2.2(e)
fig. 2.2(e)

第四章　建筑材料
Fourth Chapter: About Materials

砖

烧砖工。烧砖工（烧窑的）（图 1.17 和图 1.18）是一群责任心时有好坏的人。他们特别善于毫不吝啬地做各种好的承诺。对没有太多钱的人，这不太容易，他们需要钱购买稻草、煤和高粱秆。在起草合同的时候要特别小心，砖瓦的大小、交货时间等都要写清楚，尤其得要见主要的管事人，在付钱之前必须亲自见到。

烧窑。对于一个大工程，有一个或几个自建的烧窑是很有利的，这样能保证在工程中不会缺砖少瓦，还可以自己供暖。在这种情况下，尤其需要有一个能干又有责任心的监工在烧窑旁盯着。如果想要有好砖头，必须得有自己的烧窑，这也是唯一的办法。

Bricks

Brick Makers. The brick makers (*shaoyaode*) (fig. 1.17 and fig. 1.18) are people with a very flexible conscience. They are especially generous with wonderful promises. It is a hard job exercised by people without much capital. Nevertheless, they need cash to buy straw, coal and sorghum stalks. Take precautions, not only in drafting the contract, specifying the dimensions of bricks, tiles, etc., the time when they must be delivered, etc. Especially ask to be introduced to serious guarantors, whom you have seen yourself, before paying any money.

Kiln. For a large building, it is advantageous to have one or more of your own kilns, and to provide the fuel yourself. In this way, you will be sure not to run out of bricks or tiles in the middle of construction. That is about the only way to have well-fired bricks. But, in this case, it will be absolutely necessary to always have a competent and conscientious supervisor near the kiln.

舶来与本土：1926 年法国传教士所撰中国北方教堂营造手册的翻译和研究
Building Churches in Northern China. A 1926 Handbook in Context

大小。有 4 磅❶重的砖，也有 9 磅重的砖。这也是目前能烧制的最小和最大的砖，结合这个国家的资源来看也是比较实际的。可以在这两个重量之间烧制整个系列。

把你需要的砖的重量告诉烧砖工，可以用模子告诉他们砖的尺寸大小，还要把烧砖后的收缩系数考虑进去。要让烧砖工知道，你要的是好砖头，不是"差不多"❷的砖，明确这点后就无须担心其他的细节。你必须非常坚定、非常清楚地在一开始就起草一个很精确的合同。砖的长、宽、厚的比例是固定的，所以有了砖的两个宽度或四个厚度，就能知道砖的长度。比如，9 英寸❸长的砖，其宽度是 4.5 英寸（30 厘米×15 厘米），厚度是 2.25 英寸（7.5 厘米）。

如果从一开始就根据砖的固定模数来设计平面，而且砖是依据此模数烧制的，这种情况下，教堂的砖墙就会更容易做，也会更经济。也就是说，砖的

Dimensions. The bricks vary from four to nine pounds. This is about the minimum and the maximum of what is currently being produced, and what seems practical given the resources of the country. You can choose from the entire range between these two weights.

When you indicate to the brick makers the weight of the brick you want, and they will not ask you any other indication, you would have told them the dimensions to give to the mould, considering the firing shrinkage. No need to worry about any other detail, as long as the worker knows that you accept no 'more or less' (*chabuduo*)❹ about the weight. To obtain this: you must speak firmly, very firmly and very clearly, from the beginning, and write a very precise contract. Two widths of a brick or four thicknesses, equal one length: these are the stereotypical proportions. So, a brick of 9 inches long will be 4.5 inches wide [30×15 cm], for a thickness of 2 inches and 2.5 *fen* [7.5 cm] (the *fen* is one tenth of an inch).

The masonry of the church will be easier and more economic to do, if from the beginning the plans are designed according to an established 'module', and all the bricks correspond to this module.

---

❶ 1 磅=453.59 克。

❷ "差不多"一词作者在手册里用了好几回，作者是批评中国工匠不追求精确，只要能好看，差不多就行了。

❸ 手册原注文为"6 英寸等于 20 厘米，见后文'木料'部分"。因此 1 英寸约等于 3.33 厘米。

❹ The author often uses this expression that means 'more or less', 'about the same', when criticising the imprecision of Chinese workers who are lucky with a result that looks nice (haokan) but is not accurate at all.

第二部分　手册的内容和翻译
Part Two: The Handbook Translated

长度就是一个基本尺度，可以根据这个基本尺度算出教堂的尺寸大小。模数的具体尺寸不那么严格，随烧砖工的喜好而定，通常是 8~10 寸（寸：中国计量单位）。

这种情况下，仅仅告诉烧砖工砖的重量是不够的，还需要指出砖需要的长度，砖的长度会决定砖的重量。计算的时候，必须考虑在干燥过程和烧制过程中的收缩系数，也需要考虑砖墙的接缝（一般接缝是 3 分❶宽）。

要想烧制这样的砖，你需要自己提供模子。除了极有钱的地主，当地人是不会用什么好材料来做模子的，因为模子磨损很快，在用铁丝刮时更是如此。在这个区域，用黏土做的模子在不同时间段是有差别的，比如，在一个季节开始时做的黏土模子与在同一季节末做的黏土模子，在重量上会有 1 磅或更大的差别。如果烧成的砖厚度不一样，泥水匠在用这些砖砌墙时就会陷入尴尬，砌成的砖墙也不会那么好看。为了防止模子磨损，可以在模子的角上固定一个小铁片。但是狡猾的烧砖工总会把铁片扯掉，大家都知道，烧制 8 磅重的砖所需要的煤要比烧制 9 磅重的砖

In other words, the length of the brick will be the measuring unit [of the building]. This module is an arbitrary measure that the builder may specify to his liking, between 8 and 10 Chinese inches, and so obtain the desired dimensions for the church to build.

In this case, however, it is not enough to tell to the brick maker the weight of the desired brick, but the adopted length of the brick should be indicated. The brick's length will in this case determine the weight. In your calculations, you should take into account the shrinkage of the brick during drying and during firing. Consider also the joint of the brick masonry. This joint is about 3 *fen* (one tenth of an inch) [3.3 mm].

To obtain such bricks, you will have to provide the moulds yourself, because the local people, except for the 'moneybags', do not use such good materials. The moulds wear out quickly, when scraped with a wire, as usual. Therefore, there will be a noticeable difference between the bricks moulded at the beginning or at the end of the season, sometimes by one pound or more. Because of the uneven thickness, the masons will be embarrassed by these bricks and the masonry will be less beautiful. A small iron blade, fixed on the edges of the mould, will prevent this wear. The shrewd bricklayer, however, will find a way to break your iron blade because an 8

---

❶ 手册原注文为"1 分=0.1 英寸，相当于 0.333 厘米"。

来得少，因此要不时地检查模子，这点不容大意。

红砖。中国人不喜欢红砖。对他们而言，红砖就是没烧好的砖。从他们的角度看，他们是对的。要烧制漂亮的青砖，首先必须要烧得很好，其次紧随烧砖后还要往窑上浇水（饮窑❶），要不然砖在出窑前就变成土了。而烧制红砖，第一烧就足够了，等砖放凉后从窑中取出来即可，不需要饮窑，这样能节省很多煤。所以你必须做一个"笨洋人"，让他们烧制红砖。

混色砖和过火砖。砖头在没有烧坏的情况下，会多多少少有些红色或杂色。这是因为没有做"饮窑"这道工序。要设法把这些砖折价卖出去，因为用这些砖砌的墙会很难看，当然了，可以把它们用在地基部分或用在砖墙里面。

过火砖，即条砖，烧制时多放在窑口处。应该把这类砖用在靠近地面墙体的最下面几层，因为它们能防止由硝酸盐作用而引起的碱化。❷也可以把它们用在需要承压的地方，比如用在圆柱的

pound brick will cost less fuel than a 9 pound one. It is therefore imperative to check the moulds from time to time.

Red bricks. The Chinese do not like red bricks. For them a red brick is a badly fired brick. From their point of view, they are right. To have a beautiful blue brick, it must above all be well fired, otherwise it will not support sprinkling (*yinyao*) after firing; it will fall to dust before leaving the kiln. On the contrary, to make red bricks, the first firing is sufficient: let cool and remove from the kiln. No water to add, no bad colouring to fear, and lots of fuel saved. Therefore you would be considered a 'stupid foreigner' (*yangren*) to require red bricks.

Blended bricks and overburned bricks. Bricks, however, can be a mixed red, without being badly fired. The reason is the defective influx of water at the end of the firing process (*yinyao*). These bricks are sold at a discount because the wall would be really ugly if built with these motley bricks. They can be used for foundations or inside the masonry.

There are also the overburned bricks (*tiaozhuan*). These bricks were the most exposed to heat, near the door of the oven (*yaokou*). These bricks should be reserved for the first exterior layers of the walls at the ground level; they will

---

❶ 饮窑就是给窑顶渗水，使砖变成青色。现在还有这道工序。

❷ Saltpetre efflorescence is an important damage in construction in Northern China. This chemical process is caused by rising water. This point is developed further in chapter 8 of the handbook.

第二部分　手册的内容和翻译
Part Two: The Handbook Translated

顶板处。要在砖出窑时就把这些条砖挑出来，不要和其他砖混在一起。必须强迫砌墙工人把这些条砖用在正确的地方，他们不喜欢条砖，因为这些条砖会割伤他们的手指，也会损坏他们的工具。

烧砖土。现在我们必须说一下用来烧砖的土，尤其是那些自己建砖窑的人更要知道这些。不是所有的土都能用来烧砖，用来烧砖的土不能太肥也不能太贫瘠。土太肥的话，烧制时就会翘曲、变形和裂缝；土太瘦的话，在烧制时就会太容易被火烧化。

如果土太肥，可以加一些细砂；如果土太瘦，就加一些塑性土。这样的问题就留给专业人员去做吧，只需要记住一点，不是任何地方都能建窑的。

花砖。❶ 尤其是教堂，需要一定数量有时候需要大量的特别形制的砖来做线脚和装饰。做这类花砖要花很大的

never disintegrate from saltpetre. See p156❷ They should also be used at places that have to bear great pressure, for example above the abacus of the columns. These overburned bricks (*tiaozhuan*) should be sorted right out of the oven, otherwise they will be lost amongst the others. You must supervise the masons to force them to use these bricks at the appropriate places. They do not like to handle these bricks, because they cut their fingers and break their tools.

Brick clay. Now we need to speak about the clay used for bricks, especially if you have you own kiln. Not all earth is appropriate for making bricks. The common clay for bricks should be neither too rich nor too poor. When earth is too rich, the products warp, deform and crack during firing. In the second case, the bricks obtained would vitrify too easily and melt during firing.

If the clay is too rich, add some fine sand; if too poor, add some fire clay. This question should be left to the judgment of professionals. It is enough to know that you cannot build a kiln in just any field.

Flowered bricks [profiled bricks].❶ Especially for a church, one will sometimes need a considerable number of bricks of a special shape, in order to

---

❶ The French word 'brique à fleurs' (flowered brick) is not of common use neither in French nor in English. The author means moulded profiled bricks and modelled bricks, or perhaps bricks with decorated flower motives. The 'flower brick', however, is something different: earthenware pottery in the form of a brick with holes at the top for picking flowers. Such brick-shaped vases appear in the eighteenth century in the context of growing tulip fashion and are often decorated with imitations of Chinese motives (Dutch Delft pottery, for example).

工夫，也会花费很多的工钱。建筑师在设计图纸中必须提供这些砖的形状。应该先算一下所需要的花砖数量，然后针对每一种形状制作特殊的模子。如果不想出现意外的话，必须烧制比所需数量更多的花砖，这些花砖因为有凹进去的部分很容易破碎。那些没有被用在正确地方的花砖可以用在砖墙里面（见手册图纸 Fig. 1~ Fig. 40 中的花砖模型，图 2.3~图 2.7）。

花饰、柱头及其他的部分构件没有固定模子，需要铸模工用好的陶瓷土混合切碎的大麻来做这样的模子。等模子足够干燥后再开始烧制，因为它们很容易破碎，因此不管烧制前还是烧制后，应该准备3倍的所需量。在这个国家找到好的铸模工很容易，他们连那么复杂的龙头和中国屋面装饰都能做，做这类模子对他们来说是小菜一碟。

如何给柱头铸模的建议。如果没有人指导的话，要做一个好看的柱头模型，不是件很容易的事情。记住，那些用来装饰柱头的叶子是需要放在花篮里的，花篮是柱头的一个重要组成部分，柱头底部不能比柱子本身大。差的铸模工会做一个底部很大的柱头，那是非常糟糕的。图 2.18、图 3.33 和图 3.37

make mouldings and other ornamentation. Cutting these bricks by hand after firing requires a lot of hard work and considerable expense. The architect's plans must provide the shape of these bricks. One should calculate the quantity, and then make special moulds for each type of profile. These bricks must be made in much more quantity than actually needed if you do not want to come up short, because they are more fragile than ordinary bricks due to the necklines. Furthermore, any leftover could be used elsewhere within the masonry (see models of these bricks, pl. 1-5, handbook Fig. 1-Fig. 40) (fig. 2.3 to fig. 2.7).

The finials, capitals and other similar pieces cannot be moulded. They must be modelled by a modeller, with good pottery clay, mixed with chopped hemp. After enough drying, one can start firing. Because of the considerable number of 'breakages', either before or after firing, one should prepare about three times the amount needed. It is quite easy to find good moulders in the country, among those who can do dragon heads and other ornaments of the Chinese roof.

Recommendation for modelling a capital. Modelling a beautiful capital is not easy without some guidance. Remember that the leaves that adorn a capital, are expected to be applied on a basket [the central part of the capital], but their base merges with the basket, so that the bottom of the capital may not be larger than the column itself. The great

显示了好的柱头做法和差的柱头做法。

建议把柱头边上的叶子做成弯曲状，这些弯曲的叶子是中空的。差的铸模工通常会做很直的叶子。一个有着曲叶的柱头，即使这些叶子只有很简单的细节处理，也会给人很好的印象。相反，那些细节很多、比例又很差的柱头会让人觉得很不舒服。

挑砖。烧砖工如果有足够良知，他们就会知道你会挑砖，而那些不合格的砖是要算在他们头上的，这点在合同里必须写清楚。挑砖时可以用一个大钉或一个其他铁件，好的砖头与铁件碰撞后会发出金属般的声音，而差的砖头会发出"沙哑"的声音（图1.17）。

数量。通常砖头数量是按每立方米计量的。如果有设计方案图纸，计算起来就很容易。对于一般的房子，中国工匠知道每间需要的砖量。砖的尺寸和"间"的尺寸都需要考虑在内。

mistake of unskilled modellers is making capitals too big at the base, which looks awful. See pl. 15, Fig. 76, good capital; Fig. 76 bis wrong capital. (fig. 2.17, fig. 3.33 and fig. 3.37)

Also, instruct the modeller to 'curve' the leaf that goes to the corner, that is to say to hollow the curve sufficiently. Usually the clumsy modellers make leaves that are too straight. A nicely curved capital always makes a good impression, even when the details of the leaves are very basic. In contrast, a very detailed capital that is badly proportioned, will always be unpleasant to see.

Sorting bricks. Given the elastic conscience of brick makers, they must be told that you will choose the bricks, and that any rejected bricks will be at their cost. The contract should stipulate this point. For sorting bricks, use a large nail or any other piece of iron. A good brick will be recognized by the metallic sound it makes. The other bricks ring 'cracked' (fig. 1.17).

Quantity. Obviously the quantity of bricks needed for a building is calculated per cubic meter. If the plans are well made, this calculation is easy to establish. For ordinary buildings, Chinese masons know the quantity by bay (*jian*). Both the dimensions of the bricks and those of the bays should be taken into account.

瓦

八砖[1]。八砖主要用在中国屋顶（图1.18和3.36）。它们有不同大小，通常都是方的。像砖一样，用铁件敲打八砖，好的八砖会发出清脆的声音。八砖厚度通常是1英寸，用不着把它们烧制得很薄，它们的重量对于大木架来说不算什么。八砖必须烧得很好，要不然过了两三年，说不定就会有瓦片掉下来砸到你，如果这些坏的八砖没有从两椽之间掉下来，也没有从15英尺的高处砸下来，你就是幸运的了。我就见过有些坏瓦像飞机碎片一样穿过屋顶掉下来。

要烧制比所需量更多的八砖，许多八砖可以用来做边饰和造型。

屋瓦（瓦）。瓦都是和砖一起烧制的，烧制时瓦应该放在窑的一角，因为瓦比较薄，如果用烧砖一样大的火烧制往往会烧过。烧瓦时要有"转瓦的"定时翻转瓦，如果自己建窑的话，"转瓦的"应该区别于烧砖工另外雇用。

瓦不能过小，要能够防风，也要能

---

[1] 八砖是边长八英寸的方砖，铺在椽子上。

Tiles

Square tiles (*bazhuan*). These small slabs are used primarily for the Chinese roof (fig. 1.18 and fig. 3.36). They are of different dimensions, such as bricks, but always square. Like bricks, they must make a clear sound when struck with a hard object. The usual thickness of these square tiles is 1 inch. No need to make them too thick, their weight would unnecessarily charge the timber frame. They should, however, be well fired; if not, after two or three years, pieces will fall on your head while you are happily at the table! You will be lucky if bad square tiles (*bazhuan*), don't come off in blocks from between two rafters, falling on you from a height of 15 feet. I have seen some crash through the ceiling like a bomb shell!

Count more square tiles (*bazhuan*) than necessary for the roof; a lot is used for borders and mouldings.

Roof tiles (*wa*). The tiles are generally fired together with bricks, in a corner of the furnace. Because of their thinness, when exposed to the same heat as the bricks, they are usually well fired. Tiles are turned (*zhuan*) by specialists (*zhuanwade*) that should be hired apart from the brick makers, if you have your own kiln.

The tiles should not be too small so

防止乌鸦钻进来。

瓦有两种。第一种是筒瓦，半圆柱形，通常用在宝塔上，过去也用在皇宫上。中国人觉得筒瓦好看，但可以肯定的一点是，它们不能很好地遮盖屋顶。因此，即使筒瓦再好看，也不要在教堂上用。另外一种瓦是一般的瓦，1/4的圆柱面。

中国瓦工计算瓦的数量是按每平方英尺来计算的，比如每平方英尺用20瓦。一"间"如果梁长是16英尺，就需要3000瓦。采用这种方法，就能计算出所需要瓦的数量，比如，梁长20或30英尺所需要瓦的数量。在做这样的计算时需要结合具体的屋顶种类来计算。在屋顶部分我们再细说。

很令人惊奇的是，据说希腊人和罗马人用的瓦，与中国人今天及几千年前用的瓦是一样的。在今天的意大利，与中国同样做法的屋顶还在使用。3厘米厚的瓦片放在橡子上，之间的空隙用灰浆填上。在很厚的土层上铺瓦，这应该是孔夫子之前久远年代的做法。

as not to be too easily disturbed by wind or by ravens.

There are two kinds of tiles. The first kind, the hollow tiles (*tongwa*), are half cylindrical. They are used almost exclusively for the pagodas, and in the past, for the imperial palaces. The Chinese say they are beautiful to see. What is certain, however, is that they do not cover the roof well. Therefore, you should never use them for a church, under the pretext they are nice looking (*haokan*). The other kind is the ordinary tile, a quarter of a cylinder, like a truncated cone.

Chinese masons count 20 tiles per square foot. A common bay (*jian*), with a 16-foot beam, requires about 3000 tiles. With this data, one can calculate the amount of tiles necessary for another known surface; for example if the beams are 20 or 30 feet. In this calculation you also will take into account what kind of covering you chose. We will develop this when speaking about the roof.

Amazingly, it is said that the Greeks and Romans used the same kind of tiles thousands of years ago that the Chinese use today. In Italy, the same method of roofing as in China is still in use. 3 cm-thick slabs are placed on the rafters and the joints are filled with mortar. On this kind of paving, one arranges the tiles, probably on a thick bed of mud, as has been done here since long before Confucius. (Barberot, Construction civiles, p. 455).

除了普通瓦，还有用在屋檐处的瓦，被称为滴水和勾檐。烧制这些瓦需要用特殊的模子，这些模子要比其他的模子贵很多。

木料

我们的传教区位于直隶东南部，这里唯一的木材是杉树。天津还有其他的木料，与南方相比，天津的木料要贵很多。建议专门找一个购买木料的专家，不然的话，你就会买到半腐蚀的木料或很粗糙的树干，根本不能用来做木板，还可能会碰到其他情况。木料购买商一旦选好了他要的木料，他就会控制木料的质量，在木料运上船的时候甚至运木料途中进行检查，把不符合质量的木料换掉。预定木材要留足时间，因为相互间的沟通往往会很慢，而且木头砍下来后需要进行干燥，这样才能保证在使用时木头都是干燥的。

在我们传教区的南边，有很多杨树和榆树，这些树料可以用来做屋架，质量一样好，而且比天津便宜很多。但是不要用这些木料来做门、窗和家具，即使在木料很干燥的情况下，这类木料也很容易变形（图 2.22）。

千万不要用柳树来做任何重要的结构，不然几年后都会被虫蚁吃尽而化为土。

Apart from ordinary tiles, do not forget the tiles that end the lower part of the roof, called gutters (*dishui*) and covering (*kouyan*). These tiles are moulded in a special way and are more expensive than the others.

Timber (muliao)

In the north of our Mission of Southeastern Zhili, the only timber is fir, but other timber can be found in Tianjin. It is very expensive in comparison with the timber that is found more to the south. We recommend that experts purchase the timber, otherwise, you will receive half-rotten wood, or gnarled trunks that cannot be used for planks, etc. When the buyer has made his choice, he will have to watch that there are no substitutions made when loading the boats, and even during transit. We also recommend ordering timber early, because communications are often very slow. Then timber will need to be cut, to allow for drying before use.

In the south of our Mission, there are many poplar trees (*yangshu*), elm trees (*yushu*) etc. As lumber these trees are as good as, and cheaper than timber from Tianjin. Doors, windows and furniture, however, will be hard to make with such local woods that warp too much, even when perfectly dry (fig. 2.22).

You must avoid using willow for any serious construction. After a number of years, it becomes worm-eaten and

在砍树之前，提前几个月先把树皮剥掉，这样会增加木料的强度，据说会增加六分之一的强度。必须在树木最旺盛的时候剥树皮，比如，在五月剥树皮，在九月砍伐。

如果在砍伐和锯料之后把木料放在石灰水里浸泡几天，之后再对木料进行涂刷，经过这样处理的木料可以无限期保存。

不管怎样，必须在树浆开始流动之前砍伐木头，也就是在冬天，十月到二月之间。

木料的自然干燥过程中，一方面要避免木料被日晒雨淋，另一方面还要把木料放在空气流通的地方。

屋顶结构的大构件，稍微的弯曲变形几乎不会有什么影响。绿色木头（即未干燥的木头）都可以用作这些构件。

用绿杨料做檩条，直径可至 6 英寸（0.20 米），长度可以达到 10 英尺（3.3 米）。不能无限夸大檩条的长度：对于不固定在山墙里的檩条，11 英尺是最大长度，如果是绿色木头的情况下更是如此。用榆木做檩条，如果木料没干透，11 英尺长的檩条就会在中国屋顶的压力下变弯。

通常情况下，椽子为 2 英尺（6.6~7 厘米）见方，长约 4 英尺。❶ 当然，椽

falls into dust.

Stripping the bark from the trees a few months prior to felling will increase by one sixth the strength of the lumber. This must be done when the tree is in full vigour, in May for example, and then cut in autumn.

Lumber can be conserved indefinitely, if, after having been stripped and sawed, the pieces are soaked for a few days in lime water, and accurately coated.

In any case, timber must be felled before the sap rises, during the winter, from October to February.

The natural way to dry wood consists of protecting it from direct sun and rain, while letting it dry outdoors.

For the big beams of the roof structure, 'warp' is not a problem. Almost green wood can be used for these elements of the timber frames.

The dimensions of the purlins (*lin*), made of green poplar, are 6 inches (0.20 cm) in diameter by 10 feet (3.30 m) in length. We must not exaggerate the length of the purlins: 11 feet (3.35 m) is a maximum for the purlins that are not fixed in the walls of the gables, especially if it is green wood. 11 foot purlins, made of green elm, would bend under the weight of the Chinese roof.

Usually, the rafters are 2 inches square (6.6 to 75.08 cm), for a length of

---

❶ 更多关于梁的尺寸在教堂项目草案 1 里的剖面图里有提到（图 2.28）。

子的长度还取决于檩条之间的距离。在订购椽子之前，要参考设计图纸，了解檩条间的距离，这点很重要，要不然就会花冤枉钱买了材料还用不上。

每间的椽子数量是可以通过八砖尺寸来计算的。通常是每间每排 12 根椽子，而排数取决于房子的面阔。

石灰

石灰有纯石灰和差石灰之分。纯石灰是碳酸钙或经过强热处理过的石灰岩，这两者在煮沸之前基本都有一样的外观。给这些石头加上水，就会变热、融化和膨胀，体积会膨胀很多，也会产生很多热量，这样就变成了熟石灰。生石灰就是加水之前的烧石灰，中国人称之为生灰。熟石灰和水混合形成浆料，是砂浆的一个组成部分。拎灰是指稀释石灰并通过细网把石子和鹅卵石刷掉。经过这样处理过的石灰被存放在石灰坑里，再用一层干燥的土把石灰坑盖上，这层盖土很快会将多余的水分吸走。这层盖土很重要，它把石灰与空气隔开，能防止它与碳酸发生化学作用。如果石灰坑被盖得很好的话，熟石灰能一直存放，直到它被用作砂浆。为

about 4 feet (1.22 m).❶ The rafters' length, of course, depends on the spacing of the purlins. Before ordering the rafters, is important to know this exact spacing, according to the plan, to avoid paying for wood of the wrong dimension.

The number of rafters by bay is calculated from the dimensions of the square tiles (*bazhuan*). This is usually 12 rafters by bay and by row. The number of rows depends on the length of the building.

Lime (*shihui, dahui* or *just hui*)

We distinguish pure lime (fat or rich lime) and poor lime (lean or meagre lime). Pure lime is calcium carbonate or strongly heated limestone, that retains almost the same appearance as before burning. If one wet these stones, the lime heats, fuses and expands, that is to say that its volume expands a lot and generates considerable heat. So it becomes slaked lime or hydrated lime. Quicklime or dry lime is burnt lime before being mixed with water. Chinese call it 'uncooked lime' (*shenghui*). Slaked lime in powder is called *tanghui* / *tangwu* or *shushihui*. Slaked lime mixed with water forms a binding paste that will become a component of mortar. *Linhui* means to dilute lime and filter it through a wire mesh to extract stones and pebbles. Such lime, thinned and cleared of foreign

---

❶ More dimensions of beams are mentioned on the section of the church project 1 (fig. 2.28).

了防止霜降，要盖一层厚土。新的熟石灰没有老的熟石灰好。

在熟石灰里加更多的水，就会变成石灰水，也叫石灰乳，它可以用来刷白。

差石灰。石灰岩含有不相关的东西使得石灰水合性很差、膨胀也少。如果与水混合，浆料的黏性很差。这样的石灰视为差石灰。纯石灰和差石灰都不能放在水里。

如果要形成能放在水里的水硬性石灰，石灰岩得至少含有 12%~20%的黏土。

直隶及其周边地区的石灰基本都是纯石灰，当然也有些差石灰，不过在这个区域我还没有发现有水硬性石灰。

纯石灰砂浆。石灰加上两倍或三倍体积的砂子（好的河砂），就会变成纯石灰砂浆。因此必须准备好砂子。砂浆硬起来很慢，所以要避免建得过快，要留足时间让下面的砂浆变硬，这样才能

bodies, is stored in pits called *shihuikeng*. It is covered with a layer of dry earth, as soon as the earth has absorbed the excess water. This covering with earth is important to deprive the lime of air and prevent it from combining with carbonic acid. If well covered in this pit, slaked lime can be stored indefinitely until it is used as mortar. One should also avoid frost; therefore cover with a thick layer of earth. The freshly slaked lime is not as good as older slaked lime.

When adding more water, slaked lime becomes water lime (*shihuishui* or *shihuiru*). It can then be used for whitewashing (*choashihui*).

Poor lime. Limestone containing inert foreign matter provides lime that hydrates more slowly and expands less. When mixed with water, it forms a less binding paste. Such lime is called poor lime. Both rich lime and poor lime do not set under water.

To obtain hydraulic lime, which is able to set under water, limestone containing from about 12% to 20% clay is required.

Our Zhili limes and those that come from neighbouring provinces are mostly pure limes; some are poor limes. I have not found any hydric lime.

Pure lime mortar. When adding two or three times its volume of sand (good river sand), this lime becomes pure lime mortar. You must therefore prepare sand. This mortar hardens slowly. Therefore you should avoid building too

承受上层部分的重量。因此不建议在第一年里完成一个大教堂的所有墙体，这点我们在后面也会提到。比较理想的做法是，在第一年的秋天将墙体砌到窗子高度就停工，然后到春天再开工。

购买石灰。购买合同里必须写清楚石灰里不能有石子。给石灰装车或装船的时候，商贩可能会故意加进一些石子，在拎灰时你就会发现这些小把戏了。

要尽量避免在夏天运石灰，否则你买到的就只会是尘土，而且还要为运料途中吸入的水分付钱。途中吸收的水分绝不能忽略不计。春天是买石灰、运石灰的最好时节。但是在山区，往往是冬天焚烧石灰石。

石灰运到后，就要加水做成熟石灰，然后倒进石灰坑里。至少要有一个工人知道怎么做这件事情。

千万不要用碱水，烧砖时不要用，溶解石灰时更不要用。

麻绳

这方面的花费也挺可观，做预算时应该把麻绳的费用考虑进去。

quickly and foresee enough time for the lower layers of the masonry to harden in order to support the weight of the upper parts. Therefore, we will repeat this later, we do not recommend completing the walls of a large church during the first year of construction. The ideal is to stop at the height of the windows in the autumn of the first year, then resume in the spring.

Purchase of lime. The contract of sale must stipulate that stones will not be accepted! When loading lime on the carts or boats, retailers may be tempted to add rough stones. You will only discover this fraud when filtering the lime (*linhui*).

Avoid carting lime in the summer. You will only get dust, and you will pay the water absorbed on the journey the price of the lime. This amount of water is far from negligible. The spring is the best time to buy and transport lime. Furthermore, in the mountains, burning limestone happens especially in winter.

As soon as the lime is delivered, it must be slaked and poured into the pits that I mentioned earlier. At least one of the workers should be capable of this manipulation.

Do not use brackish water, either for manufacturing bricks, or for dissolving lime.

Rope (*sheng, masheng*)

The expense for this article is considerable. It must therefore be included

第二部分　手册的内容和翻译
Part Two: The Handbook Translated

　　当地工人搭的脚手架，根本没什么艺术感可言，就是一个没有任何形式的组合，还经常是不稳定的，建钟塔时搭的脚手架就是一个例子❶（图 3.21~图 3.23）。搭脚手架无一例外几乎都是用麻绳。这些麻绳和脚手架本身一样，都没有什么形式可言，都是由小工现场编织的。

　　我们必须会分辨两类绳子。第一类是进行拖拉的绳子，借助臂力、滑轮等进行拖拉。这类绳子需要专业的织绳人（即"打绳的"）来编织。他们用各种麻来织绳，这些麻有不同的名字，麻的名字和它来自哪个地区相关。

　　通常，麻论磅出售，非常贵。它们不能用来做脚手架，因为风吹雨淋的话，麻很快就烂掉了。

　　第二种是用来做脚手架的绳子，这类绳子是用建麻或槿麻织成的。

　　这类绳子就是很简单编成的，不像"打绳的"织好绳那样编织的。小工会根据不同用途编织出不同长度的绳子。并不是说这类绳子不能用在滑轮上。如果整个建造过程需要两年的话，这类绳子暴露在外，经受雨淋受潮

in the original estimate.

　　The scaffolding of local workers is not artistic at all. It is a rather shapeless assembly, and often too unstable in the case of a bell tower for example❶ (fig. 3.21 to fig. 3.23). It is held together almost exclusively by ropes. These ropes are even more misshapen than the scaffolding itself, because they are woven on the spot by any unskilled worker.

　　We must, however, distinguish two kinds of ropes. First, the ropes used for traction in all its forms: arm strength, pulleys, etc. These ropes are carefully made by professionals, rope makers (*dashengde*). They are manufactured with a hemp that receives different names, depending on the region: *haoma*.

　　Usually, they are sold by the pound and are quite expensive. They are not used to secure scaffolding poles because they rot quickly if left exposed to the rain.

　　Second, the ropes that hold the scaffolding are made with *pennma* or *k'iouma*, also called *jingma*.

　　These ropes are simply braided; they are not twisted as rope makers do when making good ropes. The unskilled workers braid these ropes in varying lengths, depending on their use. It goes without saying that they cannot be used for any pulley. They can remain exposed

---

❶ Implicit reference to the tower's scaffolding in Daming, which needed to be reinforced, as explained in: Pierre Mertens. Une grande ville chinoise qui s'ouvre à la foi. Épilogue: achèvement de la cathédrale de Tai-ming-fou [J]. Les Missions Catholiques, 1920, 356-357, 367-369.

也不会烂掉。建造工程完成后，它们就会被剁碎和石灰混合在一起，用来抹灰。这种情况下，麻也被叫做"麻袋"或"麻刀"，一起混合的石灰叫做麻刀灰。

to rain and moisture for the duration of construction, two years if necessary, without rotting. Afterwards, they are finely chopped and mixed with the lime used for plastering. In this form, that hemp is called *madai* or *ma dao* and the lime to which it is incorporated is called *madaohui*.

脚手架（杆子）

这是另外一项大的开销，在做预算时可能也会被忘记。如果是建一座不太重要的房子，可以借杆子和板子。但这也不好，因为租金非常贵，会因为租金损失不少钱。如果你自己买材料做脚手架的话，买的小木块还可以循环使用，在装天花板时还可以用作内楞子。

在天津周边区域只有小冷杉（也叫"沙杆"）能用。可以在铁路和河流附近找到便宜的沙杆。其他地方的小杨树，用作椽子太牵强，倒是可以用来搭脚手架，而且价格还便宜（图1.19）。完工后，还可以把它们再卖掉，比小冷杉好卖。

如果杆子要埋在地面下好长时间，比如建大教堂的话可能需要一年或两年，这种情况下就需要事先把埋在地面下的那段杆子涂上一层沥青。

Scaffolding poles (*ganzi*)

This is another large expense you might forget when establishing the cost estimate. For a smaller building, you can often rent poles and planks. This is, however, not advantageous. The rental is quite expensive and you will lose the money of the rent. On the contrary, the smallest piece of wood you will purchase can be recycled, even as lattice for plastering.

Near Tianjin, only small firs are used (*shagan*). Some can also be found at good prices around the railways and rivers. Elsewhere young poplars, too weak to be used as rafters, cost less and can be used for scaffoldings (fig. 1.19). After construction, you can re-sell them more easily than the small firs.

If the poles must be planted in the ground for a long time, one or two years for example, when building a large church, it is suitable previously to coat the buried section with a layer of tar.

第二部分　手册的内容和翻译
Part Two: The Handbook Translated

第五章　砖石工程
Fifth Chapter: Masonry

地基

在直隶平原和它的周边地区，没有必要为了寻找岩石层而进行勘探。不用担心，好的工匠了解他们所在的区域及其地下土层。他们勘察时不要去妨碍他们，即使他们进行得很慢，这也是事实，但最终他们会在一个很牢靠的地基上建起最漂亮的塔。这是通过夯土来实现的。这个过程叫做打夯。用来打夯的工具叫做夯，有木夯、石夯和铁夯（图1.20 和图 1.48）。如果有需要，他们会用混凝土和木桩。让他们决定吧。

除了土层的坚硬性，砌筑基础是决定建筑结构稳定的最重要因素。我们把地基部分的墙体称为砌筑基础，底部最宽然后逐层递减。底部宽的砌筑基础将建筑结构的重量分摊到一个更大的平面，使得每单元面积的平面上受到的垂直压力降低，这样的话，建筑即使是建在不那么坚实的地面上也无妨。中国工匠知道这个原理，并能恰到其位地运用

The Foundations

In our plain of Zhili and its surroundings, it is not necessary to practice borings for finding rocky lands. No need to worry; good masons know their country and the subsoil. If you do not obstruct them in their investigations, even if they are slow, I must admit, they will erect the most beautiful tower on a solid foundation. This will be done by ramming the earth with a rammer. This operation is called *tahang*. The rammer is called *hang*. There are wooden rammers (*muhang*), stone rammers (*shihang*), and iron rammers (*tiehang*) (fig. 1.20 and fig. 1.48). If necessary, they will use concrete and pilings. Let them decide.

After the nature of the soil, the footing is the most important factor of the construction's stability. The footing is the layout of the foundation walls, wide at the base and becoming narrower layer by layer until the ground level. The footing spreads the weight of the construction on a broader surface, reduces the vertical pressure per surface unit, and allows building even on less resistant soil.

~ 169 ~

它。放手让他们去做吧（图 2.8 中的 Fig. 41）！

如果承包商同意只用少数把式来挖、夯地基，会省很多钱。实际上，这项重要的工作可以由小工们在两三个把式的监督下完成。

这些小工首先要挖地基槽（也叫修基槽），至于要挖多深，这取决于土壤的坚硬程度。他们会根据事先用小木桩和绳子做的标志来挖。夯土时往往会喊号子，伴随着喊号子进行的夯土工程，为期两周或一个月，之后开始砌砖墙，这个时候我们如果看到墙体在夯土层外差不多 2 英尺，或者只有一半墙体立在夯土层上，你就需要注意了，这样做的地基会很不牢固。你需要进行监督，不监督的话，他们会胡乱匆忙地再夯几下，当然只是最基本的一些打夯，这样做只是为了避免受责备，然后他们就继续砌墙了。

在一块完整的土地上给建筑边界划线不难，但是根据所划定的线来挖好几英尺深的地基槽，却是很困难的事情。挖了几下后，木桩就看不见了，挖出的土堆在地基槽外面也改变了原来地面的外观。所以，木桩要足够高，还要位离要挖的地基槽外一些距离。工人时不时可以从一个木桩望向另一个木桩以保证挖的地基槽在一直线上。坑完

The Chinese worker knows this principle and uses it in a very rational way. Just let him do it ! (handbook Fig. 41) (fig. 2.8).

It will save a lot of money, if the contractor agrees to use only a few master masons (*bashi*) to dig and ram the foundations. This important work can be done by unskilled workers under the supervision of only two or three masons.

These unskilled workers first dig foundation trenches (*xiujicao*) to a variable depth, depending on the nature of the soil. Usually they do not dig very accurately (*chabuduo*), following the lines that have been previously traced on the ground with pegs and cords. After having rammed and shouted a lot, during fifteen days or a month, when beginning the masonry, one will notice that the walls are 2 feet outside the rammed area, or are only half based on it. If you do not closely supervise this stage, they will hastily do some additional ramming and then start the construction. But this additional ramming, of course, will be minimal, because it is only intended to avoid blame.

Tracing the limits of the building on the still intact ground is not difficult, but digging trenches, several feet deep, according to those traced lines, is indeed quite difficult. After the first blows of the pickaxe, the pegs disappear; extracted and rejected earth changes the appearance of the terrain. Therefore marker stakes, high enough, should be placed at some distance, outside the field where

全挖好、清理、弄平后，在打夯之前，应该准确地给地基槽重新划线，将木桩放在地基槽底部，以保证直角不会变成80°或100°。

不用说，所挖的沟槽必须足够宽，能放得下砌筑基础，而不能仅仅是地面层墙体厚度那么宽。

打夯

打夯是项讨厌的工程……对付钱方来说，他们总是很急切想看见建筑拔地而起。建议不要从一开始就紧张兮兮。对所有人而言，除了东家之外，打夯就像是赶集，小孩也参加。有好吃的，每天有馍馍，时不时犒劳（有酒有肉），大家唱歌，有时候有些……粗俗；但大多数人几乎不工作，他们只打号子。如果东家很有钱，他的各门亲戚都会远道而来，如果他人缘不错，全村的人都会来帮他用晚餐。

you are digging. From time to time, workers can take a look from one stake to another and check if they are still in line. Afterwards, when the trenches are fully opened, cleared and levelled, and before ramming, one should accurately retrace the trench, and place the stakes at the bottom of the trenches, in order to accurately verify that the right angles did not became 80 or 100 degrees.

It goes without saying that the trenches must be wide enough for the footings and not only for the walls as they will be at the ground level.

Ramming (*dahang*)

Ramming is a most tedious operation... for the one who pays, especially if he is anxious to see his construction rise quickly from the ground. It is, however, essential not to be impatient at the beginning. For everyone except the boss (*dongjia*), ramming is like a fair. The children come and take part. Good food and white bread (*momo*) is eaten every day, even small feasts with meat and brandy (*kaolao*) from time to time; people sing a lot, it's sometimes a bit… bawdy; but mostly people are working very little, despite or because of lots of shouting. When an individual person is building, especially if he is a rich man, relatives (*qinqi*) of all ages come from far and near, and if he is well-liked, the whole village will help eat his good dinners.

你应该感谢这些人来，他们带着热心，吃着你的饭，督促着这项运动，通过一些额外的犒劳给打夯工人们胆量，你会看到你的教堂不会一分为二，就像某个教堂那样。

但是不要无限度地容忍懒人，应该让所有人知道这一点，最好不要在现场说某人懒或和个人说什么。就跟承包商说某人有些累了。这个懒人就会自己改正或他会以其他的什么理由被除名，这样做，所有人的脸面都会得到保留。

要想看看土壤是否被完全夯实，可以看它是否会发出一种干燥的声音，几乎是金属般的声音。承包商必须做这样的检查。如果是建一个教堂，特别是钟楼，它的地基部分还要加入夯实的混凝土。先把砖头捣碎到核桃般大小，然后把这些砖屑用足够浓的石灰水来混合，以 1∶5 的比例，即 1 份体积的石灰配 5 份体积的砖屑。要留足一天或两天时间让石灰水彻底渗透到砖屑里。

夯土的时候，将砖屑和石灰水的混合物倒入地基沟槽里，再加上土混合，之后再加水调匀，再夯，一开始用小夯，之后再用重的夯。如果夯得很好的话，就会得到和坑底形状一样的一整块，很硬实。如果要在上面挖洞的话必须用铁杆才行。它就像是一块巨大的石板承托着墙体的整个压力，将其分散传

You should make your mind up early, to thank these good people who came to spend their energy and your food, to encourage them, to give 'guts' to the rammers by some extra *kaolao*, and you will see that your church never will split in two, such as the church of X!

No reason, however, to tolerate any lazy workmen. But, let it be said once and for all, it is better to say nothing on the spot or personally. Report to the contractor which workman is 'tired'. The sluggard will either correct himself or will be fired under another pretext, with no loss of face for anyone.

In order to check if the soil is sufficiently rammed, it should make a dry sound, almost metallic. The contractor must note this finding. For a church, especially for a bell tower, add pounded concrete. Broken bricks are crushed to the average size of a walnut. These crushed bricks are watered with fairly thick lime water in the proportion of one volume of lime for five volumes of crushed bricks. Mix the whole and let the lime penetrate the brick during one or two days.

When ramming, throw this mixture of crushed bricks and lime in the trenches of the foundations, add earth and mix, then add a little water, equalize and ram again, first with little blows, later with heavy blows of the rammer. If this work is well done, one obtains a compact block that has the precise shape of the bottom of the trench. An iron bar

递给地基沟槽的整个范围。50 米高或更高的漂亮钟楼立在上面完全可能，不用有任何担心。在这种情况下，建议做一个更厚的混凝土层，一层一层更仔细地夯实。建议不要在一年里完成整个钟楼的建造，要先让建了一半的钟楼在冬天自然下沉以压实地基，要用土覆盖地基部分以免其遭受霜冻破坏。

砖石工程

砂浆。我们在手册 26 页上谈过做砂浆要准备的材料。砂子在这里起着惰性材料的作用，从经济角度它增加了砂浆的体积。砂子非常有用，因为它把石灰分开，这样更容易与空气相渗透，由于它吸收碳酸的能力也使凝固变得容易。中国泥水匠喜欢使用纯石灰，仅仅是因为它更好看。不知多少次，对他们而言，好看比强度要重要！

然而，这种纯石灰砂浆只有在露天才能变硬。因此，不要急于填埋地基，在填埋之前让它们尽量干透。要反复搅匀混合砂子与纯石灰，使其能吸收尽可能

will be necessary to drill a hole into it. It will be like a huge slab that will support a uniform pressure from the walls, spreading it over the entire extent of the foundation trench. It will be possible to erect on it, without any fear, very pretty bell towers of 50 meters height or more. If that is to be the case, we recommend making a thicker concrete layer, and pounding it more carefully, layer by layer. Do not complete the construction of such a tower during the first year, but let the slab pack down over the winter by covering the foundations in order to protect them from the frost.

Masonry

The Mortar. We talked about it on page 26, about the materials to prepare. The sand here fulfils the role of inert matter; it economically increases the volume of the mortar. Moreover, it is useful because it breaks up the lime, making it more permeable to air, and facilitates hardening thanks to its capacity to absorb carbonic acid. Chinese masons are thus wrong to use pure lime under pretext that it is more beautiful to look at (*haokan*). How many times, for them, a nice appearance takes precedence over strength!

However, this pure lime mortar only hardens in the open air. Therefore, do not be too hasty to 'bury' the foundations, but let them dry as much as possible before covering with soil. The pure lime

多的碳酸。此项混合工作要非常认真地做，要保证砂浆的完全均匀性。

　　墙。我们前面已经说过地基墙的砌筑基础。多层建筑或教堂墙体的砌筑基础，在地下土层坚硬的情况下，也必须是 0.7~0.8 米厚。

　　在讨论地面以上的墙体之前，我们先说一下怎么准备砖。做好砖墙砌体的秘诀是要用近固态砂浆和湿的砌墙材料。但总有人不那么做。我们的泥水匠更喜欢在砖头或石头上狂撒砂浆，而不是把它们弄湿，建筑师也允许他们这样做，这是不对的。干的砌墙材料，尤其是干砖会很快把砂浆中含有的少量水分吸走，这样的话，砖和砂浆都变干了。含有极少量水分的砂浆几乎就是粉末，在砂浆水分没有完全被砖头或石头吸干的部位，所预期的粘合性只能部分实现(Barberot, 民用建筑，第 887 页)。

　　8 磅或 9 磅的砖能吸收近 2 磅的水。其吸收过程相对很慢。所以光给砖浇水或将其浸泡水中一会儿工夫是不够的。它们应该湿透以让水充分渗透（即饮透了）。为了达到这个目的，做

and sand may be mixed repeatedly, in order to facilitate the absorption of the most carbonic acid possible. After, one softens it by adding water, as and when required. But the mixture should always be done with great care, to ensure perfect homogeneous mortar.

The Walls. We have spoken about the footing of foundation walls. The footing on good soil, for a multi-storey house or for ordinary walls of a church, must be 0.70 m to 0.80 m thick.

Before discussing the walls in elevation, that is to say above the ground level, let us say a word about how to prepare bricks. The following precept expresses the secret of good masonry: firm mortar and wet materials. Nevertheless, the opposite is always done. Our masons prefer flooding mortar rather than wetting brick or stone, and the builders allow this, and are wrong. Dry materials, especially bricks, have quickly absorbed the little amount of water contained in the mortar, then both dry separately. Mortar with less water is almost powder, and the aimed cohesion or adhesion happens only partially, where for any reason water has not been completely absorbed by the brick or the stone (Barberot, Constructions civiles, p. 887).

A brick of 8 to 9 pounds can absorb nearly 2 pounds of water. But absorption is relatively slow. Therefore, it is not enough to simply water the bricks, or plunge them an instant into the water. They should be soaked, to let the water

第二部分　手册的内容和翻译
Part Two: The Handbook Translated

小型结构的话，可以用一口锅或一个大瓮。做一个大建筑的话，就需要挖一个小水池，在水池底部涂上胶泥以防止水渗透下去，接着铺一层砖，这样水就能保持干净。将要用的砖放在这个水池里，在水池里放上水，浸 20~30 分钟。已经湿透的砖如果不马上用，就需要把它们从池塘里拿出来，这样好准备其他的砖，由于时间紧，在砌墙的时候应该再给这些砖浇水。泥水匠或小工，在砌下一层砖前，总会给刚刚垒好的一排砖浇水（图 1.21）。

不要用碱水给砖浇水。砌墙的砖吸收如此大量的水，砌成的墙确实会在很长一段时间里都很湿。这就是为什么在春天施工比在夏天好。这也是为什么有中国俗语这么说："一个新房子，第一年给你敌人住，第二年给你朋友住，第三年自己住。"

中式墙。地方泥水匠至少知道四种墙，我们且不说他们知道的细的分类。级别最低的是打土墙，它有两种做法，一种有木板一种没有木板，它们是打墙和垛墙。接着是坯墙，这也是细分的一类。有的书提到这种建造方式可追溯到很古的时代。这点我可以相信，不

penetrate thoroughly (*yintoule*). For this purpose, in the case of a small construction, a pot (*guo*) or a large jar (*weng*) is enough. For a large building, dig a small pond, cover the bottom with clay (*jiaoni*) to prevent water from filtering into the ground, then place a layer of bricks so that the water remains clean. Arrange the bricks that will be used in this basin, fill the basin with water, and let soak for 20-30 minutes. If the soaked bricks are not used immediately and have to be removed from the basin to prepare other bricks, because of a rush, they should be watered again just before they are used. Then the mason or an unskilled worker will water again the row of bricks he just has laid, before building a new layer (fig. 1.21).

Do not use brackish water for this operation. The walls, it is true, will remain wet for a long time, after having absorbed such a large amount of water. This is why it is better to build in the spring than in the summer. This is also why the Chinese proverb says: "*In a new house, the first year, let an enemy live in it; the second year, a friend, and the 3rd year, yourself.*"

Chinese Walls. Local masons know at least four kinds of walls, not to mention the subcategories. At the bottom of the scale, there is the clay wall, which is built in two ways, with or without boards. These are the 'hit wall' (*daqiang*) and the 'piled wall' (*duoqiang*). Then the rammed earth wall (*piqiang*) that also

~ 175 ~

要枉费心力去教老百姓建这类"建筑"。这些房子预期寿命也就20年。

接着是少数有钱人做的墙：里生外熟，就是说里面是夯土外面是砖头。事实上，有两座墙，或者说是一座真的夯土墙外面包了砖头以防雨水。

"墙面"的做法有很多种。最经济的做法，即墙体外包砖（所谓"表砖"），这种墙面相对而言不那么坚固。那些想建造更坚固墙面的人（但是他们是少数，因为谁又能预测将来呢？）会把所有的砖以"卧砖"砌筑。我们也称之为"片砖"。如果整个墙面都采用这种做法，就被称为"片砖到顶"。

这种双层墙的最大缺点是两层墙会各自产生不同的沉降。夯土和连接夯土墙与片砖的厚泥浆，有着比砖头和只用一层薄砂浆砌成的普通墙更高的沉降系数。过不了多久，墙就会凸，这样的过程会反复发生。如果这样的墙体做好的话，可以持续四十年。

knows subcategories. Books mention that this way of building dates back to very ancient times. We can take their word. Useless to teach the average peasant about this kind of 'architecture'. These houses are expected to last twenty years.

Then comes the wall of the moderately wealthy or even very rich man: raw inside, fired outside (*lishengwaishu*), that is to say rammed earth inside and bricks outside. In reality, there are two walls, or rather there is a real rammed earth wall, clad with a brick facing against the rain.

There are many ways for doing the 'facing'. The cheapest, and not surprisingly the less solid, is the wall faced with bricks, laid on their small side (*biaozhuan*). Those who want build more solidly, but they are the exceptions, because who can predict the future? Lay all bricks flat, as stretchers (*wozhuan* or *pianzhuan*). If the whole wall is clad in this way, it is called 'flat brick until under the roof' (*pianzhuandaoding*).

Compaction is the most important disadvantage of such double walls because it is a double risk. Rammed earth, and even more the thick earth mortar that binds the two parts of the wall, have a much higher packing coefficient than the fired bricks, the regular surface of which being joined by a thin layer of mortar. After a short time, the wall will become convex, and will need to be rebuilt. Nevertheless, when such a wall is well done, it can last forty years.

第二部分　手册的内容和翻译
Part Two: The Handbook Translated

最终我们做的是"里生外熟"的墙。我们在这里也只讨论这个。遗憾的是，我们的泥水匠不太习惯做这种墙，所以必须很仔细地监工，特别是在整个工程以一个固定总价外包的情况下。他们可能会按照做双面墙的做法做，这样做的话能砌得更快。他们会首先建一座很厚的墙，像他们过去做的夯土墙一样厚。此墙有两面：外面是砖里面也是砖。这之间的缝隙可能用任何东西填缝，特别是在监工转过身去的时候。在这个被遗忘的地方❶，他们有时会乱弄一通，如果他们对你给的犒劳不满意的话，他们甚至都不会花时间把砖屑、土、泥和纯石灰均匀弄平。他们或许会在每 6 或 8 块顺条砖之后放一块丁砖，用来遮住那被遗忘的地方，这样你就会看见一个纯砖墙。这种情况可能会在整个工程以一固定总价外包的情况下发生，那么一座夯土墙，梁下面是用砖砌成的柱子。那就是一个被堵塞的真的烟囱。

Finally, there is 'our' wall with bricks inside and outside (*lishengwaishu*). We will only talk about that one. Unfortunately our masons are not used to building such walls and they must be supervised closely, especially if they work for a set price. They will prefer the principle of their double walls, especially because in this way masonry rises much faster. So, they will first build a very thick wall, as thick as the rammed earth walls that they are used to doing. Such a wall will be double sided: one brick facing outside and another inside. The gap between will be filled with anything, especially when the construction supervisor has his back turned. Thus, in this 'oubliette',❷ they will stack haphazardly, without even taking the time to even out the crushed bricks, earth, mud, and baskets of pure lime, if they are not happy with your *kaolao*. Then, on 6 or 8 courses of stretchers, they will place one course of headers, to hide what is in the 'oubliette', so you will think you have a pure brick wall. I have seen buildings like this, of course for a flat rate: a wall with brick columns supporting the beams, set in a rammed earth wall. It was a real blocked chimney.

---

❶ "oubliette"一词英法文皆一样，通常用来指中世纪城堡塔楼下的地下监狱。关在地下监狱的犯人会被遗忘。该手册的作者将双面墙之间的缝隙比作地下监狱中的犯人，是要说明这空缝隙是被人遗忘的地方，但是此缝隙又关系到墙体的牢固性。

❷ This French word is used in English to designate an underground prison located under the most important tower (the dungeon) of a medieval castle. There prisoners were literally 'forgotten' (in French 'oublier'). The author of the handbook compares the empty spaces inside the double brick wall with 'oubliettes' because all the filling would be 'forgotten', but will be a tread to the solidity of the wall.

你需要核实一下泥水匠在砖石砌合方面的知识（图1.21）。在结构上，砖石砌合是指建筑中处理石头或砖头的细节。对不同尺寸的石块进行凿石琢边，因为在与砖体组合之前要准备好结构所需要的形状。这会是石头雕刻师的工作，但通常是先由一个工头勾出轮廓、剖面等，然后再由石头雕刻师来做。在这个地区，为做好大教堂墙体的一些部位，我们需要这些石头雕刻师。在这种情况下，我们也需要找到懂得砌筑这些石块雕刻的专家。一块好的石头如果砌筑不好的话会形成一座糟糕的墙体。砌筑石块的最好方法是：先把石块放好位置，在石块四角附近垫上木楔，用5~10毫米厚的砂浆填之间的空缝。其后，我们将石块抬起，抬起1/4，也就是说抬起一边，涂上一层比木楔厚一点的砂浆，将石块放下，用大铁锤敲打石头直到砂浆压实使石块和木楔接触上。当砂浆开始变硬时，将木楔子拿走。

砖的砌合。如果都是标准的砖，事情就比较容易做。在墙体倾斜和相交的地方才需要做砖块砌合，要根据墙的厚度来做。

最简单、最经济的做法是中国的表砖做法。我们用它做隔墙，其中抹上石灰的板条往往不够牢固。这个时候必须

First, you should check the masons' knowledge of bonds (fig. 1.21). In construction, 'bond' means the details of the arrangement of stones or bricks in a building. 'Dressing' a freestone is to prepare in advance the shape of the stones before composing the masonry. This will be the work of stone carvers, led by a foreman who draws the sketches, the sections, etc. We will need them, even in this country, for some parts of walls of a large church. In this case, masons, expert in assembling freestones should be found too. A good stone that is placed incorrectly, can only result in poor masonry. The best way to lay the stone is to position the stone on the place it must occupy and put wooden wedges at its four corners. Adjust the wedges in function of the thickness of the mortar that will form the joint, 5 to 10 millimetres. Then raise the stone on one quarter, that is to say one side only. Spread a layer of mortar a little thicker than the wedges, place the stone back in place, struck the stone with a mass until the mortar flows back and the stone touches the wedges. When the mortar starts hardening, remove the wedges.

Brick Bonds. For standardized bricks, things are easier. The bond consists only in the way bricks are placed and crossed on the wall, following its thickness.

The simplest and cheapest bond is the Chinese way of placing bricks on their small side (*biaozhuan*). It is used

第二部分　手册的内容和翻译
Part Two: The Handbook Translated

用木柱来加固隔墙，木柱之间砌砖，木柱与木柱之间的距离最远不能超过 2 米（图 2.8 中的 Fig. 42）。

为了使隔墙更坚固，砖必须沿长边方向卧放，在砖墙相交处其处理要小心谨慎（图 2.8 中的 Fig. 43）。在中文中，躺着的砖叫做"卧砖"。

一砖厚的墙，24~25 厘米，可以根据不同的砌合方式来做。

（1）丁砖（图 2.8 中的 Fig. 44），砖长就是墙厚。

（2）顺丁相间（图 2.8 中的 Fig. 45），就是说，每一层由丁砖和顺条砖交替使用砌成。

（3）一顺一丁（图 2.10 中的 Fig. 48）。

大名地区新教徒的圣殿，他们叫做礼拜堂，❶ 就是由一砖厚的墙建成的。这些墙只需要承受压型钢板屋顶的重量。❷

for partition walls, when the plaster lath is not strong enough. One would then have to consolidate this partition with wooden posts, filling posts, spaced at a maximum distance of 2 meters each (handbook Fig. 42) (fig. 2.8).

For a more solid partition wall, brick should be placed flat and lengthwise, taking care, of course, to cross the joints (handbook Fig. 43). In Chinese, lying bricks are *wozhuan*.

The wall of one brick thick, about 0.24 m to 0.25 m, can be built according to different bonds (fig. 2.8):

(1) headers (handbook Fig. 44), the brick making the entire thickness.

(2) headers and stretchers (handbook Fig. 45), that is to say each course is formed of alternating headers and stretchers.

(3) alternating courses of headers and stretchers (handbook Fig. 48).

The Protestant temple at Daming, ❸ the 'tabernacle' as they call it, is built with walls of one brick thick. It is true that these walls have only to support a roof of galvanized sheet metal. ❹

---

❶ 1919—1949 年期间美国新教徒在大名建了宣圣会的建筑，见：R.Gary Tiedemann. Reference Guide to Christian Missionary Societies in China from the Sixteenth to the Twentieth Century [M]. Armonk, 2009, 149; L.C. Olsborn. The China Story: The Church of the Nazarene in North China, South China and Taiwan [M]. Kansas City, 1969.

❷ 这样的屋顶比中国式屋顶要轻很多，或者是普瓦片的西式屋顶。

❸ Pentecostal Church of the Nazarene, American Protestants established in Daming from 1919 to 1949. R.Gary Tiedemann. Reference Guide to Christian Missionary Societies in China from the Sixteenth to the Twentieth Century [M]. Armonk, 2009, 149; L.C. Olsborn. The China Story: The Church of the Nazarene in North China, South China and Taiwan [M]. Kansas City, 1969.

❹ Such a roof is much lighter than a Chinese roof or a Western roof with tiles.

图 2.10 和图 2.11 中的 Fig. 47~ Fig. 53 显示砖墙砌体可以有一砖半厚、两砖厚或更厚。

地方上的泥水匠，除非他为欧洲人干过活（教堂、铁路等），必须学会做里生外熟的墙。本能地他会抵制这种做法，因为这是新做法，而且他不能用任何碎砖，此外还有其他各种原因。

墙的厚度。我应该承认，墙厚的问题也是写此书的最初原因。许多传教士将其微薄预算中最好的一部分钱埋在墙的厚度里了。我希望能帮助他们避免这个错误。

我已经说过大名地区新教徒的礼拜堂，它的墙在底层是一砖半厚。大名的教堂是由一位建筑师设计的，它也是一砖半厚的墙（图 2.10 中的 Fig. 52）。只有近 50 米高的钟楼的第二层墙是两砖厚。我来告诉你，有那么几十个谷仓教堂，❶ 没有任何个性，也没有任何样式，梁下距离也不高（最高 25 英尺，而大名的教堂的梁下距离有 40 英尺高），它们用一个简棚保护罩（比教堂其他部位都难看）替代钟楼来遮盖 30 千克重的钟，这么重的钟没有足够的力量根本敲不响。这些谷仓教堂的

Fig. 47-Fig. 53 will show the bonds that can be given to a wall of one and a half brick, and two or more bricks (fig. 2.10 and fig. 2.11).

The local mason, unless he has worked for Europeans (church, railway, etc.), will have to learn how to build a wall with bricks inside and outside (*lishengwaishu*) as just described. Instinctively, he will oppose it because it is new for him and, because he will not be able to use up any crushed bricks, etc.

Thickness of the walls. I must confess that the question of what thickness to give walls was what first inspired these pages. Many missionaries lose the biggest part of their meagre budget in the thickness of the walls. I would like to preserve them from this error.

I have already mentioned the great 'tabernacle' of the Protestants in Daming with its walls, one and a half brick thick at the ground floor. The Catholic Church of Daming, the plans of which were designed by an architect, also has walls of one and a half brick thick (handbook Fig. 52) (fig. 2.10). Only the bell tower, nearly 50 meters high, has walls of two bricks at the first floor. I could mention dozens of barn-churches,❶ without character, without any style, as well as without high beams, maximum 25 feet, while the church of Daming has beams at 40 feet. They have a shelter, uglier than the

---

❶ 'Prayer barns' and 'barn churches' refer to the rural architectural type of the barn used for poor churches because of their low costs and relatively vast inner space.

立面墙,竟然有 4 英尺甚至 5 英尺厚,相当于 1.65 米厚,厚墙加上扶壁,大教堂才值得这么做。边墙有三砖厚,至少 0.8 米,里外都有扶壁,是为了防止屋顶结构掉到教堂里面来。建这般坚固的纪念物的家伙好像就想告诉基督徒他们是多么在乎结构的坚固性啊。

这是真的事情,当然那个时候 1 磅好的砖头的价格还不到 1 方孔钱。那是过去好的日子。但从那以后,我们的津贴一年比一年少,而基督徒的数量在增加,两倍、三倍地增加。所以要把我们的墙做薄,让我们做分量更轻的梁。

我先不谈建筑师如何确定墙厚的技术过程。

去向他们咨询,按照他们指示的厚度来做墙。这是 Barberot 就建造一个七层房屋给出的数据。

地基低处:0.75~1 米,3~4 块 0.26 米宽或 8 磅重的砖。

地下室顶部:0.65~0.80 米,两块半至 3 块 0.26 米宽的砖。

底层低处:同上。

rest, which is used as a bell tower for a bell of 30 kilos that cannot always be rung with full force. The walls of the facades of these barn churches, however, are 4, even 5 feet, 1.65 m, thick, with buttresses worthy of a cathedral. The side walls are in keeping with the rest: 3 bricks, at least 0.80 m, with buttresses both inside and outside, to prevent the roof structure from falling into the church. Those who erected these monuments were showing Christians their concern about the solidity of the building.

It is true that at that time, bricks cost less than one cash (*sapèque*) for one pound of good bricks. Those were the good old days. But since then, our allocations have decreased year by year, while the number of Christians has only increased, doubled, tripled. Conclusion: let us put the courses of our bricks closer together and let us make the beams a little less heavy.

I leave aside the technical processes by which architects establish the thickness of the walls.

Ask them and use the thicknesses that they indicate. This is the table that Barberot gives for a house of six floors:

At the lower foundations: 0.75 m to 1 m, thus 3 to 4 bricks of 0.26 m, or 8 pounds.

At the top of the cellars: 0.65 m to 0.80 m, thus 2½ to 3 bricks of 0.26 m.

Lower ground floor: the same.

二层：0.45~0.55 米，两块 0.26 米宽的砖。

三层：0.55~0.45 米，一至一块半 0.26 米宽的砖。

四层和五层：0.25~0.40 米，一至一块半 0.26 米宽的砖。

六层：0.40 米，一至一块半 0.26 米宽的砖。

七层：阁楼。

这与谷仓教堂不同，它的立面墙屋顶层有五砖厚，边墙有三砖厚，还用双扶壁加固。这样的扶壁只应该用在有着一块半厚墙砖的大教堂上。它们可承受木结构的所有重量，它们不会被由八砖、土和瓦等材料建成的沉重的中国式屋顶压垮。这是经过经验证实的。

"灌浆"是不是中国的发明？我不太清楚。根据我掌握的资料书籍，里面没有任何相关说明。以下是过程。

浆是一种非常稀的混合物，由石灰、砂子和水组成，由一名或两名小工不停地在石灰坑里搅拌而成。当一层甚至两层砖砌好的时候，他们就把浆浇到上面。浆会流进砖墙的每一个缝隙里面，抹平，然后再砌新的一层。也许这种方法一开始是为了省钱，即使在当时也不是很贵。很显然，这样做的效果很好，建筑师也不反对这种做法。但是像其他所有事情一样，他们灌浆时你必须

First floor: 0.45 m to 0.55 m, thus 2 bricks of 0.26 m.

Second floor: 0.55 m to 0.45 m, thus 1 to 1½ of 0.26 m.

Third and fourth floors: 0,25 m to 0,40 m, thus 1 to 1½ of 0.26 m.

Fifth floor: 0.40 m, thus 1 to 1½ of 0,26 m.

Sixth floor: attic.

This is very different from the façade of the barn church with its 5 bricks at the roof level and side walls of 3 bricks, reinforced with a double buttress. Such buttresses should be only used for a large church, with walls of 1½ brick. They will bear all the load of the timber structure and will not be crushed by the very heavy Chinese roof with its slabs, earth and tiles. This has been tried, and works.

Perhaps 'gorging with porridge' (*guanjiang*) is a Chinese invention? I do not know. I do not find any reference in the books that I have available. This is the process:

The *jiang* is a very liquid mixture of lime, sand and water that one or two unskilled workers stir constantly in a basin dug in the ground. When the builders have completed one or even two courses of bricks, they pour full buckets of *jiang* on it. They will let it flow into every interstice of the masonry, then equalize, and start building a new course. Perhaps this technique was first intended to save mortar, even when it was not expensive.

第二部分　手册的内容和翻译
Part Two: The Handbook Translated

监工，尤其是当整个工程以一价格整体外包的情况下。

In any case, it gives very good results, and architects are not against it. As in all things, however, you should closely supervise the procedure, especially if the construction is being built at a flat rate.

隔碱层

Saltpetre (*jian*), Insulation

在中文里，正确说法是硝而不是碱。但是中国人总说墙成碱墙了或砖碱了。法语里我们把它翻译成墙变硝墙了。砖一旦与土壤接触，土壤中的盐分就会渗透到砖里，我们暂且先不对土壤中的盐分进行化学分析。直隶在古代是海，所以其土壤里有丰富的氯化物、硝酸盐和各类盐。如果建筑不是用石头建的话，砖头有很多孔，所有这些土壤里的物质都会渗透到砖表面，然后引发极易蔓延的侵蚀，会对所有结构产生严重的损害作用。我们知道海绵能吸收很多水。带有盐的湿物通过毛细管从土壤里上升，在空气和阳光作用下蒸发。水蒸发后，盐留下了。它们作为结晶物在砖孔内增加，然后以一小层一小层凸现，总是从外面一层开始。砖就会剥落，如果砖没有被烧透的话，剥落速度会更快。釉面砖、条砖或琉璃砖几乎都抵抗不了这种侵蚀。石头本身如果又软又有孔的话，最后也会被侵蚀殆尽。

In Chinese, the right word for saltpetre its *xiao* and not *jian*. The Chinese, however, say that the wall is attacked by the *jianqiang* or *zhuanjianle*. In French we translate: 'the wall is saltpetred' (*le mur est salpêtré*). Let us leave aside the chemical analysis of all the salts that will infiltrate the bricks as soon they are in contact with the ground. It is obvious that Zhili, an ancient seabed, is very rich in chlorides, nitrates, and all kinds of salts. All this comes back to the surface and causes damp and blight, a deadly wound for any construction, if it is not made of stone. Brick is very porous. We know the amount of water that this sponge can absorb. Humidity loaded with salts that rises from the soil by capillarity, evaporates under the action of the air and the sun. Of course, only water is evaporated out. The salts remain. They accumulate as crystals within the pores of the bricks and shatter in small layers, starting with the outer surface. The brick will flake and fall to dust, even quicker if it is under-fired. Glazed bricks, *tiaozhuan* or *liulizhuan* withstand the damp almost indefinitely. Stone itself, if soft and porous, will eventually succumb.

因此，必须设隔碱层，用来阻止湿气由地面上升通过毛细管侵蚀建筑的高处。否则，在硝达墙顶之前，整个结构都会由于底部地基的腐蚀而倾倒。对硝碱不闻不问就像是结构领域犯了一个不可饶恕的大罪。❶ 此罪往往是由于缺乏经验而犯的。

可以用不同的隔离材料来对付砖碱。

中国人用各种碱草来对付砖碱：芦苇、高粱秆、各类秸秆，甚至枯叶。这些材料的主要缺点是它们很容易下沉。这样会使墙开裂，如果是重要建筑物的话，就更容易产生结构问题。在有柱子的教堂，柱子几乎不会下沉，但是墙会移几厘米。

第二个缺点是各种碱草的存在时间短。10 年后，可以说，30 年后，最好的碱草也烂了。有钱人倾向于用木头隔离。木头确实能防止硝碱上升，也不会有下沉，但是木头终究也会烂掉。

A barrier, therefore, must be installed against the humidity that rises up from the ground and invades by capillary to a certain height in the buildings. Otherwise, the whole construction will collapse by erosion of its base, before the saltpetre reaches the top of the walls. Neglecting to defend against saltpetre is committing a mortal sin in the field of construction. ❷ This sin, however, is committed primarily by lack of experience.

There are numerous insulating materials that protect against the rise of saltpetre.

The Chinese mostly use various herbs against saltpetre (*jiancao*): reeds, sorghum stalks, various straw, and even dead leaves. The first drawback of these materials is their considerable settling. The walls split, especially if the building is large. In a church with columns, you will have almost no settling under the columns, but the walls will move several centimetres.

Their second drawback is their short duration. After ten years, or at most thirty years, the best herbs (*jiancao*) become rotten. Rich people prefer using wood as insulation. Wood, indeed, insulates against the rise of saltpetre, and there is no settling, but wood will eventually rot too.

---

❶ 基督徒定义不同的罪，轻罪或可饶恕的罪无关轻重，但是不可饶恕的罪，是更坏的，很难被上帝原谅。手册的作者如此说法是为了说明砖碱的严重性。

❷ Catholics have defined a hierarchy of sins: 'venial' or forgivable sins are the less important ones, but 'mortal sins', the worse ones, are hard to be forgiven by God. The author of the handbook uses this expression to stress the seriousness of the issue of saltpetre.

第二部分　手册的内容和翻译
Part Two: The Handbook Translated

用石头隔离是有钱人用的唯一解决方法。石头无疑是非常棒的隔碱物，也是这个国家能用的最好的隔碱物。不幸的是，很多地方没有石头，或者石头价格由于运输费而承担不起。另外，正如我们所说的，只有紧密的石头才能抗拒湿气。这就是为什么在欧洲我们的大教堂建造者用厚铅板来保护石头防止其硝碱，这些铅板是连续焊接而成的。"它们是青的，只适合乡巴佬。"❶ 而这正说明了做隔碱层的重要性。

涂有沥青的纸板也有被用来做隔碱层的。沥青是隔离砖碱的一种最好的材料。欧仁·维奥莱-勒-杜克❸在一些大教堂的地基里发现过一些这样的做法，那些沥青看上去就像刚刚放上去一样。❹沥青纸板在天津、上海等地有售，叫做"铺沥青"，卖得不贵。每卷宽度不一样，纸板的厚度也不一样。

此外，传教士可以制造他自己的沥

Using stone is considered the best solution by wealthy people. Stone undoubtedly is an excellent insulator, the best in the country. Unfortunately, in many areas there is not any stone available or its price is exorbitant because of the cost of transport. Moreover, as we have said, only fine grained stone resists. This is why in Europe our cathedral builders protected their stones against saltpetre with thick lead sheets, made continuous by good welding. *"They (grapes) are unripe and only fit for boors."* ❷ This shows the importance of insulating against saltpetre.

Felt plates soaked with asphalt are also used. Bitumen is one of the best insulators. Viollet-le-Duc found some in the foundations of cathedrals, which were in as good a condition as if it had just been installed. ❸ Bituminous cardboard is sold in Tianjin, Shanghai, etc. : 'Asphaltina', 'Maltoïde' etc. It is not expensive. The rolls are of different widths, and the cardboard itself comes in different thicknesses.

Moreover, the missionary can

---

❶ 此说法摘自法国诗人杨·德·拉·方登的寓言《狐狸与葡萄》，该寓言于 1668 年出版，用来批评那些吃不到葡萄说葡萄酸的人。在解释了中世纪建造者怎么隔离盐碱的做法后，手册的作者在这里引用寓言中的说法是为了说明在中国不能用同样技术的无奈。

❷ This sentence comes from The Fox and the Grapes, a short tale by the French poet Jean de la Fontaine (1621-1695), first published in 1668. It criticises people who speak with disregard of things that they cannot reach. After having explained how the medieval builders insulated cathedrals against saltpetre, the author of the handbook refers to the tale in order to express his frustration of not being able to use the same technique in China.

❸ 欧仁·维奥莱-勒-杜克（1814—1879）是一位著名的法国建筑师，他修复了大量的中世纪伟大的建筑物，他也撰写了关于中世纪建筑的参考书。

❹ The French architect Eugène Viollet-le-Duc (1814-1879) was a famous restorer of medieval monuments and the author of reference books on medieval architecture.

青纸板，价格更便宜，也同样好用。

用厚的黄纸（叫草纸），用一个大刷子刷上一层沥青，放在太阳底下晒干。三层刷过沥青的草纸叠在一起，隔碱效果和"铺沥青"一样。

用来隔离盐碱的材料还有席子。它们可以编得和许多墙一样宽。除此之外，可以用两卷席子互相叠压以达到需要的宽度。

放好之后在两面涂上沥青，这是很好的隔碱材料，不容易被压实，不容易腐烂，能存在很长时间。

在哪里做、怎么做这些隔碱层？ 要做好的隔碱层，最好做两层保护层：第一层隔碱层与外地面齐平，另一层隔碱层稍微高一点，放在建筑一层的内墙底部。通常唯一的隔碱层做得更高，在地面以上 1 英尺或 2 英尺，或者甚至在底层。这样的做法会导致地面以下的所有砖层都被碱化。

然而隔碱层最大的缺点是它的延续性。我们应该还记得中世纪的伟大建造者们是如何焊接铅板的。我们也看到中国人是如何仔细地放上同墙一般厚的碱草，在顶部还要宽出几毫米。但还是有很多传教士建造者觉得他们在外

manufacture his own bituminous cardboard. It is much cheaper and just as good for the insulation we have been discussing.

Take thick yellow paper, called *caozhi*. Spread a layer of tar with a large brush, and let dry in the sun. Three layers of such paper equal any 'Asphaltina'.

Another insulating material against saltpetre are the small mat bands (*xizi*) that are used for tying the heaps of grain (*liangshi dui*). They are braided to the width of many walls. Besides, one can use them double layered, and obtain exactly the desired width.

Tar-coated on both sides when put in place, they are an excellent insulator, not subject to settling and rot-proof at least for a very long time.

Where and how to place the insulating body? For an efficient insulation, it is better to place two protecting layers: one at ground level and the other a little higher, placed at the inner level of the ground floor. Often the only protecting layer is placed much too high, one or two feet above the ground floor, or even at the level of the rooms. This unnecessarily sacrifices all the brick courses that are underneath.

The biggest flaw, however, would be any interruption in this insulating layer. Remember how the great builders from the Middle Ages welded their lead plates. Also see how carefully the Chinese cover the entire thickness of the wall with their herbs (*jiancao*) and let

墙做一排好看的雕刻过的石头就已经很好了，对内墙却完全没有做任何处理。他们说："内墙不暴露在地面湿气里。"地面湿气就像异教徒的神一样，总是直着往前走，而不会去找什么边门，你以为一道影壁墙能阻挡得了他们？

注释。我们相信在这里讨论空心墙是没有用的，做空心墙的话就只有一层空气来隔离雨水和寒冷。中国北方的干燥气候确实能避免做隔碱层带来的巨额花费。尽管大名的新教医院❶就是以这种方式建成的。那确实是一个医院。在底层，外墙一砖厚（30厘米），内墙半砖厚，内外墙之间距离相当于 1/4 砖厚。

装窗框和门框

可以在建筑完工后安装门窗。通常门窗是在砌墙时一起装的（图

them extend a few millimetres beyond. Yet many missionary-builders think they have done well by placing a nice row of beautifully carved stones on the outer half of the wall, while doing absolutely nothing to insulate the inner half. They say: *"It is not exposed to humidity from the ground."* Humidity from the ground should thus be like the spirits (*shen*) of the pagans, who are always going straight ahead and do not find the side door, when a screen wall (*yingbiqiang*) is there to stop them?

Note. We believe it useless to talk about hollow walls, with an insulating layer of air against the rain and the cold. The dry climate of northern China seems to dispense with this wealth of precautions, even if the Protestant hospital in Daming is constructed in this manner. ❷ It is true that this is a hospital. At the ground floor, the outer wall has a thickness of one brick (30 cm) and the inner wall half a brick; the space between is equivalent to a quarter-brick.

Installing Door and Window Frames

One can wait to install these frames when the building is completed. Usually they are put in place gradually, as the

---

❶ 1919 年期间美国新教徒在大名建了宣圣会的医院，见：R. G. Tiedemann. Reference Guide to Christian Missionary Societies in China from the Sixteenth to the Twentieth Century [M]. Armonk, 2009, 149.

❷ Hospital of the Pentecostal Church of the Nazarene, established in Daming 1919. R. G. Tiedemann. Reference Guide to Christian Missionary Societies in China from the Sixteenth to the Twentieth Century [M]. Armonk, 2009, 149.

1.22）。这样能够节省劳工，安装也会更牢靠。为了防止在完工后安装门窗时可能会产生的一些严重隐患，在这种情况下，还要做一些必要的措施。

最重要的是，必须考虑到墙的沉降问题。如果没有考虑到这一点，窗框的垂直边就会被压碎或会翘曲，这样就不可能安装门窗了。你必须采取一些措施，首先在窗框下方的横木下面做必要的措施。不然它们就会在窗下墙体的推力下被提升。为了防止这种墙体的推力，有经验的工人会在窗下建一个倒拱形。但只有有经验的工人才能做出这样的拱。对于其他工人，建议让整个窗框保留有来回浮动的余地。在给墙抹灰时窗框就会被永久固定住。这不会对牢固性造成任何影响。

需要找到好的泥水匠来做门窗上方的拱（图1.22）。中国人只使用"过木"，即便在漂亮的宝塔和北京的皇宫里也是如此。这将是如何把中国样式适应性应用到基督教堂中要讨论的一点，❶ 不过没关系。

做一个普通拱的话，可以先做一个

walls go up (fig. 1.22). This way saves labour and the installation is more solid. Yet in this case, some precautions must be taken to avoid the serious drawbacks that can happen when the frames are not installed at the end of construction.

Above all, the settling of the walls must be foreseen. Without this precaution, the uprights will be crushed or at least will warp, making it impossible to place the doors and the windows. You have to allow a bit of play, first under the lower crosspiece of the window frames. Otherwise, they will be lifted by the thrust of the wall under the window. To prevent this considerable thrust, skilled workers build a reversed arch under the window. But only skilled workers can make such arches. For other workers, have them leave some space around the whole frame. It will be fixed definitely when plastering the wall. This will not be to the detrimental to overall solidity.

You need good masons too for making the arches above the doors and the windows (fig. 1.22). The Chinese only know their old wooden lintel (*guomu*), even in the most beautiful pagodas and imperial palaces in Peking. This would be a good point in the discussion about that famous adaptation of the Chinese style to our Christian churches.❶ But never mind.

To bend a common arch, make a

---

❶ The author considers wooden lintels as less quality than brick arches. Therefore he add this technical aspect of Chinese construction to his critics of the Sino-Christian style (see chapter 3 of the handbook).

简单的石膏模型，然后在上面垒砖头或石头。不要急于拆除石膏模型，等到砖墙稳定后再拆不迟。

在这里我们没必要讨论不同形状的拱，如半圆拱、扁圆形拱、三角尖形拱、柳叶刀形拱、扁圆尖形拱、都铎拱、提篮形拱等。这是做设计方案的建筑师的事情。在任何情况下，都要考虑推力的问题，尤其是在有拱顶的情况下（地下室、教堂等）。此外，地下室已经几乎不用拱顶；它们需要足够的高度来做门廊，也需要做很厚的砖墙。现在它们已经被铁桁架加钢筋混凝土代替了。在后面我们会提到。辅助拱会缓解门窗上方拱的负荷（图 2.12 中的 Fig. 54）。这些拱在稍微高一点的地方，会将墙的重量转到拱墩上。

在砌墙时同时安装窗框和窗子的固定部分，除了以上谈到的沉降问题之外，还会有另外两个缺点。首先，木头由于没刷油漆无法防水。其次，因为碰撞而引起的损害（当小工搬砖头、石头、梁等材料的时候会发生碰撞）。当然这个可以通过用木板条保护窗框的垂直边而得以避免。

最好是在泥水匠完工离开后再做这些木工。要把窗框永久地固定在砖墙上，可以用这些方法：窗框每边用两个或三个铁构件固定，用砂浆、水泥等将

simple plaster casting and lay the bricks or the stones on it. Do not demolish this casting too quickly, but wait until the masonry hardens.

No need to mention here the various forms of arches: round arch, segmental arch, ogee arch, lancet arch, segmental ogee arch, Tudor arch, basket handle, etc. This is the business of the architect who will design the plans. In any case, take into account the thrust, especially if it is a vault (cellar, church, etc.). Moreover, for cellars, vaults are nearly no longer used; they require a great height for doorways, and very thick masonry. They have been replaced by iron floors in reinforced cement. We will talk about this later. Discharging arches relieve the arches of doors and windows (handbook Fig. 54) (fig. 2.12). These arches are placed a little higher and will transfer the load of the wall to the imposts.

If you install the frames and the fixed parts of windows together with the masonry, there will be two more disadvantages, besides the already mentioned settling. First, the action of rain on wood that is not protected by oil or varnish. Second, degradations resulting from knocks, caused by unskilled workers carrying baskets, stones, beams, etc. You can prevent this latter drawback by protecting the uprights with wooden laths.

Often one prefers doing the carpentry after the masons are gone. The frames should be permanently fixed in the masonry, with two or three hooks at each

铁构件固定在墙里，再用铁构件把窗框的垂直两边钉住，或者在墙体部分镶入木板，这种做法更好。

中国木工经常在过梁上省钱，用的木头没有达到与其长度相对应的厚度。"已经够牢固了"，他们会告诉你。但必须承认，这确实不好看。至少它的外侧部分，过梁的整个下沿部分必须是直的，而不是锯齿不齐的。

如果从一开始在砌墙的时候就装门窗框，泥水匠就会做到差不多就好了，这样就不仅仅只是"不好看"的问题了。为了避免压碎窗框的垂直边，在过梁和门窗框上方的横向构件之间会留有至少5英寸的空隙。为了安慰你，他们会说这样做能使冬天取暖时的烟气散发出去。这毫无疑问。但是在一个教堂里呢？还有北风怎么办？还有麻雀怎么办？

从这点来看，在最后安装门窗框更好。

如果你不想冬天在教堂里或你的房间里被冻僵的话，或者不想整年被麻雀烦个不停的话，在屋顶建造时你就应该监督那些工匠的"差不多"的做法。墙的顶部会有椽子和八砖，会"差不多"做

side. Seal the hooks in the walls and screw the uprights. Using wooden blocks, encased in the walls at the right places, would be even better.

Chinese carpenters usually save money on the lintels, using wood that has not the required thickness to be squared on the full length. "It is still solid enough," they will tell you. Admit, however, that this is truly not nice looking (*buhaokan*). At least the edge should be complete over the full length at the outer side.

If one installs the frames from the beginning, together with the walls, the masons will allow themselves to work approximatively (*chabuduo*), which will be more than not nice looking (*buhaokan*). Under the pretext of avoiding crushing the uprights, they will leave a space of at least 5 inches between the lintel and the upper transversal piece of the upright. To comfort you, they will point out that the smoke will go out that way when you heat in the winter (*kao-kao*). No doubt, but in a church? And what about the northern wind? And sparrows?

From this point of view it would be better to install the frames at the end of construction.

If you do not want to freeze in your room or in your church in the winter, or be invaded by obnoxious sparrows all year, you should supervise even more closely the approximate work of the local workers (*chabuduo*) when they install the

好。通常装好梁之后，应该加砌几排砖，差不多能支撑土檩（檐檩），这些土檩在地方建筑中是由砖墙支撑的。把椽子用钉子钉在土檩上，再把八砖放在椽子上，往往不去考虑墙和屋顶之间的空隙问题如何解决。可是，我们从外墙部分看，墙和屋顶之间的空隙部分会有椽子厚度那么大，两三英寸（正好麻雀能穿过！），里墙部分则会有差不多 2 英尺的空隙。

"我们后面会做，在抹墙的时候做"，砖工们会这么说。不幸的是，外墙部分是不会抹灰的，至于里墙部分，砖工会忘记抹灰一直抹到墙顶。就是因为这"差不多"，冬天的北风会呼呼地刮进来，这何止是为了排散烟雾啊。

因此我们必须要求把墙完全做好，在做下一道工序之前，八砖下方必须封严实。这项工作很简单，没有理由去推迟它。

roof on the walls. The top of the walls, where the wall meets the rafters (*chuanzi*) and the square tiles (*bazhuan*) of the roof, will be completed very approximatively (*chabuduo*). Once the beams are in place, they will add a few rows of bricks, to support approximatively (*chabuduo*) the lower purlin (*tulin*), which is supported in turn by the masonry in this local art of building. The rafters (*chuanzi*) are immediately nailed to the purlins (*tulin*). The square tiles (*bazhuan*) will be attached to the rafters (*chuanzi*) without worrying at all about the interruption between the wall and the roof. However, at the outer side of the wall, this interruption will have exactly the dimensions of the thickness of the rafters, two to three inches (joy for the sparrows!), and inside, it will be about 2 feet.

"*We will finish later, when plastering*" will say the masons. Unfortunately, the outer side of the walls will not be plastered, and inside, the masons will forget to plaster up to the top. It will be (*chabuduo*) and in wintertime the north wind will enter more easily than smoke.

One must therefore require that the wall be completely finished and sealed hermetically, up to the bottom of the square tiles (*bazhuan*), before starting to place those tiles. This work is very easy and there is no reason to postpone it.

第六章　屋顶
Sixth Chapter: The Roof

中式屋顶

比起其他部位，地基和屋顶是一所房子最重要的部分。不幸的是，好的屋顶，至少是很理想的屋顶还有待发现。此外，我们的传教士（本手册就是写给他们看的）还没有发现金属屋架和好屋顶。他们甚至不知道平瓦（也叫做机械瓦）。不管你喜欢与否，你应该用筒瓦，早在希腊文明之前亚洲人就开始使用它了。

我们先说一下直隶平原的中国式屋顶吧。它们有不同的做法。

（1）土平顶。如果这种平顶是架在土屋上的，那整个房子就叫土平房。它的确是平的。不要用艺术规范去要求它。随便问一个你碰见的农民，他们都会做。对我们传教士而言，这种土平顶并不好，因为它需要有人去维护。每两年或三年应该加一层泥，此番地方工艺叫做"泥房子"。

The Chinese Roof

Foundations and roof are, more than any other, the vital parts of a building. Unfortunately, a good roof, or at least an ideal roof is still to be found. Moreover, those for whom these pages are written have no access to metal roof frames and complex covering systems. They do not even have interlocking roof tiles, also called machine-made roof tiles. Like it or not, you must make up your mind to use the round tile, which Asians have used long before the Greek civilization.

Let us just concentrate on the Chinese roof as it is built in the plain of Zhili. There are several construction techniques.

(1) The earthen flat roof. When it is placed on an earthen house, the whole is called 'flat earthen house' (*tupingfang*). It is, indeed, flat. No need to define the rules of the art. Ask the first peasant you meet. For us, missionaries, such earthen roofs are bad, because nobody will maintain them. Nevertheless, one should add a layer of earth (*nifangzi*) every two or three years.

（2）平顶，上面覆盖有瓦片或砖墁顶（即砖勾缝做得很好）。由于砖的多孔性，在连续下雨的天气，这种屋顶必然会漏雨。

（3）石灰锤顶，又叫灰顶。此种屋顶如果做得好，是很好的屋顶。很多工匠做不了。所以在分配任务之前，必须保证他们有此项专业技能。灰顶的最大敌人是雪。雪融化后的水渗进灰顶表面的孔里，晚间天气降温，渗进的水分就会结冻然后就会摧毁石灰，就像碱摧毁砖头一样。

如果是私人住宅的话，住户会在雪停之后马上就把雪扫掉。但是在一个基督教社区里又会有谁不惜牺牲自己爬上属于整个教区的屋顶去扫雪呢？

（4）瓦房。对我们的传教士而言，这是唯一可行的屋顶。不过它有个很大的缺点，就是非常重，需要一个屋架去支撑它。但是如果建造质量好，这个缺点也不算很大的问题：首先，如果建造得好，它会很坚固；其次，如果建造得好，屋顶还会防水；不管何种情况，与薄金属板屋顶和石板屋顶等相反，这种屋顶下的室内会冬暖夏凉。这最后一条优点对于像我们这种只有一层的房子（即教堂中殿）来说价值很大。

(2) The flat roof, covered with tiles or bricks with good pointing (*zhuanmanding*). Proper pointing is hard to do. But above all, because of the very high porosity of the brick, these roofs leak infallibly when the rain is persistent.

(3) The macadam roof (*shihuichuiding* or simply *huiding* or *zhending*). This roof is very good, when it is well done. Many masons cannot do it. One must therefore be certain about their knowledge, before assigning them the task. Snow is the great enemy of the macadam roof. Water from the melting snow penetrates into the pores of the surface, and at night, when the temperature drops, the frost shatters the lime like saltpetre shatters the bricks.

In private houses, someone will sweep the snow off as soon as it has stopped falling. But who, in a Christian village, will volunteer to climb on a roof that belongs to the community?

(4) The tiled roof (*wafang*). This is the only practical roof for the missionaries for whom we write. It has the great disadvantage of being very heavy and requiring a frame in relation to that weight. This disadvantage, however, is offset by real qualities: first, the strength, if it is well built; then being waterproof, also on the condition it is well built; in any case, contrary to metal roofs, slate roofs, etc., it keeps cool in the summer and preserves heat in the winter. This last advantage is of great value for one-storey buildings, like ours.

做瓦顶有以下几种方法。

（1）最不推荐的做法，但有着"好看"的美誉，就是用筒瓦做的屋顶。我们已经说过了，这是宝塔和皇宫所用的屋顶。在民国之前，清朝时期，普通人是不能用这样好看的瓦来盖他们的房子的。绝不要屈服于基督徒的恳求而把这种瓦用在我们的教堂上。兖州❶的主教教堂就是用了这种瓦。它们是很漂亮的琉璃筒瓦，有着皇家颜色（黄色），与北京故宫类似。但每次碰到暴雨天气屋顶就漏雨漏得厉害。我想那个亮光闪闪的屋顶应该已经被改建了。至少，20年前就有要改的决定了。

（2）仰瓦屋顶，即所有的瓦将其凹面朝上。在其他地区，这种做法叫"反毛鸡"（图2.11中的Fig. 55）。在直隶地区的南部，来自广平府❷的工匠做这个最出名（图1.25）。事实上一直有人想知道为什么这样平列而放、没有叠压也没有注浆的仰瓦屋顶不会漏雨。确实，即使是暴雨天气，这样的屋顶也不

There are several ways of doing a tiled roof:

(1) The least recommended way, which, however, has the reputation of being *haokan*, is the roof made with semi-cylindrical roof tiles (*tongwa*). We have already talked about it. This is the roof used for the pagodas and the imperial palaces. Before the Republic, at the time of the Qing Dynasty, common people did not have the right to cover their houses with such beautiful tiles. Never yield to the entreaties of Christians to put such tiles on our churches. The cathedral of Yanzhou❶ was covered with such tiles. They were very beautiful vitrified *tongwa*, of imperial colour (yellow), similar to the palaces of Peking. At each heavy rain the roof leaked miserably. I think that that whole glittering roof had to be changed. At least, twenty years ago, the decision was taken to change it.

(2) On the *yangwa* roof, all the tiles have their concave side facing upwards. In other regions it is called the 'hen with upside down feathers' (*fanmaoji*) (handbook Fig. 55) (fig. 2.11). In the southern part of Zhili, the workers from Guangpingfu❷ are the most renowned for this work (fig. 1.25). In fact, one wonders why tiles that are just juxtaposed without overlapping and without grouting, do not

---

❶ Yanzhou (Yenchow, Romanised as Yen-tcheou-fou) is a city and prefecture in Jinan province (previously in Shandong province). The catholic vicariate apostolic of 'Southern Shantung' was created in Yanzhou in 1885 under the supervision the German missionaries of the Society of the Divine Word. It became the vicariate apostolic of Yenchowfu in 1924 and a diocese in 1946. The cathedral still exists.

❷ Guangping (Hebei province), located 34 km northwest from Daming.

会漏一滴雨，风也不会进来，整个表面就像一整块板似的。然而，要做好如此杰作的屋顶，必须有人能踏在瓦片上行走，同时还不会踩坏或弄坏任何瓦片。好的工匠能这么做。说实话，他们需要很长的时间：三排瓦，六个工匠一起做的话，也需要一天才能完成。他们也会要求很高的犒劳，要不然屋顶就会漏雨。

（3）最后一种方法是用扣瓦（图2.12中的 Fig. 56），也叫大陇。这种屋顶类似前面那种，是由形状为四分之一锥形的瓦铺成。这种瓦有不同的尺寸。现在流行把瓦片做得越来越小，大概是出于经济和生活成本的缘故。小瓦当然没有以前古时候的大瓦那样坚固。另外，你还可以自己定尺寸，定好后告诉制模者和烧瓦工。在这个传教区的北部，"扣瓦"很流行。就像我们所看到的，在那种情况下，会用双排瓦。一排凹面朝上，另一排凸面朝上，交接部分用灰浆抹上，整个都是用很好的灰浆抹填的，在长时间饮足水分之后双排瓦就完全融为一体了。

相反，仰瓦干着铺，不用石灰水。这三种做瓦顶的方法，都需要准备同样的"床铺"：在椽子上铺上八砖。它们不是

leak. But so it is; when raining in torrents, not a single drop will go through a well-made roof. The wind has no grip either: the entire surface is like a block. However, for making this roof a little masterpiece, one must be able to step on it, without breaking or even disturbing any tile. Good workers succeed. To tell the truth, they need time: three rows of tiles a day for half a dozen men. And they need lots of good food (*kaolao*), otherwise the roof will leak.

(3) The last way is the roof with overlapping tiles (*kouwa*) (handbook Fig. 56) (fig. 2.12), also called 'big gutters' (*dalong*). This roof, as with the previous one, is made with tiles that have the shape of a quarter truncated cone. They have varied dimensions. There is a trend to make them smaller; question of economy and cost of living. They are, of course, less solid than the large tiles from the good old days. Moreover, you can impose the size of your choice on the modeller and the fireman. In the north of the mission, *kouwa* roofs are in fashion. One sees, in that case, a double row of tiles. One row is turned on its concave side and the next, turned on its convex side, covering the joints. The whole is thoroughly grouted with very good mortar, after the tiles have been completely soaked with water by a long 'dip' as for bricks.

On the contrary, the *yangwa* tiles are laid dry, without lime. For these three ways of tiling roofs, a same kind of

用钉子钉的，而是在四角用很好的灰浆来固定。需要事先将八砖在水中浸湿，这样灰浆才会与八砖相粘。这样的八砖屋面，本身已足够应付地方小雨，再在上面铺上一层土草混合物（图1.24 和图 1.26）。放着晾干。这样房子就不用担心大雨了。只有等"床铺"干燥了，才能开始铺瓦。铺瓦时，首先涂上一层混合着干草和石灰水的泥浆，接着压实，这样就将瓦片嵌在这种结实的泥浆里了（图 1.25）。

瓦顶的"床铺"还有一种比较经济的做法。这种做法不适用于教堂，但可以用在学校、住宅等的屋顶上。这种屋顶的预期寿命至少有三十年。

这种经济的"床铺"不是用椽子或八砖做成的，而是直接在檩条上放高粱秆或芦苇秆，不要像土平顶那样自由铺开。它们事先被捆成一小把一小把的。放的时候，这些草把通过相对较长的木钉而连接在一起，木钉沿着水平方向从一边穿到另一边。木钉的放置应该有变化：如果一开始木钉在草把两头穿过，下一根木钉就要从草把中间穿过，以此类推，一直不停交换变化。在脊檩上，多用几杆草，当它们被很好地用木钉钉好后，折弯它使其压到屋面另一坡上。这样，它们就会钩在脊檩上了。

'bed' is prepared: small square slabs (*bazhuan*) are placed on the rafters. They are not nailed but fixed with good mortar on the four edges. One must first soak the slabs in water so that the mortar adheres. On this slab roof, which is already sufficient in itself to stop a light rain, spread a layer of mud mixed with straw (fig. 1.24 and fig. 1.26). Let dry. Now the building is protected against a 'two fingers' rain. Start placing the tiles only when the bed is dry. First place a thick layer of mud mixed with straw and lime. Than press, virtually 'cast', the tiles in this substantial layer of mud (fig. 1.25).

There is an economical way to make the 'bed' of a tiled roof. It would not be good for a church, but it can be used for a school, a dormitory, etc. The life expectancy of such a roof is about thirty years at least.

It is not made with rafters (*chuanzi*) or square slabs (*bazhuan*). Sorghum stalks and reeds are directly placed on the purlins (*lin*). They are not freely spread, like on a clay roof. The stems are first tied in small bundles. When they are laid, the bundles are joined together by relatively long wooden pegs, which pass through the bundles from side to side in the horizontal direction. The location of these pegs should be alternated: if pegs traverse the first two bundles at both extremities, the next one will be pegged in the middle, and so on, always alternating. On the ridgepole, let a few stems extend over it, and then, when they are well

接着抹上泥浆，就像在抹在八砖上一样的做法，然后像往常一样铺瓦。

高粱秆的底端不应该超过屋面的下檐，就像做平顶那样，但是要把墙一直盖到房檐。从外面看，它像普通瓦顶一样整洁。不做房檐的话，下排的瓦就会滑落。

屋面"床铺"可以用苇笆做。任何一个中国人都能告诉你怎么做。这种情况你可以不用八砖，但椽子必须要有。

欧式屋架

屋架一词是指用来支撑屋顶的整体木构架。屋架是一个有着3米或4米间距的木结构，是屋顶结构的构成部分。檩条将屋架连接在一起。在中文里，屋架又叫房架、木料架或木梁架。没有一个专门的词来翻译屋架，人们只对组成屋架的每块木头命名，比如"檩条"。

屋架有两种：中式屋架和欧式屋架（图2.12和图2.13中的Fig. 57~ Fig. 59）。

pegged, fold them over to the other slope of the roof. So, they are hooked to this purlin.

Then spread the mud, as on the slabs, and place the tiles as usual.

At the bottom the sorghum stalks should not exceed the edge of the roof, as for flat roofs, but the wall is built up to the cornice (*fan yan*). Seen from the outside, it is as neat as a roof with common tiles. Without the cornice, the lower tiles would slip.

The 'bed' of the roof can be made with large reed mats (*weipa*). Any Chinese can tell you how to make it. In this case you will save on the square slabs (*bazhuan*) but not on the purlins (*chuanzi*).

The European Roof Frame

The word 'roof trussing' refers to the entire wooden structure which supports the roof. A truss is a wooden framework that, at intervals of 3 or 4 meters, strengthens the roof trussing. The purlins link the trusses together. In Chinese, roof trussing is 'support the home' (*fangjia*) or 'wood-support' (*mu liaojia / mu liangjia*). There is no specific word for translating 'truss'. People just name each piece of wood that composes the truss. A purlin is named *lin*.

There are two kinds of roof trusses: the Chinese truss and the European truss (handbook Fig. 57-59) (fig. 2.12 and fig. 2.13).

中式屋架（图 2.13 中的 Fig. 59）有大梁、二梁，有时有三梁，梁之间用小柱子（也加"高柱子"）连接，这样的屋顶很重而且也贵。中式屋架下不可能有开间很宽的建筑，即不能做像建筑师说的"跨度很大"的建筑。柱子在中国建筑以及有着中式屋架的教堂里使用很频繁。

欧式屋架（图 2.12 中的 Fig. 57, Fig. 58）轻得多，也没那么贵。最重要的是，欧式屋架得益于阻力点的组合，能产生比中式屋架更大的跨度（图 2.12，图 1.27 和图 1.28）。

特别是欧式屋架的中柱，它像"联络官"。中柱与梁组合(Fig. 60~Fig. 62)，再加上主椽(Fig. 63)，共同作用下的结构体系能防止梁由于承载过多小支柱而产生的中间弯曲（图 2.13 和图 2.14）。因此，中柱不会给梁以负担，相反，还能对梁有一些支持。

用这样的欧式屋架，就不需要用柱子，即使是跨度 10 米的教堂，也不需要过大的中柱和主椽。

建筑师需要提供欧式屋架的图纸。他也需要计算材料的强度并确定屋架每个构件的强度。

The Chinese truss (handbook Fig. 59) (fig. 2.13) consisting of a long beam (*daliang*), a second beam (*erliang*) and sometimes a third beam (*sanliang*), connected with small columns (*gaozhuzi*), is heavy and expensive. It does not allow for very wide buildings, 'span long distances' as the builders would say. This explains why columns are so frequently used in Chinese buildings and churches that have this kind of truss.

The European truss (handbook Fig. 57-Fig. 58) (fig. 2.12) is much lighter and therefore less expensive. Above all, thanks to the combination of resistance points, it can span much wider distances than the Chinese trusses (fig. 2.12, fig. 1.27 and fig. 1.28).

It is the king-post which works as a 'liaison officer'. The king-post, when assembled with the tiebeam (handbook Fig. 60-Fig. 62) and maintained by the principal rafters (handbook Fig. 63), prevents the tiebeam from bending in the middle, if it were overloaded by the struts (fig. 2.13 and fig. 2.14). It is, therefore, evident that the king-post should never load the tiebeam, which it is intended to support.

With this truss, no need of columns, even for a 10 m wide church, as long as you do not give excessive dimensions to the tiebeam, or to the principal rafters.

The architect shall provide the plans of the trusses. He has to calculate the strength of the materials, and determine

中式屋架则只能依赖于好的地方工匠的经验了（图1.47）。

但愿这本手册附的图纸能为一些建造工程提供帮助（图 2.28，图 2.29，图 2.32，图 2.34，图 2.36 和图 2.38）。

欧式屋架种类。

主体部分如下。

主椽(A)。它决定了屋顶坡度并支撑着檩条。主椽底部嵌在梁上的凹槽中，以完成与梁的连接，主椽顶端则通过榫卯和中柱连接。

梁(B)。它的主要作用是防止椽子移动，不需支撑任何重量，而中国的梁则需要独自承载屋顶过重的重量。因此欧式屋架的梁没有必要有像中国的梁一样的强度。它甚至可以由两根平行件构成，称之为双梁，双梁与主椽及中柱的连接是用螺栓完成的(图 2.13 中的 Fig. 58bis）。梁(C)与梁(B)平行，减轻了主椽的负重（图2.13 中的 Fig. 57）。

中柱(D)。它起着拉力作用，前面已经说过。它不需要给梁以负荷，只与梁连接，主要通过以下方法连接：a)用一个穿过梁的螺栓及中柱的榫头，b)用一根木锚栓（把中柱伸到梁里），c)用一个穿过梁的铁构件和两个螺栓把中

第二部分　手册的内容和翻译
Part Two: The Handbook Translated

the solidity of each piece of the roof frame.

For the Chinese roof frame, you can rely on the experience of good local workers (fig. 1.47).

The plans annexed to this book can help in a number of cases (fig. 2.28, fig. 2.29, fig. 2.32, fig. 2.34, fig. 2.36 and fig. 2.38).

European truss type.

The main parts are:

Principal rafter (A). It dominates the roof slope and supports the purlins. At its lower end, the rafter is assembled by slit and tongue joint with the tiebeam (B). In the upper part, it is assembled with the king-post (D) by mortise and tenon.

Tiebeam (B). Its main function is to prevent the rafters from moving, rather than supporting any load, while the Chinese beam (*liang*) has to bear alone the crushing weight of the roof. Therefore it is needless to give to the European tiebeam the strength of the Chinese beam. It may even consist of two parallel pieces, called double members, which clasp the principal rafters and the king- post (handbook Fig. 58 bis). The collar (C), parallel to B, relieves the principal rafters (handbook Fig. 57) (fig. 2.13).

King-post (D). It works in traction as already mentioned. It must not load the tiebeam, but be connected with it by: a) a bolt which passes through the tenon of the king-post, or b) a wooden rod (in that case the king-post exceeds under the

柱和双梁固定连接（图 2.13 和图 2.14 中的 Fig. 60~ Fig. 62）。中柱上方与主椽的连接，将主椽榫头嵌入中柱上方的榫槽之后再用钉子钉住（图 2.14 中的 Fig. 63）。

斜撑(E)。它通过中柱减轻了主椽的负载。斜撑必须与主椽垂直，且尽可能靠近檩条。它的另一端和中柱连接。一个欧式屋架上，一根中柱有四个斜撑，两斜撑用以减轻主椽负载，另外两斜撑用以支撑屋脊梁。

小支柱(G)与梁相连接减轻了主椽的负载。

撑脚(H)把梁和二梁连接在一起。

"系梁"(I)。

屋脊梁(J)。这是屋顶的上端部分，其上安装有椽子和檩条。顺着屋顶坡度将椽子钉在脊梁上（图 2.14 中的 Fig. 64）。

檩条(M)，依于主椽，将屋架架在建筑两侧的山墙上。檩条尽端必须长出 0.33 米，这样可以架在墙上。把檩条架在山墙上的做法是很好的。中国工匠熟知这些，他们通过燕尾结把檩条互相组合。就放手让他们做吧。

tiebeam), or c) by an iron stirrup that passes under the tiebeam and is bolted by its two upper branches to the king-post (Fig. 60, Fig. 61, Fig. 62). The assembling of the upper extremity of the king-post with the principal rafters is done with a pegged mortise and tenon joint (handbook Fig. 63) (fig. 2.14).

Brace (E). It relieves the principal rafters by relying on the king-post. The brace must be placed perpendicularly to the principal rafter, as close as possible to the purlins. By its other end it is assembled with the king-post. On one truss, the king-post receives four braces; two relieve the principal rafters and two others support the ridge (handbook Fig. 63).

Strut (G) is a small piece that relieves the principal rafter by relying on the tiebeam.

Crown post (H), it connects the tiebeam to the collar beam.

'Blochet' (I).

Ridge or ridge plate (J). This is the upper edge of the roof, on which lean the rafters, such as the other purlins. It is adapted to the roof slope in order to nail the rafters (handbook Fig. 64) (fig. 2.14).

Purlins (M) *lin*, link the trusses together, by relying on the principal rafters and, at the ends of the building, on the two gables. The purlins at the extremities must be at least one foot longer, 0.33 m, than the others in order to be placed on the wall. It is good to anchor the purlins in the gables. Chinese carpenters know

檩托(N)，垫木以防止檩条滑落。在实际操作中，就简单钉在主椽上，没有凹槽。

椽子(O)钉在檩条上，来承载中国屋顶的八砖。八砖的大小取决于檩条之间的距离。椽子通常是 5 英尺长，至少 2 英寸粗。对于压型钢板屋顶，很显然不需要那么大的强度。

土檩(P)。这是最下端的檩条，与墙面平行而放。它承接椽子的底部。它可以由多段木料拼接而成，因为它不需要承重。

飞椽(R)（又叫飞子），半人字形，上方斜切。它放置在檐部的突出部分，这样可以缓解屋顶挑檐处的倾斜度，同时也能使雨水下落尽可能远离墙面。通过这些飞椽，中国建筑师能把他们的屋顶做成上翘的鼻子形状。

中式屋架（图 2.13 中的 Fig. 59）：

（1）大梁。
（2）二梁。
（3）下梁。
（4）高柱子。

and still practice the assembling of purlins with each other with 'dovetail joints' or 'swallowtail joints'. Just let them do it.

'Échantignole' or 'Chantignole' (N), a bracket that prevents purlins from sliding. In practice, simply nail down on the principal rafter, without slit and tongue joint.

Rafters (O) *chuanzi* are nailed to the purlins and support the square slabs (*bazhuanzi*) of the Chinese roof. Their dimensions depend on the spacing between the purlins. Usually for rafters 5 feet long, the sides are at least 2 inches thick. For a galvanized sheet roofing, obviously less strength is needed.

Wall plate (P) *tulin*. This is the last purlin, placed horizontally on the wall. It receives the lower end of the rafters. It can be in several pieces, since it carries nothing.

Sprocket (R) *feichuan*, half of one rafter, chamfered on top. It is placed on the cantilevering part of the entablature in order to soften the roof pitch from where it sits on the cornice, and also to project rainwater as far as possible from the facade. With these sprockets, the Chinese architects give their roofs the shape of an upturned nose.

Chinese roof trussing (handbook Fig. 59) (fig. 2.13):

(1) Main beam *daliang*.
(2) Secondary beam *erliang*.
(3) Secondary beam *xialiang*.
(4) Small columns *gaozhuzi*.

（5）明柱。

其他构件及其名称与欧式屋架一样。

屋面坡度

在研究屋顶的时候，建筑师首先要考虑屋面的坡度问题。屋面坡度很大程度上取决于所用的建筑材料。从经济可行的角度，总是倾向于最小的坡度，当然前提是最小的坡度尚够。

关于这个区域的屋面：用砖或碎石做成的平屋顶坡度只有5°；压型钢板屋顶的坡度有26°；扣瓦做法的屋顶坡度也是26°；仰瓦做法的屋顶坡度有35°。中国工匠有一套约定俗成的方法，不用度数，他们也懂。他们称之为"正五斜七"，与屋面坡度为45°的一样。比如，梁长的一半，或者a、b之间的距离是水平的，即为"正"，是5英尺；屋面坡度a、c之间的距离是斜的，即为"斜"；那么主椽应该是7英尺。

也许"正五斜七"的坡度做法是为了排水，这个国家有时候会下暴雨。确实，有这样的坡度，雨水会很快被排走。但是瓦片很容易滑落（尤其是在有乌鸦的情况下），即使在没有下暴雨的情况也是如此。

(5) Columns *mingzhu*.

The rest has a similar layout and the same names as for the first, so-called 'European', timber frame.

Roof Pitch

The first thing that should concern the builder in the study of a roof trussing is the pitch to give to the slopes. This pitch will be mostly determined by the available roofing materials. By economy, one will always chose the minimum pitch, on the condition that it is still sufficient.

Regarding the roofs in the region that concern us: a flat roof covered with bricks or macadam can have a pitch of only 5 degrees; a metal sheet roof, 26 degrees; a *kouwa* tiled roof, 26 degrees too; a *yangwa* tiled roof, 35 degrees. Chinese workers do not understand degrees so they use stereotyped formulas instead. They say *zhengwuxieqi*, which is the same as a 45 degrees pitch. Half of the beam or the distance a-b is horizontal or straight (*zheng*), let us say 5 feet; the roof slope a-c, is oblique (*xie*); the principal rafters should thus be 7 feet.

Perhaps this *zhengwuxieqi* pitch is calculated for the torrential rains that sometimes fall in this country. In fact, with such slopes, the water certainly will flow. But the tiles will slide easily also (especially if the crows get involved), not

35°的屋面坡度已经足够了，这也是最大的坡度，尤其对朝北的房屋而言。北屋，其屋顶北坡可有40°坡度，其屋顶南坡可有35°坡度。为了更好地对称，在我们的区域，40°坡度是仰瓦屋顶的最佳坡度。这种情况我们就需要用比市场上卖的大一点的瓦。

counting the storms.

The pitch of 35 degrees has been successful, but it is certainly the maximum to be used, especially for the slope of the roof facing north. For a north building (*beiwu*), one could give 40 degrees to the northern slope and 35 degrees to the southern slope. For a better symmetry, in our regions, 40 degrees seems the best slope for a tiled roof (*yangwa*). One should then produce slightly larger tiles than those currently on the market.

第七章　屋面装饰
Seventh Chapter: Roof Ornamentation

用整章内容来讨论屋顶装饰这样的事情在欧洲是不会发生的。但我们既不是在欧洲也不是在美国，在美国用"屋顶装饰"来做题目会更糟糕。再次声明：写这章内容就是为了阻止你按照你周围的人所希望的那样去装饰屋顶。

中国建筑师装饰他们的屋顶，仅仅是为了装饰而装饰。当他们不能改善任何建筑的结构性能的情况下，他们至少能赋予这些建筑以某种特点。

Devoting an entire chapter to roof ornamentation would be inappropriate in Europe. However, we are neither in Europe nor in America, where the title 'roof ornamentation' might sound even worse. Be reassured: this chapter aims to prevent you from decorating your roof like the people around you might wish.

The Chinese architect adorns his roof, because he needs something to decorate. When people cannot give architectural quality to any construction, they try at least to give a certain character.

中国建筑被定义为：一个过于沉重、装饰过于华丽的屋架，被搁在大木柱上。❶ 这个定义不仅适用于北京的天坛，也适用于皇宫建筑。

在我面前有一张慈禧太后的宫殿——颐和园的照片，❸ 在那里或许有着一些阴森可怕的记忆。慈禧太后当然有足够的钱去建造一座宏伟壮观的宫殿，以彰显她有着统治四亿人民的至高无上的权力。但是这里还是像其他地方一样，一个非常重的、装饰过度的屋顶被架在大木柱子上。如果你想亲自体验一下，用不着去北京。去任何衙门，❹ 你就能理解那些皇宫的房子是怎么建的。

只有简单的一层，没有二层。立面由木雕隔板组成，有走廊。走廊本身是屋顶的延伸部分。玻璃是否已经替代窗纸（那些窗纸经常被粘附在隔墙上）？这张照片没有显示这点。这应该是因为慈禧多多少少相信装玻璃的话会有穿堂风。

建筑师不用担心后墙和两边山墙的门窗。所有的空气和光线都来自南边（这些房子一直都是朝向正南）。

Chinese construction has already been defined: a very heavy and decorated roof, placed on large wooden columns. ❷ This definition does apply to the Temple of Heaven in Beijing, as well as to the imperial palaces.

I have in front of me a photograph of the palace of Empress Cixi,❸ of dismal and malicious reputation. Sure, she certainly had enough money for erecting a grandiose mansion, worthy of the undisputed sovereign of 400 million people. Yet here, as elsewhere, it is just a very heavy and decorated roof, set on large wooden columns. No need to travel to Beijing if you want to experience this yourself. Go to any *yamen*, ❹ you will understand how all the houses that are called imperial palaces are built.

There is no upper floor, it is a simple ground floor. The façade is a carved wooden partition, sheltered by a veranda. This veranda is formed by the extension of the roof. Has glass replaced the paper, which is usually glued to the partition wall? The picture does not show it. It must have depended on Cixi's more or less firm conviction of whether or not glass lets the breezes through…

The architect had not to worry about doors and windows at the back wall and the two side gables. All the air and all the

---

❶ 作者在第三章"样式"部分有此定义。
❷ The author refers to his long note on 'style' in chapter 3 of the handbook.
❸ Summer Palace in Beijing, built by empress dowager Cixi or Ts'u Hsi who ruled China from 1861 to 1908.
❹ A yamen is a government office building of a governor or a mandarin in Imperial China.

第二部分　手册的内容和翻译
Part Two: The Handbook Translated

屋子里面没有拱券，只有层层叠压的木材。我就不描述了。看看你之前看过的第一个大塔。天坛和夫子庙的塔都是同样的做法。

建筑师很失望，找不到东西去装饰，至少砖头没法装饰，于是他们就尝试用瓦及与瓦相关的东西去装饰屋顶。与他的外国同事相反，他从脊开始做屋面装饰。在脊檩上，他砌了一个真墙，四五英尺高，有时候有 6 英尺高。想象一下屋架要承受的重量吧！作为整个房子最显眼的部分，这座墙几乎完全被装饰。首先，用几排砖头形成线脚，接着在两边放上大量中空外有莲花的花砖；紧接着再放上一些砖做装饰线脚；最后放上不同形状的瓦，摆放出各种好看的样子。脊的两端是脊兽。这里通常会放大龙头，没有任何美观性可言。对大学者来说，这些张着大口的龙头事关整个区域的风水。普通老百姓家的龙头是不能张开口的。

等脊塑造好、装饰好、涂上各种颜色后，对两边山墙的脊的装饰就开始

light come from the southern side (these buildings are always strictly oriented to the south).

Inside, no vaults, but an incredible accumulation of wooden lumber. I will spare you the description. Have a look inside the first large pagoda you come across. The Temple of Heaven and the Pagoda of Confucius are not made any differently.

In desperation, the architect finding nothing to adorn, at least with his bricks, turns to adorning the roof with tiles and 'derivative products'. Contrary to his foreign colleagues, he begins with the ridge (*ji*). On the ridge piece (*jilin*), he raises a real wall 4, 5, sometimes 6 feet tall. Imagine the weight on the roof trusses! Being the most visible part of the whole building, this wall is also the most decorated. First, there are several layers of bricks, forming mouldings; then massive cast pieces, hollow inside, with lotus flowers, etc. on both sides; then, other moulded bricks; finally, tiles of various shapes, cleverly arranged to give a feeling of nice looking (*haokan*). Both extremities of the ridge end in two crest heads (*jishui*). These are huge dragons that are anything but aesthetic. For the important 'literati', these dragon heads have a gaping mouth that catch the *fengshui* of the whole neighbourhood. For common people, the dragons may not open their mouth.

Once the ridge wall is well fixed, well decorated, and sometimes well

了。这里的装饰有莲花，还有猴子、小马、潜头鱼、蝙蝠和做祈福状的和尚。

最漂亮的是在屋顶高处用铁制的鹿角来装饰（叫做"钢叉高引"），但不是每个人都能这么做的，必须是很高级别的官员才行。塔除外，因为神有一切权利。

等这些装饰玩意儿放好后，就开始铺瓦。我们看到的那些古迹，屋面瓦是琉璃瓦，就像陶器和瓷器，是黄色的。一些绿瓦或蓝瓦用来放在屋顶中央形成大字，如"福""卍"等。铺瓦是从底端开始的。第一排瓦用大钉子钉住以防滑落，大钉子穿过第一排的每一片瓦，这些钉子是用特殊方法铸造的。这些钉子上有用泥铸的中空的人形，形成一排祈福的小和尚。

连续铺瓦以形成凹形弧状。丑的屋顶是做一个直坡。

以前，在一个基督徒村庄，工头没有听从牧师。牧师要求木工和砖工把教堂的屋顶做成直坡。当牧师外出传教时，工头就让工匠们做了一个凹形屋顶。工匠们就把屋顶做成凹形，非常凹

daubed, the ornamentation of the two ridges of each gable can start. Here again, there are lotus flowers and a whole menagerie of monkeys, small horses, diving fishes (*qiantouyu*), bats, praying bonzes, etc.

The most beautiful is when one has the right to put iron deer horns (*gangchagaoying*) on top of the roof. But everyone does not have that right. A very high rank in the mandarin hierarchy is required, unless it is a pagoda: the spirits (*shen*) have all the rights.

When all these toys (*wanyier*) are in place, the tiles are laid. For the monuments that interest us, the roof tiles are glazed and vitrified in the kiln, like earthenware and porcelain, and are yellow. Some green or blue tiles are used to draw huge characters in the middle of the roof: 'happiness' (*fu*) or 'ten thousand (*years*)' (*wan*). Laying the tiles begins from the bottom. The first row is protected against slippage by huge nails, which go through each of the first tiles, moulded in a special way. These nails are capped by hollow figures, in clay, forming a row of little praying bonzes.

Laying the roof tiles continues in a concave arch. Making a straight slope would be the height of ugliness.

In that time, there was a Christian village whose directors did not obey their pastor. He had asked the carpenters and the masons to make the roof of the church with a straight slope. When he left to go to his missions, the directors

第二部分　手册的内容和翻译
Part Two: The Handbook Translated

的那种。在经历第一场大雨后，这个新屋顶漏雨了，在经历第一场大风后，上面的瓦滑落了。屋顶一到大雨天就经常漏雨。椽子和檩条开始腐烂。后来，把屋顶凹面处的瓦，一半都揭掉，再加上去差不多 2 英尺的土，然后再铺上瓦。它就不再漏雨了，但是檩条的强度被高估了。在经历第一场大雪后，有几根檩条就断了，后来加上了一些应急的檩条。很丑，一点也不好看。最后，20 年后，整个屋顶包括屋架都被拆了，重新建了一个直坡屋顶。这个新屋顶再也没有漏雨。当时花了 7000 串（那个时候中国的钱币）。教堂屋脊好的装饰见图 2.14 中的 Fig. 65（图 2.14，图 3.11，图 3.23 和图 3.24）。

小尖塔。我们拒绝把教堂屋顶变成佛教天堂，拒绝教堂屋顶有佛教圣徒，但同时我们又想使教堂屋顶具有一些地方风情，于是传教士就有了基督教式的发明——更恰当地说，是他们使用了基督教的发明。教堂的外侧有一排小尖塔，每个人都觉得好看。传教士就在屋顶放了小尖塔，放了很多，甚至有点太多了（图 1.30 和图 1.31）。在"谷仓教堂"❶的四角放有小尖塔，在钟楼

asked the workers to make a concave roof. So the workers made it concave, extremely concave. At the first big rain, the new roof leaked, and at the first big wind storm, the upper tiles slid. The roof often leaked, basically at every heavy rain. Rafters and purlins began to rot. So half of the tiles, where the roof was concave, were removed, about 2 feet of earth added and the tiles replaced. It no longer leaked, but the strengths of the purlins had been overestimated. At the first big snow, several purlins broke, and emergency purlins were added across. It was very ugly, not nice looking (*haokan*) at all. Finally, after twenty years, the whole roof including the trusses was removed and rebuilt with a straight slope. This new roof did not leak. Cost: seven thousand ligatures of Chinese coins, at that time. See Fig. 65 a good decoration of the ridge of a church (fig. 2.14, fig. 3.11, fig. 3.23 and fig. 3.24).

The pinnacles. Refusing to have the Buddhist paradise and its saints on the roofs of their churches, but wanting beautiful roofs with a local flavour, the missionaries made a Christian invention, or rather they applied a Christian invention. [In Gothic architecture], the outer side of a cathedral is a forest of pinnacles that everybody finds beautiful. Missionaries placed pinnacles on the roofs, many pinnacles, even a little too many (fig. 1.30 and fig. 3.10). First, at the four cor-

---

❶ "谷仓教堂"是指那些农村的谷仓建筑被用作教堂，因为谷仓建造廉价且有着大的内部空间。

舶来与本土：1926 年法国传教士所撰中国北方教堂营造手册的翻译和研究
Building Churches in Northern China. A 1926 Handbook in Context

的四角放有小尖塔，为了使人相信塔有两层，通常就放八个小尖塔。连每个扶壁上都有小尖塔，就像 13 世纪教堂做的那样。

小尖塔的做法是成功的：有钱的基督徒在他们的大钟楼上放上成排的小尖塔，异教徒们也采用了这种新样式。我们可以在他们的大门、围墙和校堂上看见小尖塔。小尖塔一定是有感染细菌的（手册作者是说小尖塔作为一种时尚是很容易传染的）。

然而，小尖塔的形状很像"睡帽"，过分使用就会显得品味极差，也没有任何意义。就像那些立面上雕刻成鸽子、葡萄等样子的昂贵雕塑一样，小尖塔很贵，还会造成屋顶漏雨。有台风时，小尖塔因为没有与屋顶很好地连接或由于计算不当，就会从教堂屋顶掉下来摔到路面上。愿上帝保佑，这些小尖塔不会在节日有很多人时掉落。我听说在一个基督村庄就在十字架苦路上发生过这样的事故：幸运的是，是在第十四站苦路。❶ 小尖塔就像一颗流星从钟楼顶部掉下来，庆幸的是所有参加仪式

ners of these 'prayer barns'. Then, on the four corners of what is said to be a bell tower. Often eight pinnacles were placed to give the impression of a tower with two storeys. Pinnacles were also placed on top of each buttress, like thirteenth-century cathedrals.

Pinnacles became popular: rich Christians placed rows of them on top of their towers, and pagans in turn, adopted that new style. Pinnacles can be seen on pagan gates, enclosure walls, and schools (*xiaotang*). There must be a pinnacle virus, it is so contagious…

This profusion of pinnacles resembling 'nightcaps', however, is in utter bad taste, with no rhyme or reason. Like pigeons, grapes, etc. sculpted at great expense on the facades, pinnacles are expensive and make the roof leak. When a typhoon comes through, the pinnacle, poorly attached or badly built, will fall through the roof onto the floor of the church. May the good angels see that this does not happen on a holy day, during High Mass. I know a Christian village where such an accident happened during the Stations of the Cross: fortunately, it was at the fourteenth station. ❷ The pin-

---

❶ 苦路是指天主教的一种模仿耶稣被钉上十字架过程重现的宗教活动，也称之为"拜苦路"，主要进行于四旬期间。17 世纪时方济会士圣利安纳（St.Leonard of Port Maurice）开始宣扬拜苦路的敬礼，最后在 1731 年教宗克勉十二世（Clement XII）确定 14 处地方为苦路地方和敬礼仪式，已故教宗若望保禄二世曾重编拜苦路内容，重编内容包括了纪念主的复活。

❷ Series of 14 images depicting Christ on his way to crucifixion. Every church has such a series of images, usually hanging on the walls of the aisles. Christians are stopping in front of each image and remember Christ's passion with specific prayers. The fourteenth and last station is located close to the sanctuary. Therefore, when happened the

的人都聚集在圣坛附近。

每一处装饰，至少在建筑上，一定要有理由，它必须有其有用的一方面。不要为了装饰而装饰。❶

举个例子，我们想把通向教堂侧道上方的屋顶的旋转楼梯遮挡住：我们把它藏在一个非常小的转塔里（图3.34）。如果拱符垛的扶壁需要支撑很大的重量，以抵消砖拱顶或石拱顶的推力，建筑师就会找来一块花岗岩石块，将它放在那里。"既美观又实用"❸，石头的诗。但是为什么谷仓教堂要去学那些大教堂呢？不用像寓言里那样模仿青蛙。❹

nacle fell like a meteor from the top of the bell tower above the facade; luckily, all the attendance were grouped near the altar.

Every ornament, at least in architecture, must have a reason; it must have its useful side. Never ornament just for ornament. ❷

For example, if you want to hide a spiral staircase giving access to the roof of the aisles of a church: hide it in a nice little turret (fig. 3.34). If a heavy weight is needed on top of the buttresses of the flying buttresses in order to counteract the pressure of stone or brick arches, the architect will search among the granite blocks for exactly the right one. "Combine business with pleasure" (*miscuit utile dulci*);❺ the poetry of stones. But why give to a prayer barn the allure of a cathedral? Do not imitate the frog in the fable!❻

――――――――――――――――（接上页 continue）

accident the author described, the people were far from the tower.

❶ 这种说法符合那个时代的理性视角。手册作者是否知道当时欧洲关于建筑的争论讨论，阿道夫·鲁斯的作品《装饰和罪恶》1908），还有战后现代建筑对装饰的排斥。

❷ This statement fits in the perspective of Rationalism. Did the author knew the architectural debates in Europe, the work of Adolf Loos (Ornament and Crime, 1908-1913), and the radical refusal of ornament by the modern post war architects?

❸ 这句拉丁谚语选自 Ars Poetica 的第 343 句诗，由罗马诗人 Horace（公元前 65 —公元前 8 年）写成。完整的句子是"Omne tulit punctum, qui miscuit utile dulci"，意思是："他得到了所有的选票（大家一致的共识），他将有用和甜蜜结合在一起了。"

❹ "青蛙和牛"，是由法国诗人 Jean de la Fontaine（1621—1695）写成的短片故事，最早发表于 1668 年，批评那些社会底层的人模仿上层社会。手册的作者用这个故事来表明：祈祷谷仓，即没有风格的穷酸教堂不应该学大教堂的奢侈装饰。

❺ This Latin proverb comes from Ars Poetica (verse 343) written by the Roman poet Horace (65 BC-8 BC). The full sentence Omne tulit punctum, qui miscuit utile dulci means literally: "He got all the votes (unanimous consensus), the one who mixed the useful with the sweet." In other words: "who is able to combine business with pleasure, reaches perfection."

❻ The Frog and the Ox, a short tale published in 1668 by the French poet Jean de la Fontaine (1621-1695), that

I forgot another useless and costly ornament: the battlements (*duokou*); they load the walls and cause roof leakage. The battlements have their rationale for the rich people, who are obliged to defend themselves against brigands. But for a rectory or a small church? About this question of battlements, it is hard to have the last word with the administrators. They would gladly put battlements on the kitchen, or anywhere, especially when it is paid with money from the Father. Therefore give precise orders and supervise closely.

差点忘了,还有另外一个没有任何用途且极其昂贵的装饰:城垛,它们加重了墙体的重量,还会造成屋面漏雨。城垛对于有钱人来说合情合理,因为他们必须对付山贼保护自己。但对于神父住宅或小教堂,就算了吧。说到城垛的细节,我们说了不算数,这要看管理员。他们总是很高兴地把城垛建在厨房甚至其他地方,如果这些钱是神父出的话,更是如此。因此需要对装饰(花销)有更精确的指示说明,还需要到现场进行监督。

第八章 建筑细节的处理
Chapter Eight: Miscellaneous

屋顶形式

Roof, various forms

在中国的这个区域,大多数屋顶都是双坡屋顶,这也几乎是这个区域唯一的屋顶形式。双坡屋顶两边坡度一样,山墙一直攀升到屋脊用来承托脊檩。这种建造方法最经济,从屋顶结构的稳固性来说也是最好的(图 2.14 中的 Fig. 66)。

中国人不喜欢欧洲的四坡屋顶(图

The main, almost the only form of the common roof in this area of China, is the saddle roof with two equal sides, with gables rising to the ridge and supporting the ridge purlin (jilin). This is the most economical construction, as well as the best for the roof's stability (handbook Fig. 66) (fig. 2.14).

The Chinese do not like the Euro-

(接上页 continue)

criticises people of a lower social rank who imitate the upper class. The author of the handbook refers to the tale in order to state that a prayer barn, that is to say a poor church without style, should not imitate the richness of a cathedral.

2.15 中的 Fig. 67）。我也不认为这种屋顶有什么好处。比起双坡屋顶，它花费更多，特别是在木材极其缺乏的地方。这种屋顶也更容易被风刮落。

滴水石

滴水石（图 2.15 中的 Fig. 68）是整栋建筑里最重要的细节，它位于屋顶与墙上方相交的地方。教堂两边侧道上方的屋顶与教堂中殿的墙体相交的地方也要用到滴水石。如果圣器室屋顶和教堂后殿的屋顶都比教堂中殿低的话，这两个地方也会用到滴水石。

滴水石是檐口的一个组成部分

它由石头或砖头做成，凿成如图 2.15 中 Fig. 68 的形状。也就是说在石头或砖头的底端有一个沟槽，1 英寸深，1 英寸宽，横向切凿而成。这个小沟槽的作用是隔断雨水使其不会向内沿着墙壁流下来，而是使雨水像滴水一样或像眼泪一样垂直滴落。

为了使滴水石充分发挥作用，必须精确地算出它的位置。这需要做一些精密计算，这相当不好做，因为在砌墙时就需要把滴水石嵌入墙里，也就是说在做屋顶之前就要完成。如果计算错

pean 'hipped roofs' (handbook Fig. 67) (fig. 2.15). I do not see the benefits of this type. It certainly costs a lot more than the first type, especially in a country where wood is scarce. The wind has more grip on such a roof.

Dripstones

The dripstone (handbook Fig. 68) (fig. 2.15) is an extremely important detail throughout the whole building, at the place where a roof meets a higher wall. This will be the case with the roofs of the aisles of a church, at their junction with the main wall of the nave. Dripstones will also be used for the roofs of the sacristy and of the apse, if the latter is lower than the nave.

The dripstone shaped cornice

It consists of stones or bricks, cut as shown in Fig. 68. That is to say that these stones or these bricks have, at their bottom side, a small groove or waterdrip, 1 inch deep, and 1 inch width, cut into the direction of the width. The aim of this small groove is to prevent rainwater from flowing down the wall, and force water to fall vertically as drops or 'tears'.

For this cornice-dripstone to fulfil its function, you must accurately calculate its location. This is quite tricky to do, because it must be done during the construction of the wall where the drip is located, that is to say before making the

误，屋顶就会遮住檐口，或者檐口高于屋顶：这两种情况下滴水石是没有用的。等檩条、梁的位置完全确定后，就应该考虑教堂侧道上方屋顶和圣器室屋顶等所用的材料。滴水石必须至少要比屋顶下檐高出 1 英寸，必须挑出墙体 5 英寸。

如果教堂耳堂低于中殿，屋檐口滴水石部分就需要遵从耳堂屋顶的布局。应该逐步提升以保证滴水石一直比与中殿墙体交接处的瓦高出 1 英寸。

广义上来说，滴水石是指那些墙垂直线外的凸出部分。有滴水石的情况下，雨水滴落的地方就会离墙有一段距离，看看我们大教堂的那些雕刻精细的怪兽，那些实际上就起着滴水石的作用。

这类滴水石可以用石头或砖头做成。如果是石头，可以做成如图 2.15 中的 Fig. 69 那般样子：A 点处是滴水。如果是砖头，屋檐口就做成像图 2.15 和图 2.16 中的 Fig. 70~ Fig. 72 那样。

窗沿处的滴水处理也很重要，里外都要有，就像之前一个教堂在它的双层和三层大窗户的平面图上标出来的那样，都有同样的一处凹槽。

roof. If the calculation is wrong, the roof will cover the cornice, or the latter will be too high above the roof: in both cases the dripstone will be useless. Therefore you must take into account all the materials that will form the roof of the aisles, of the sacristy, etc., after setting exactly the location of the first roof plate (*lin*) of this roof. The dripstone must be at least 1 inch higher than the roof, and must spring about 5 inches out from the wall.

If the transept is lower than the nave, the cornice-dripstone should fit with the layout of the roof of that transept. It should therefore gradually raise with the roof in order to always remain one inch above the tiles that touch the wall of the main nave.

By extension, any projecting element from the wall's plumb, made for stopping rainwater, is called 'dripstone'. In this case it is intended to shed rain water at a suitable distance from the wall's base, like the oddly sculpted gargoyles on our great cathedrals.

This kind of dripstone can be made of stone or brick. If it is stone, it is given a shape as on Fig. 69: the tip A serving as the waterdrip. If it is brick, the cornice will be built as shown on Fig. 70, Fig. 71, Fig. 72 (fig. 2.15 and fig. 2.16).

It is also important to place dripstones at the window-sills, both inside and outside, as mentioned on the plans of large double and triple church windows. They also have a little waterdrip, as

第二部分　手册的内容和翻译
Part Two: The Handbook Translated

做窗户的细木活时，一个好的窗户在它的可动部分和固定部分的下方都要有"喉咙"，也要有凹槽（图2.16中的Fig. 73A和B）。

拱顶

拱顶能使教堂有更大空间和更有特点吗❶？

"不，必须微带愤怒地回答那些纯粹主义者，考虑一下你的财政情况吧，要做一个真正的有着雕花拱顶石、连排扶壁、石柱和奢华柱头来支撑的石头或砖头拱顶，这样的想法是愚蠢的。于是你用木板和石灰做一个拱顶仿制品，你准备用这去欺骗谁？如果你不能做真的，就什么也别做。"❷

然而，那些寒酸的仿制拱顶，那样的假拱顶，却有很多优点。这点不容置疑，那就是它使教堂更有特点，更有样式。至少，除了更有样式外，它会更整洁；你看不到那些木结构的裂缝裂纹。它也更干净（很抱歉不能讲太多细节），因为有着欧式拱顶的教堂，其室内没有燕雀，那可是这个地区连最好的

mentioned on the same plans.

In joinery, a well-made window will have a 'throat' practiced both in the lower part of the mobile window frame and of the fixed frame, as well as a groove (handbook Fig. 73 A and B) (fig. 2.16).

Vaults

Is it appropriate add vaults to our monumental churches, to give them more character [style]?

"*No, answer the purists with a little indignation, since in the state of your finances, it would be folly to think of a real stone or brick vault, with its carved keystones, with its forest of buttresses, with stone columns and their rich capitals that would bear it. So, who do you mean to fool with your pastiche vault, made of boards and lime? Do nothing if you are not able to do something true.*"

However, this poor pastiche vault, this sham vault, offers many advantages. Without wanting to fool anyone, it will still give more character [style] to the church. Aside from the issue of character, it will look neater; you will not notice all the little flaws of the wooden timber structure. It will be cleaner too (sorry for this detail) because the sparrows, who

---

❶ 手册作者在这里用"特点"一词，实质意为"风格"。这里提出这样的问题是因为大多数中世纪的西式教堂都有拱顶，因此，建于中国的一座真正的哥特风格的教堂应该有拱顶。

❷ 英国建筑师A.W.N. Pugin（1812—1852）对哥特复兴式建筑的真实性问题进行过讨论，一些最重要的相关见解可参考《对照（1836）》和《尖顶建筑或基督教建筑的真实原则（1841）》这两篇文章。

传道者也感到害怕的竞争对手。有了拱顶，它们哪里还有地方能筑巢呢？此外，拱顶使得教堂内夏天更亮冬天更暖。尤其是在那用普通板条架做的拱顶做得很好的情况下，教堂的音响效果将彻底被改善。

怎样用木板、板条和石灰做一个木制拱顶呢（图 2.18 中的 Fig. 74-Fig. 76）？

可以在地上画出要做成的拱的弧度。接着就按照这个弧度来切割木板。这些木板用钉子钉在一起，交接的地方要轮换变化。木板之间的交接边不能方方正正地切割，而是应该沿着弧的半径方向切割。在用钉子钉之前，要将木板塑形，像图 2.18 中 Fig. 74 中左边图示那样。

假的哥特式拱用木头做成，有两种：大的那种是与教堂中轴线平行或垂直的（墙拱和横向拱）（图 3.40~图 3.43）；小的是交叉拱（拱顶肋骨）。这些交叉拱相交于"间"中心处的拱顶木上（图 3.44）。这种拱顶木是用单块圆形木做成的，有着与拱一样的线脚（见图 2.18 中 Fig. 75 中的虚线）。在拱顶木的底面钉上木板，这些木板事先被雕刻成具有装饰性的花朵形状、字母图样等。

can drown out the best preachers in these regions, will not be at home under this… European vault. Where could they build their nest? Moreover, it will be cooler in the summer and warmer in the winter. And furthermore, especially if the vault, made of common lathing, is well done, the acoustics of your church will be completely changed for the better.

How to build a wooden vault with planks, lath and lime (handbook Fig. 74-Fig. 76) (fig. 2.18).

You should trace on the floor the curve of the arches to be obtained. Then boards are cut according to this curve. These boards are then nailed together, alternating the joints. These joints should not be cut straight, but following the radius of the curve. Before nailing the boards you should mould them, as shown on the left side of the drawing.

False Gothic arches, made of wood, come in two models: a large model for arches parallel or perpendicular to the axis of the church (wall arches and transverse arches) (fig. 3.40 and fig. 3.43); a small model for diagonal arcs (ribs of the vault). These diagonal arches are assembled by a wooden key at the place they cross each other, that is in the middle of the bay (*jian*) (fig. 3.44). This key is made of a single round piece of wood, moulded with the same mouldings as the arches that converge in it (see dotted line on the drawing). On the underside of the key one may nail a wooden board, that has previously been carved

第二部分　手册的内容和翻译
Part Two: The Handbook Translated

在拱顶木与拱交接的地方，做一个榫槽，使得拱的榫头能插入。榫槽要雕刻一定的线脚，使得它看上去像是拱顶木本身的一部分。教堂侧道上方的拱顶木，尺寸会小一点（举个例子，如果中殿的拱顶木是 1 英尺，那么侧道的拱顶木是 8 英寸）。但是它们应该在同一高度而且要有同样的线脚。如果有足够大的车床，可通过旋转来做柱顶石的线脚。

众多交叉拱通过与拱顶木相交形成一个牢固的肋骨结构，用来支撑由板条和石灰做成的拱顶（图 1.34 和图 1.35），无需附加任何东西到屋架上。

在钉板条时，应该稍微往上弯曲一点，这样拱顶板就会形成更牢固的拱顶（图 1.33）。如此做法只用在拱顶板上端，这样渐渐变得水平从而形成天花板。

要想给拱顶面抹上石灰，必须在板条之间留够足够空间以保证石灰能够穿过。一个工人站在拱顶上面，拍打石灰使其渗过板条。

通常，要在教堂的纵向剖面图上把拱的中点标出来。原则上，拱顶木必须至少与其周边的横向拱和墙拱的拱尖在同一高度（图 2.32 和图 2.38）。同样，那些形成拱顶肋骨的交叉拱（那些

with an ornamental flower, an initial, etc.

At the point where the arch reaches the key, one should make a mortise for receiving the tenon of the arch. The latter should be carved to match the mouldings of the key. For keys of the vaults of the aisles, the dimensions will be smaller (for example 8 inches if the key of the nave is 1 foot). They must, however, keep the same height and the same moulding. You should turn the mouldings of the keys, if you have a large enough lathe.

The arches assembled by the key must form a perfectly solid support for the panels of the vaults made of lath and lime (fig. 1.34 and fig. 1.35). No need to attach anything to the roof frame.

When nailing the lath, one should give them a slight curve upwards, so that the vault panels themselves form a more solid vault (fig. 1.33). This precaution should only be taken in the upper part of the vault panels, which is horizontal and forms the ceiling.

To coat the vault panels with lime, space the laths enough so that the lime could pass through. A worker, standing above the vault, would spread this lime over the laths.

Usually, the centre of the arch is marked on the 'vertical section' drawing of the church. In principle, the key must be placed at least as high as the top of the transversal and wall arches that frame the vault (fig. 2.32 and fig. 2.38). In princi-

与拱顶木相交的拱）必须是圆拱（半圆形），不是所谓的尖形拱。在哥特样式里，墙拱和横向拱都是尖形拱。兰斯教堂[1]是为数不多的这样的做法：那里的交叉拱也是尖形的。

如果拱顶肋骨做成圆拱的话，拱顶石就会太低（见上述第一条规则），可以通过使圆拱的中心高于柱头所在的水平线而加以补救。只有在这般加高法变得过于夸张时，才应像兰斯大教堂那样做尖形拱。

如果想要做一个非常牢固的天花板（拱顶面板），两层板条应该互相交叉安放：第一层板条与墙拱和横向拱垂直，第二层板条与第一层板条十字交叉。可参考教堂方案三的平面图（图2.32）。

砖拱

要建造这样的拱，应该准备好做各种拱的拱券砖。拱券砖应该装饰成馒形式以便做线脚（图2.19中的Fig. 79）。

ple too, the diagonal arches forming the intersecting ribs (those that intersect with the key) must be round arches (semicircles) and not what is commonly called ogees (pointed arches). In the Gothic style the transversal and wall arches are pointed. The cathedral of Reims is one of the few exceptions to this rule: there, the diagonal arches are pointed.

If, when using the round arch for the diagonal ribs, the keystone is too low (see the first rule above), one could remedy this by fixing the centre of these round arches higher than the line at the level of the capitals. Only if such heightening would become too exaggerated, one should do as at the cathedral of Reims and break the top of the pointed arches.

If you want a very strong ceiling (vault panels), two layers of crossed lath should be placed one over the other: the first perpendicular to the transversal and wall arches, the second crossing the first one. See also the Church project 3, ground plan (fig. 2.32).

Brick arches

To build such arches, you should prepare arch-stones appropriate to the different arches. Keystones will be well adorned with a torus, that would form a moulding (handbook Fig. 79) (fig. 2.19).

---

[1] 兰斯教堂，1991年被列为世界遗产，是13世纪法国哥特式建筑的代表作。

要确定适用于每种拱的拱券砖的形状，就必须在木板上画出 1∶1 大小的拱。那些半径稍微不一样的拱可以使用同样的拱券砖。

建造拱时，在它中心点钉上钉子并系上绳子，用来表示拱的半径。工人在每一次砌砖时都必须拉紧绳子以保证砖头离中心的距离准确合适，所砌之砖的中轴线在半径方向上。砖拱的剖面见图 2.19 中的 Fig. 79。哥特拱的顶端总有一个接缝（图 2.19 中的 Fig. 80），而不是柱顶石（图 2.19 中的 Fig. 81）。

彩色玻璃窗

说到被传教国家的彩色玻璃窗，这里尤其要重申的是前面已经提到的石拱和砖拱："太嫩了"！❶ 确实，即使在大教堂里，有着故事情节和场景或只有人物图像的彩色玻璃窗，也许会招来批评："这是不合适的"。

然而，如果中国传教区目前的财政状况强迫我们放弃（不管愿意与否）这样壮观的教堂装饰及通过图像的形象教育，我们可以认真地考虑有没有可能做一个漂亮的单色玻璃窗。❷ 即使在这

To define the form of the arch-stones that are appropriate to each arch, one must draw the arch on a board, life-size. Arches that have a slightly different radius can use the same arch-stones.

To build an arch, fix a nail at its centre and a cord with a knot, which indicates the radius of the arch. Each time the worker lays a brick, he must measure with the cord to ensure that the brick is at the right distance from the centre, and that its axis is in the direction of the radius. Full section of a brick arch: see Fig. 79. The top of a Gothic arch must always present a joint (Fig. 80), not a key (Fig. 81) (fig. 2.19).

Stained glass windows

Stained glass in mission countries! Here especially should be repeated what has been said about the arches of stone and brick: 'too green'! Indeed, even in a cathedral, figurative stained glass with scenes or just individual figures, probably would be criticized: "*this is not appropriate*" (*non erat hic locus*).

However, since the current state of finances of the missions in China force us to renounce willingly or reluctantly such grandiose decoration of our churches, as well as the education of the faithful by image, you could still rea-

---

❶ 手册作者意思是说"太糟了"。用绿色来指没有成熟的水果。中国的教堂建造还没有到能做彩绘玻璃窗的成熟阶段。

❷ 灰色玻璃是指只绘有灰色阴影的玻璃。

种情况下，也应该放弃欧洲某些杰作的做法。而是要根据"手头所有的"，即这个国家的资源，来做好工作。当然你的作品会被烙上"中国制造"❶的标签。这又有什么关系？因为它又不是用来出口的（图 3.45~图 3.47）。

无疑可以有其他的做法。我们可以很容易地在天津和其他地方找到铅块。同样也能找到彩色玻璃。对一个大教堂而言，从欧洲进口彩色玻璃更好。然而，需要预定小尺寸的玻璃条，建议要求很仔细的包装，完全不是"中国制造"那种，要不然的话，你会收到一车厢的碎片，当然你可以用这些碎片为圣诞木屋做很漂亮的岩纸。❷

接着你需要有一台用来做铅槽的小轧制机。我认识一个传教士，他有一台很好的小轧制机。他会把它借给那些不太富裕的邻居。对于那些存在时间长、区域大的传教区的主教住宅来说，要求有一台配套齐全的轧制机是比较容易的事情。这类花费比较有限。去

sonably consider the possibility of a beautiful grisaille. Yet even in this case, you should abandon the project of having a masterpiece composed in Europe. You should make a good job with 'what is at hand', that is to say the country's resources. Obviously your composition will bear the mark 'Made in China'. What is wrong with that, since it is not an export item that you want to produce (fig. 3.45 to fig. 3.47)!

Here is, amongst others, one way to proceed. It is easy to find ingots of lead in Tianjin and elsewhere. There is also coloured glass. It would be cheaper for a large church, to import glass from Europe. However, you should order glass sheets of small dimensions, and ask for careful packaging, not 'made in China' at all. Otherwise, all you would receive are broken bits and pieces, with which you could, at best, make beautiful rock paper for a Christmas Nativity scene…

Then you will also need a special little rolling mill for pellets grooves. I know a mission that has a very good one, and lends it kindly to less fortunate neighbours. The episcopal residence of the oldest and most extensive mission could easily acquire such a rolling mill with all its accessories. The expense

---

❶ 手册作者说这句话的背景是在 20 世纪 20 年代，而不是当今的 21 世纪，手册作者意思是说"地方当地制造"：中国制造的东西质量差不适宜于出口，只适用于中国当地。

❷ 圣诞节是为庆祝基督诞生的节日，天主教徒在 12 月 25 日庆祝（东正教教徒在 1 月 7 日庆祝。在教堂里、家里，甚至是公共场所，都会有圣诞节木屋，里面有基督、圣母玛利亚、约瑟夫、牧羊人等人物雕塑，以纪念基督在伯利恒的一个破羊圈里诞生的事情。岩纸可以用来模仿岩石。圣诞节木屋的传统可追溯到 13 世纪并由此产生了很多艺术座屏。圣诞节的神学意义是上帝选择了人类的模样。圣诞节木屋旨在使这种化身尽可能真实。

找或培训至少一个好的工人。这样的"聪明家伙",他们的十个手指非常灵活,这种人在这个国家并不难找。

你具备所有成功的要素,战争所需的命脉。❶ 然而,我要重申的是,如果彩色玻璃的使用局限于一定比例,既不会有任何风险也不会过度夸张,且你负担得起一座又漂亮又大的教堂的费用,你可以在不铺张浪费的情况下加上这项开支。

最后说一点:灰色玻璃如果有一定的细节会更好,就是说,构成它的玻璃是小尺寸的。当然工人需要花费更多的时间去焊接铅条去切割玻璃,但取得的效果会让你觉得所花的工夫和费用都值得。

钟楼尖顶

怎么建造它(图 2.17 中的 Fig. 77)。方形钟楼的八边尖顶是架在四排石头上的(AA,图 2.17 中的 Fig. 77),钟楼四角的石头进行叠涩,这样四边形就能变成八边形。对那些没有经验的工匠(砖瓦匠和石匠),需要用一块根据所需形状雕刻成的木片展示给他们看,这样做效果会很好。从里面

would be limited. You should find or train at least one good worker. Shrewd workers 'smart guys', clever with their hands, are not 'rare birds' (*rara avis*) in this country.

Now you are in possession of all the elements of success, assuming you have the necessary funding 'the lifeblood of war'. However, I want to repeat that if it is limited to such reasonable proportions, having stained glass is neither risky nor inappropriate. If you can afford the expense of a beautiful large church, you could add this expense without being extravagant.

A note to end on: a grisaille will work better if properly detailed, that is to say, if composed of little pieces of glass. The worker will need a little more time to solder the leads and cut the pieces of glass, but the result will reflect the extra work and expenditure.

The spire of a bell tower

how to build it (handbook Fig. 77) (fig. 2.17). The octagonal spire of a square bell tower rests on four rows of stones (AA, Fig. 77) (fig. 2.17) being cantilevered on each corner of the tower, so as to transform the square into an octagon. Show inexperienced workers (bricklayers and stonemasons) how this is made by using pieces of wood carved according to the desired shape. Seen

---

❶ 这里手册作者想通过说"钱是战争所需的命脉"来表达要做彩色玻璃窗必须要有钱。

看，这些石头像倒着的楼梯（图3.34）。从外面，它们必须切割以形成墙面。在安装时其末端要能嵌入墙里，不与墙重合，而是形成"齿状"，使其与砖咬接在一起。由于内侧部分不可见，可以切割得很粗糙。每个角需要四块9英寸的石头。所需石头的准确数量是由钟楼的尺寸大小和石头的厚度决定的，有些石头不到9英寸厚。

八边形形成后，就要用特殊的砖来建造尖顶，这种砖的一边有着同尖塔一样的坡度（图2.4和图2.5中的Fig. 6~Fig. 10）。做这类特殊砖的模块之前，先在地上画出一个1∶1的尖顶图样。尖塔里面是中空的，内壁大概6英寸厚。在砌内壁的时候，非常重要的一点是工匠不应该节省石灰，他应该将石灰抹在砖的整个表面，而不是像中国泥瓦匠习惯做的那样仅仅抹在边上。此外，应该使每块砖稍微往外倾斜。不这样做的话，雨水会渗透进砖头缝隙，整个钟楼就会湿透。

最重要的是，不要以加固的借口在

from inside, these cantilevered stones look like an inverted staircase (fig. 3.34). On the outer side, they must be cut in order to form the facing. They should be placed so that their ends, embedded in the wall, do not coincide, but form 'teeth' [or 'chain'] to intermingle with the brickwork. Because the inside part is not visible, it could be very roughly cut. You will need about four stones, 9 inches thick, at each corner. The precise number of stones is determined by the dimensions of the tower and by the thickness of the stones, which may be thinner than 9 inches.

Once the octagon is obtained, build up the spire with special bricks, one side of which having the slope of the spire (hand book Fig. 6-Fig. 10) (fig. 2.4 and fig. 2.5). Before making the mould for these special bricks, draw the spire full scale on the floor in order to determine accurately the spire's slope. Inside, the spire remains hollow, the walls being about 6 inches thick. When bricking up the sides of the spire, it is important that the worker should spread the lime generously: he should put lime on the whole face of the brick, and not only at the edges, as the Chinese masons usually do. In addition he should give each brick a very slight inclination towards the outside. Without these precautions, rain water will infiltrate the interstices between the bricks, and the bell tower will be flooded.

Above all, do not put timber in this

这种尖塔里放木屋架！

如果你将每排砖都往里砌不多不少同样的距离，建造尖顶就会是很容易的事情。为了能顺利完工，有经验的工匠会把一根长杆完全垂直地固定在钟楼中央，立在尖顶的底平面上。在长杆上端放上一块与尖塔同样形状的八边形的木板。在这个小八边形平台的八个角上系上八根绳子，然后一直接到尖顶底部的八个角上。尖顶的屋脊就是根据这些绳子来做的。

尖塔顶部必须用一块或多块总高 5~6 英尺的石头砌成，最上面的石头有花饰。这些石头从底到顶开凿有洞以固定铁十字（图 2.17 中的 Fig. 78 和图 3.35）。在精确到位地垂直安放铁十字架后，最后用铁垫片、熔铅或熔硫将铁十字架固定，不要用木头固定。

尖顶屋脊是用以下特殊砖做的：半圆形的砖，与有着同样圆的"头部"的砖一起使用，这种砖更长，能遮挡勾缝，或者用其他的特殊砖（图 2.5 中的 Fig. 12），也可以用"卷叶"状的砖（图 2.5 中的 Fig. 11 和图 3.35）。

kind of spire, under the pretext of consolidating it!

Building the spire is very easy if you carefully shift each row of bricks towards the inner side at just the right distance, no more, no less. In order to do this successfully, experienced workers fix a high pole, perfectly straight, exactly in the middle of the tower at the level of the spire's base. On top of this pole they set an octagonal board of the same form as the spire. From the eight corners of this small octagonal tray, eight cords are spanned to the eight corners of the spire's base. The brick edges follow these cords.

The top of the spire must be completed by one or more stones 5 to 6 feet high, the upper one ending with a fleuron. These stones are pierced with a hole from the bottom to the top, in which the iron cross will be fixed (handbook Fig. 78) (fig. 2.17 and fig. 3.35). After having been accurately placed vertical, the cross should be definitively fixed with iron wedges and molten lead or molten sulphur, not with wood.

The edges of the spire are made, either with rounded half bricks, mixed with bricks having the same round 'head', but longer, so as to cover the joint, or with other special bricks (handbook Fig. 12) (fig. 2.5). If 'hooks' are placed, one may use bricks (handbook Fig. 11) (fig. 2.5 and fig. 3.35).

勾缝

勾缝，或称勾墙，就是用砂浆填充建筑材料（石头或砖头）之间的缝隙，是砖石砌体的构成部分。勾缝有两种方法。第一种最便宜也最牢固，即在第一时间里完成砖石砌体。砖放上后，工匠就用瓦刀将石灰毛刺刮掉，如果有需要的话，他会洒上一些水。接着就在砖头之间的石灰上画一条线。

另外一种方法：整栋建筑完工后，在有序地拆除脚手架的时候，挖刮出 2 厘米或 3 厘米深的缝，用力磨和刷，浇上大量的水，接着用一把小的尖瓦刀用砂浆填缝。

抹泥

要等到砂浆压实后再开始抹墙。

"里外砖"只抹里侧也只用石灰抹，不用泥抹。但中国工人往往会坚持先抹一层泥再用一薄层的石灰抹。他会说那样做更经济。毫无疑问当然要考虑经济因素，但不是以牺牲强度为代价。这个国家的水和泥土里总或多或少含有硝。过一段时间，涂泥层会从墙体

Grouting of masonry

Grouting a wall (*gouqiang*) consists of filling in with mortar the joints between the building materials, stones or bricks, that compose the masonry. There are two ways to do this work. The first and cheapest way, as well as the most solid, consists of completing the masonry as you go along. As soon a brick is placed, the worker uses his trowel (*wadao*) to remove any dribbles of lime, and if necessary, he sprinkles with a little water. Then he draws a line on the lime, between the bricks.

The other way: first, the entire building is completed. Then, progressively, as the scaffolding is being removed, you dig and scratch the joints 2 or 3 cm deep, clean energetically, brush, and wet with a lot of water. Then fill the joints with mortar, using a small pointed trowel.

Roughcast or coating (*moni*)

Wait until the construction has settled before coating a puddled clay wall

A pure brick wall (*liwaizhuan*) is plastered only inside and only with lime, not with mud. The Chinese worker, however, will insist on first applying a layer of mud, then covering it with a thin layer of lime. It is more economical, he will say. No doubt economy should be sought, but not at the expense of

剥落成片掉落。

在石灰石膏里混合麻刀❶，这类抹面砂浆特别牢固。

那些做什么只图"差不多"的工人会不认真做这类抹面砂浆的。于是，泥土就会留在粗糙的表面，明显可见，这样的墙看上去很脏。

天花板部分，在板条用钉子钉好之后，对它的粉刷也是同样的做法。

和泥

对于那些用泥盖的普通房子，中国泥瓦匠有一个非常不好的习惯，就是在房子中央和泥。他们挖一个坑，把泥和水放在里面搅混。想象一下这公寓将来的卫生情况吧。他们会说他们之后会用干土填坑。但是连续几周那么多桶水不是已经渗透到土里了吗？更有甚者，之后，还会有人把水倒在房间的铺地上。他们在夏天会故意这么做，说是为了使房间凉快。他们根本不考虑风湿病，更不会考虑干净的问题。我必须提醒传教士注意这个问题。

---

❶ 相关内容见手册第4章麻绳部分。

strength. In this country, there is always some saltpetre in the water and in the soil. After a while, mud coating will detach from the wall and fall down in slabs.

Lime plaster is mixed with finely chopped hemp (*madao*). This coating is extremely solid.

The negligent worker (*chabuduo*) will smooth his coating very imperfectly. Dust will settle on these asperities and making them more visible, giving the walls a very dirty aspect.

The ceiling should be done in the same way, after having nailed the lathes.

Mix earthmortar (*huoni*)

For their ordinary buildings using earth mortar, Chinese masons have the very bad habit of mixing the mud in the middle of the room. They dig a hole, put earth and water in it, and mix. It is easy to imagine the consequences on the future hygiene of the apartment. They say they have filled in the hole with dry earth. But what about all the full buckets of water that have soaked into the ground for weeks? Moreover, later, the inhabitants will continue to throw water on the flooring of the room. They will do it intentionally in the summer to cool the room. Rheumatism is not taken into account. Cleanliness, even less. The things I have to warn the missionaries about, for whom I am writing…!

保护铺地

防止硝化。不是每个人都能富裕到能做镶木地板或普通地板的。教堂几乎总是有铺地的。硝会很快渗进铺地砖里，潮湿使得所有东西冷冰冰、不健康。有一种经济的基本做法就是用手册第 38 页所提到的黄色厚纸❶来预防硝化。准备好足够多这样的厚黄纸。铺地的时候，首先要将这个区域的地面弄平。抹上一薄层砂，盖上一张或两张黄纸。必须确保它们之间互相叠压。在黄纸上面，抹上砂浆，或者抹上泥，这种泥是用那些没有硝的好土做成的。接着铺上干砖。这种做法很便宜，这是已经过验证的。当然前提是不会有人为了少走一两步直接把茶壶里的剩茶倒在室内。对于教堂里圣器室的铺地，这种做法很值得推荐。

磨砖

我想说，那些又贵又无益的磨砖没

Protecting the flooring

of rooms against saltpetre. Everybody is not rich enough to afford a beautiful parquet floor, nor even a common floor with planks. Churches, however, are almost always paved. Saltpetre can infiltrate the bricks of the paving very quickly, and humidity makes everything cold and unhealthy. An economical and radical way to prevent against saltpetre are the sheets of thick yellow tar paper that have been discussed earlier. Prepare enough of such sheets. When laying the pavement, first equalize as carefully as possible the whole area of the apartment. Spread a thin layer of sand and cover with one or two sheets of tar paper. Care must be taken that they overlap. On the paper, spread the mortar, or just thick enough mud, made of good soil without saltpetre. Then lay well dried bricks. This process is cheap and has been proven successful. But it assumes that one would not empty his teapot in the middle of the room, to avoid taking two more steps to the door! This method is highly recommended for the sacristies of churches.

Polish the bricks (*mozhuan*)

I did not think necessary to address the expensive and harmful habit of pol-

---

❶ 相关内容见第 5 章。

有任何用。建造者这样做据说是为了使他们的教堂具有漂亮的外观。因此他们大力磨砖，做成好看光滑的墙面。其实这样做却把砖最好的那层能够防止硝的保护膜磨掉了。我是说那层在砖窑里烧制而形成的釉面层。

廊子

有些人喜欢，有些人讨厌。也许可以综合两方观点做一个不太宽而足够高的廊子。这样的话，夏天能避免过多日晒，冬天则足够暖，同时也有足够的光线进来（图 1.38）。

这是说北屋的廊子。东西厢房和南屋的廊子没有上述缺点。学校里，廊子在天气恶劣的情况下是很有用的，相比其他任何地方，学校的廊子绝对不能影响室内光线。

没有必要在这些廊子上用木构架，好像它们要承载千吨重的分量。希望木头能更便宜点！

墙角砖砌合

用墙角砖砌合把两面墙连接起来，这样它们就可以互相支撑。如果是普通房子的话，没有必要做墙角砖砌

ishing bricks. It is said that builders did it to give a beautiful look to their church. Therefore the side of the brick that will be apparent (face side) is rubbed vigorously against other bricks. By doing this, one removes the brick's best protection, its small armour, against saltpetre. I mean the thin and more or less glazed layer that has been under fire in the kiln.

Veranda (*langzi*)

Some recommend it, others hate it. One could perhaps reconcile both opinions by giving verandas less depth and more height. In this way, they would deflect direct sunlight in the summer, but in the winter, would not block heat and light (fig. 1.38).

This is about the verandas along the northern houses (*beiwu*). To the east, to the west, and especially to the north, the disadvantages that I just mentioned do not exist. In a school building, verandas are very useful in case of bad weather. But more than anywhere else, verandas should not block light entering schools.

It is not necessary to make a timber frame for these verandas as if they had to carry thousands of tons. It would be nice if wood were cheaper!

Clamping

To tie a building is to attach the walls to each other, so they could support each other reciprocally. Clamping is less

合。它适用于重要建筑,而且是有着轻屋顶的建筑。中国屋顶由土、瓦和八砖等构成,分量很重,对屋架的压力很大,有中国屋顶的建筑可以不做墙角砖砌合,对教堂也是如此。

连接墙体的方法有很多种。在我们欧洲老家,我们总能看到在墙上方有圣安德鲁十字架外形的铁构件,熟铁做的,也有 S 形和 Y 形,它在不同的高度与楼板搁栅及木屋架梁的末端咬接,把墙连接一起。建筑师会告诉你要不要做墙角砖砌合及怎么做。

我经常在地窖里看见墙角砖砌合的做法,这样做的那些人不会承认这种做法浪费钱还没有用。这里引用 Barberot 的话:"为什么用墙角砖砌合连接那些应该抵抗土壤压力的墙体呢?相反,它们应该从里到外被扶撑才对。"

除了考虑土壤压力外,对于地下室,应该首先考虑来自上方墙体巨大的压力。这种大压力下没有必要做墙角砖砌合,土壤压力也几乎不用担心。同样的道理也适用于拱券,拱券下方的柱子能承受很大的重量,比如钟楼二层的拱券等。所有的墙角砖砌合的做法都是多余的。

necessary for ordinary houses. It would be used for a more substantial construction with a light roof. Because of its considerable weight and its pressure on the roof trusses, the Chinese roof, with pavement, clay tiles and tiles, could replace clamping, even for a church.

There are many ways to tie walls. In our old homes in Europe, we have all seen on the upper levels of facades, anchors in wrought iron, having the shape of St Andrew crosses, S and Y, which connect the walls through floor joists, with the different floors and the extremities of the tie beams of the roof trusses. The architect will decide whether to use clamping and how it will be established.

I often have seen clamping in cellars, and those who had used this technique refused to admit that it was a waste of money. This is what Barberot tells: *"Why clamp walls that have to resist the earth's pressure? They should instead be buttressed from the inside towards the outside."*

Apart from any consideration of earth pressure, for cellars, one must first take into account the considerable pressure exerted by the walls erected on top. This pressure makes all clamping unnecessary, and the pressure of earth is hardly a problem. The same is true for the curvatures whose jambs support a heavy load, for example the arches of the second floor of a tower, etc. All clamping is superfluous.

灶囱

在欧洲，壁炉建造有很严格的规定，即使如此还是避免不了一半的火灾是由建造很差的壁炉引起的。

在中国，常识即规则。对大房子而言，其设计方案里必须有烟囱的所有细节处理，在工人建造烟囱的地方必须一丝不苟地进行监督。与烟囱挨着的木地板房间或阁楼，千万不要粗制滥造，要十分仔细地做；墙更应如此。烟囱经过的墙体要做隔离，这是最基本的预防措施。

楼梯

中国房子里的楼梯，如果有的话，是在塔里，是完全不能通行的。中国木工没有做楼梯的传统，因为在中国，人们对楼层和阁楼是没有概念的。因此，很重要的一点是不要让工人全凭灵感去做，要不你会追悔莫及的。你要把你所期望做成的楼梯的所有细节告诉木工。1：1 的楼梯图还是比较容易能在墙上画出来的，至少对普通房子是这样的。楼梯不是只有一种形式，它的形式变化万千。再说一次，如果楼梯形式很特殊，建筑师应该详细

Fireplaces (*zaocong*)

In Europe, the building of fireplaces is carefully regulated, which does not prevent half of the fires being caused by poorly built fireplaces.

In China, common sense replaces rules. In a large building, the plans must provide full details of the chimneys, and one must scrupulously supervise the workers at the places where chimneys are built. Never trust a coating, supposedly done carefully, to directly support an element of the floor or of the attic against the wall of a chimney; a fortiori in the wall. It is an elementary precaution that the wall, where a chimney passes, should be completely insulated at the place of this passage.

Staircase (*louti*)

The staircases of Chinese houses, when there are, like in the towers, are absolutely impassable. The carpenters have no tradition for this work, since in the real China, floors as well as cellars are unknown. It is therefore important not to leave the worker to his own imagination; or you will bitterly regret it. All the elements of the stairs you wish should be traced for him. Such drawing is easier to realize full-scale against a wall, at least for ordinary cases. Because there is more than one kind of staircase; the form can vary endlessly. Again, the

解释。

图 2.19 中的 Fig. 82 显示了做一个普通木楼梯的尺寸。图 2.19 中的 Fig. 83 显示了做旋转楼梯的踏步时，哪种做法是正确的，哪种做法是错误的。

钟架子

正如我们所理解的，钟架子是用木头做成的一个个能够支撑钟重的结构。铸钟匠应该根据钟楼的尺寸和钟的重量提供钟架子的图纸。

重量很大的钟不能直接架上钟楼的墙上，这样会损坏墙体。钟必须架在钟架子上，钟架子不能与钟楼墙体的任何一个地方（除了墙基部位）有接触。钟架子的梁应架在石头上，架在如上文所说的钟楼四角的叠涩石块上。

图 2.20 中的 Fig. 84 和 Fig. 84bis 是为四口钟所画的图纸（图 2.20 和图 3.51）。这些钟用的是"复古"的钟锤，不需要那么复杂的钟架子。要问一下铸钟匠有关钟锤的系统。这类悬挂系统并不都是一样的。

architect must explain if it is a special form.

Fig. 82 gives the various measurements to understand the meaning of the word for an ordinary wooden staircase. Fig. 83 shows right and wrong steps for a turning staircase (fig. 2.19).

Belfry (*thong jiazi*)

As we understand it here, the 'belfry' means the timber frame that forms an structure capable of supporting the weight of the bells. The bell caster should provide the plan of the belfry, according to the dimensions of the bell tower and the weight of the bells.

Heavy bells should not be supported directly by the wall of the bell tower; they would undermine it. They must rest on the timber structure of the belfry, which should not touch the wall of the bell tower at any of its points, except at its base. The wooden beams that constitute this base rest on stones that are fixed in 'cantilever' at the four corners of the tower.

See the plan of a belfry for four bells (handbook Fig. 84, 84bis) (fig. 2.20 and fig. 3.51). These bells have a 'retro-launched' clapper, that allows for narrower belfries. Ask the bell caster about the different systems of bell clappers. The suspensions systems are not all the same either.

焦油

　　焦油很容易被船夫偷走，对他们来说焦油是很贵的。最好是用油桶装，不要用木桶装。为了把焦油从桶里倒出来，船夫很容易地将木箍弄开。他们会说焦油丢了，不是他们的错。他们也会把焦油从油桶里倒出来然后重新装上水。因此，在焦油运走之前，要趁船夫在场的时候和他们核实，确保没有这般舞弊之事发生。

　　如果焦油过稠，在采取合适措施的情况下，可以对其进行加热，也可以添加质量差一点的汽油，但绝对不能加水。

排雨水

　　第一年应该仔细检查雨水的排放问题。排水沟可能会被砖屑、土等堵塞。如果你不当心，地基就会被淹。

玻璃

　　在直隶可以找到工匠和工具来做有色的"单色玻璃窗"（图 3.45）。有色玻璃可从天津、上海等地购买，或者直接从欧洲购买。这种单色玻璃窗，尤其是那些真的彩色玻璃窗，应该加以保

Tar (*jiaoyou*)

　　Tar is easily stolen by the boatmen, for whom it has great value. It is better to transport it in oil cans than in barrels. To let the tar out of the barrels, the boatmen simply loosen the barrel hoops. They will say that the barrel leaked and that it was not their fault. They can also steal the tar from the cans and replace it by water. It is therefore necessary to check for fraud, in presence of the boatmen, before taking delivery.

　　When tar is too thick, it can be heated, by taking appropriate precautions. Cheap gasoline can also be added to thin it, but not water of course.

Flow of rainwater

　　The flow of rainwater should be monitored very carefully during the first year. Gutters (*shuigou*) can be blocked with remains of bricks, earth, etc. If you are not careful, your foundations may be flooded.

Glass (*boli*)

　　One can find in Zhili, workers and tools for installing coloured and 'grisaille' glass (fig. 3.45). Just purchase the coloured glass in Tianjin, Shanghai, etc., or directly in Europe. These 'grisailles', and even more the real stained glass,

护，免得它们被冰雹砸坏，而冰雹在中国的这个地区是很常见的。可以利用当地制造的镀锌铁丝做成铁丝网罩在上面。

也可以用白色磨砂玻璃保护彩色玻璃窗。这样太阳光线不会直接射进教堂，而是透过白色磨砂玻璃漫射进来。当地的白色磨砂玻璃要比在天津或其他地方买便宜得多。

油漆

油漆匠会对家具、门窗及暴露在雨中的木结构部分（如椽子末端、廊子的柱子等）进行油漆。

油漆不要用桐油，要用亚麻子油。在阳光、雨水和风的作用下，涂在木头上的油很快就会消失。因此，需要经常油漆：从最初第三年或第四年。如果不做这类保护措施，木头会很快腐烂。同样的桐油用在家具上，在夏天很容易变成黏黏的胶质状。我想几年后它总会变成那样的。如果用的桐油是由一个不懂诀窍的工匠随便煮成的，或者商人在里面掺了更便宜的豆油，必然很快会变成那样。桐油里混有豆油会使油漆显得更有光泽，这也会诱使工匠和商人勾结。

should be protected against the hail that is so frequent in this part of China. Protection is done with locally manufactured, galvanized iron wire mesh.

You can also protect the coloured glass windows with frosted white glass. Then the sun's rays do not arrive directly in the church, but are diffused through the frosting of white glass. It is cheaper to frost ordinary white glass locally, rather than buying it in Tianjin or elsewhere.

To oil (*youqi*)

A worker applies oil or varnish (*youqijiang*) to furniture, doors and windows, and on the parts of the timber structure that is exposed to the rain: extremities of the rafters, columns of a veranda, etc.

Chinese oil (*tongyou*) used for this purpose is not as good as linseed oil. Also, the sun, rain and wind quickly eliminate it from the wood on which it was applied. Therefore, one must renew it often: from the 3rd or 4th year initially. Without this precaution, the wood crumbles quickly. The same Chinese oil (*tongyou*), applied to the furniture, easily becomes thick and sticky in the summer… I think it always becomes sticky after a number of years. But it will infallibly and quickly become sticky if it has been badly prepared by a worker who did not have the knack, or if the merchant mixed it with bean oil, which is much

如果你想保持大教堂在高度上的效果，不要试图去油漆拱顶。在这种情况下，白色会增加远距离感，其他颜色则会"拉近"拱顶。

土坯墙的梁下柱

这些柱子承受屋顶的所有重量，墙不用承重。我们必须特别保护柱子基础，防止其腐烂。很重要的一点，要防止从土壤里渗透上来的湿气与柱子基础接触，即使是墙体地基部分也要有这样的防止措施。可以用石头基础、碱草层、沥青纸板或简单通过几层黄纸❶来实现，如果不做这类预防措施的话，整个房子在10年或20年后就会有倒塌的风险。

---

❶ 见前文相关内容。

cheaper. Mixed with *tongyou*, bean oil makes paint more glossy; thus tempting the worker to be in league with the merchant.

Do not yield to the temptation to paint the vault of a large church, if you want to conserve its effect of height. In this case, white colour enhances the feeling of large space; all other colours 'bring closer' or reduce that impression.

Columns under the beams, in adobe walls

These columns support the entire weight of the roof, and thus relieve the weaker walls. You must, however, carefully protect the base of the columns against rot. It is therefore important to avoid the contact of moisture rising from the soil, even within the base of the walls. This is achieved using stone bases, at the level of the insulating layer (*jiancao*), made with bituminous cardboard, or simply several layers of tar paper. Without such precautions, the whole house will risk collapsing after 10 or 20 years.

结　语
Conclusion

最后，就把这作为这本小书的结束语吧：房子建好后必须进行维护，维护关系着房子的耐久性，就像古人所说的"它们自身发生着变化"。❶ 这是显而易见的事情，但遗憾的是，就中国人的习性，维护是令人厌恶的，无论是衣服、机器或房子的维护，他们都不喜欢。如果事关公共建筑的维护，你碰到的困难可能会是物质上解决不了的。看看那些衙门、塔、皇家陵墓、道路、桥梁、城墙。我们那穷酸的教堂也是公共建筑！如果你需要确信这点，就只需看一下窗户数一下破碎的玻璃，再看一下立面上被风刮落的小尖塔，根本没有人会想到去重新修一下，尽管人们曾经都以此为骄傲。如果教堂的屋顶尚好，砖头还在，我们就该感到高兴了。这里确实是教堂遗址开始的地方。

因此，维护好你的教堂，维护好你

Finally, and may this be the conclusion of this modest work: you must maintain your building once it is completed; maintenance and longevity of a building, 'convertuntur' as the ancients would say. ❷ You will exclaim that this is a fine example of truism. Unfortunately any maintenance seems repugnant to the Chinese temperament, be it of a dress, a machine or a building. If it comes to the maintenance of a public building, you clash with an impossibility that is almost physical! See the *yamen*, the pagodas, the emperors' tombs, the roads, the bridges, the city walls. Our poor churches are… public buildings! If you need to be convinced, just look at the windows and count the broken glass. Look at the number of pinnacles that the wind knocked down off the façade and that nobody thinks to replace. People, however, were once so proud of it! Be happy if the roof is still on and all the bricks in place. This is, indeed, where the final ruin begins.

So, take care of your church, you

---

❶ 这里的古人是指古罗马人，convertuntur 是拉丁语，意思是"它们自己变化着"。
❷ The 'ancients' are the Romans, and 'convertuntur', in Latin, means 'they convert themselves'.

第二部分　手册的内容和翻译
Part Two: The Handbook Translated

曾经为了主的荣耀而力争建的足显主威严的建筑，或许这至少能表达你对上帝信仰的一份尊敬。维护那些先辈们留给你的教堂吧，即使它们可能已经变得有点狭小。你没有必要像日本的伊势神宫那样每二十年重建一次，但至少应该保证它们整洁干净。❶

对那些希望自己做得比鲁滨逊❷建造者更好的传教士，我们推荐一本最近才出版的书：《教堂建筑的基本知识》，作者是神父迪·杜雷特，他也是佩莱尔地区沙瓦格的神学院（位于法国旺代省 Vendée）的教授。该书有一卷，八开，200 页，有 481 幅铅笔画，价格是十法郎，可转账给作者：143-81 南特❸。

关于此书的一篇书评，已经发表在 1926 年 8 月 19 日的第 33 期《教士之友》里，黄色封面，第 210 页。

杜雷特神父想编一本真正的关于

who have struggled to raise, to the divine Master's glory, a building worthy of his majesty, or at least one that shows the respect you have for his worship. Maintain and see that those churches, your predecessors have bequeathed you, are maintained. Perhaps, they have become a little too small. At least they should be kept clean, so you will not be obliged to rebuild it every twenty years like the Ise Grand Shrine in Japan. ❶

To the missionaries who would like a more detailed book than this 'Robinson builder', ❹ we would like to recommend a recent book: *Notions élémentaires d'architecture religieuse*, [Basics of religious architecture], by father D. Duret, teacher at the Minor Seminary of Chavagnes-en-Paillers (France, Vendee). One volume, in-8°, 200 pages, with 481 pen drawings, 10 French francs. Author.

This is the book review published in *Ami du Clergé* [The Friend of the Clergy], No. 33, August 19, 1926, yellow cover, p. 210:

Father [Donatien] Duret wanted to compose a true manual of religious ar-

---

❶ The author thinks that maintenance should avoid rebuilding. This material comparison proves that he does not understand the ritual and immaterial dimensions of The Grand Ise Shrine ! See footnote 18.

❷ 这个名字取自鲁滨逊·克鲁索（Robinson Crusoe），是丹尼尔·迪福（Daniel Dufoe）的小说《鲁滨逊漂流记》里面的主人公，他单独一人生活在一个小岛上，所有的事情必须亲自做。这个名字也取自一本在上海出版的手册，该手册里描述了那些在中国的传教士必须亲力亲为所有事情，见：Pol Korigan. Le manuel pratique de Robinson（鲁滨逊实用手册） [M]. Shanghai: Catholic mission, 1912.

❸ 南特是法国西北部大西洋沿岸重要城市。

❹ The name refers to Robinson Crusoe, the hero of Daniel Dufoe's novel (1719), who was alone on an island and obliged to do everything bon this own. It also refers to a handbook published in Shanghai with all kind of do-it-yourself hints for missionaries in China: Pol Korigan. Le manuel pratique de Robinson（鲁滨逊实用手册） [M]. Shanghai: Catholic mission, 1912.

宗教建筑的手册。在序言中他这样写道："关于这类主题的学术书籍并不少见。对教授来说它们是很有价值的工具手册，但有时候它们的技术特性也让售书商感到为难，这些是专门针对专家的书，而售书商希望找到一本适用于每个人的基本手册，有点像针对学生的教材。这就是写这本书的目的。"

作者完全达到了他的目的，这本书确实就是一本针对学生的教材。这是我们所能给予的最高的评价。书中内容的易懂性很值得推荐，教学内容很清晰，解释也很到位，基本包括了从古至今的宗教建筑的全部历史。

书中有近500幅铅笔速写，是作者亲笔所画，相关内容也由作者详细解说。我们相信这本书值得神学院的教士和学生读一读，特别是那些传教士建造者。对作者来说，他的写作是为了"激发对上帝之殿的谨慎甚微的激情"，让他们认识和热爱我们国家基督教的艺术精品。

chitecture. In the foreword he writes: *"The scholarly books on these subjects are not uncommon. They are valuable work tools for professionals, but sometimes their technical character baffles booksellers. These are books for the Master. Booksellers need an elementary manual for everyone, that would be like the book for the Student. This is the purpose of this little treatise."*

The author has fully achieved its purpose, and its lessons are truly for students. This is the highest praise that we can give. It is recommended for both the simplicity of its explanations, and the clarity and precision of its teaching, which embraces the whole history of religious architecture from the beginning until today.

It contains nearly 500 pen sketches, drawn by the expert hand of the author and explaining the text in a remarkable way. We believe that this book will really be of great service to the clergy and the students of our seminaries, to whom it is especially designated. According to the author, it was written *'to inspire enlightened and discreet zeal for the house of God'*, and to let them know and love the masterpieces of our national and Christian art.

第二部分　手册的内容和翻译
Part Two: The Handbook Translated

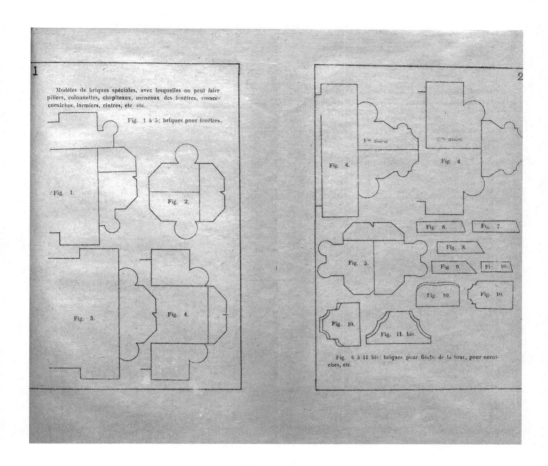

图2.3　手册图纸1：
Fig. 1 ~ Fig. 4，模型砖，用在柱子、柱身、柱头、竖框、玫瑰花窗、檐口等处。
fig. 2.3 Handbook, plate 1:
Fig. 1-Fig. 4, moulded bricks used for pillars, shafts, capitals, mullions, rose windows, cornices, etc.
(© Université Laval, Bibliothèque)

图2.4　手册图纸2：
Fig. 4 ~ Fig. 5，用在竖框和玫瑰花窗处的模型砖；
Fig. 6 、Fig. 8 ~ Fig. 11，用在尖顶、檐口等地的特殊砖。
fig. 2.4 Handbook, plate 2:
Fig. 4-Fig. 5, moulded bricks for mullions and rose windows,
Fig. 6, Fig. 8-Fig. 11, special bricks for the spire, cornices, etc.
(© Université Laval, Bibliothèque)

舶来与本土：1926年法国传教士所撰中国北方教堂营造手册的翻译和研究
Building Churches in Northern China. A 1926 Handbook in Context

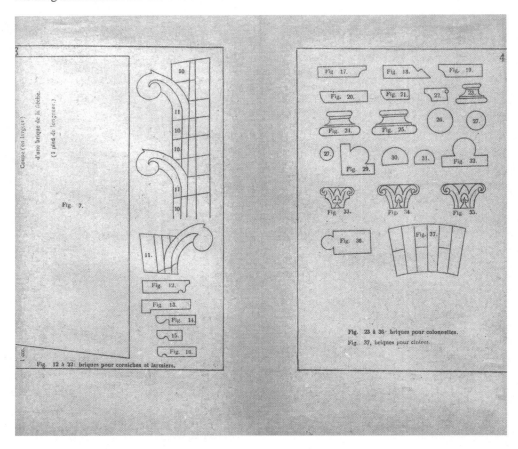

图2.5 手册图纸3：
Fig. 7, 用在尖塔处的砖的剖面；
Fig. 12~Fig. 16, 用作檐口和滴水石的砖。
fig. 2.5 Handbook, plate 3:
Fig. 7, section of a brick of the spire,
Fig. 12-Fig. 16, bricks for cornices and drip-stones.
(© Université Laval, Bibliothèque)

图2.6 手册图纸4：
Fig. 17~Fig. 22, 用作檐口和滴水石的砖；
Fig. 23~ Fig. 36, 用在小柱子上的砖；
Fig. 37, 用在拱门上的砖。
fig. 2.6 Handbook, plate 4:
Fig. 17-Fig. 22, bricks for cornices and drip-stones,
Fig. 23-Fig. 36, bricks for small columns,
Fig. 37, bricks for arches.
(© Université Laval, Bibliothèque)

第二部分　手册的内容和翻译
Part Two: The Handbook Translated

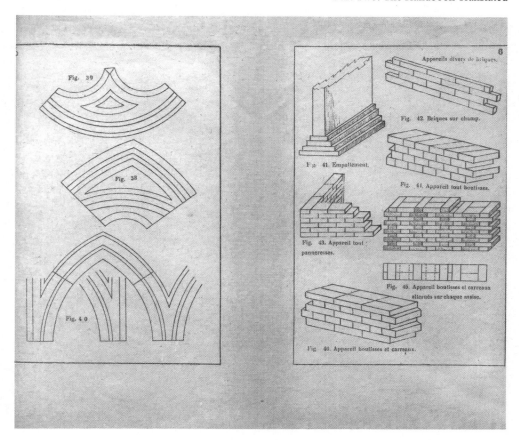

图2.7　手册图纸5：
Fig. 38~ Fig. 40，拱券和花饰窗格的细节。
fig. 2.7 Handbook, plate 5:
Fig. 38~Fig. 40, profiled elements for arches and traceries.
(© Université Laval, Bibliothèque)

图2.8　手册图纸6：
Fig. 41，阶式基础；
Fig. 42~Fig. 46，砖墙的不同砌法。
fig. 2.8 Handbook, plate 6:
Fig. 41, stepped foundation,
Fig. 42-Fig. 46, brickwork, different types of bonds.
(© Université Laval, Bibliothèque)

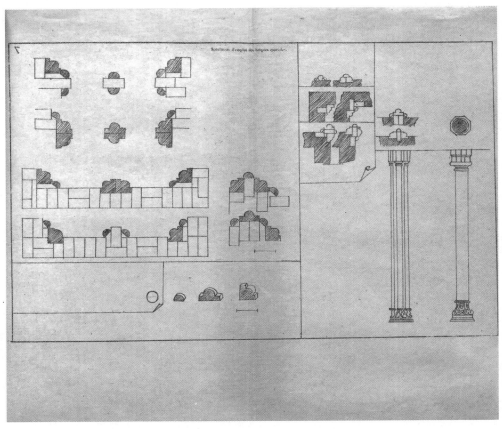

图2.9 手册图纸7：需要用到特殊砖头的部位。
fig. 2.9 Handbook, plate 7: examples of special bricks.
(© Université Laval, Bibliothèque)

第二部分　手册的内容和翻译
Part Two: The Handbook Translated

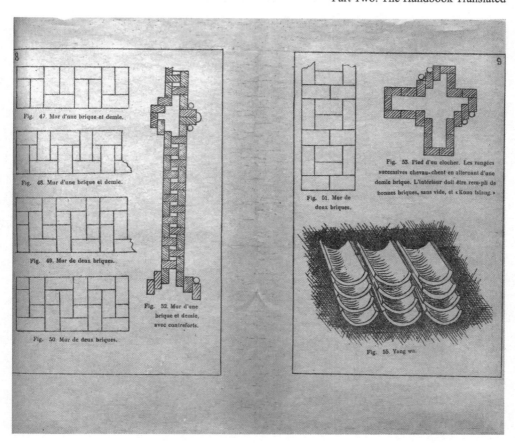

图2.10　手册图纸8：
Fig. 47~ Fig. 50，砖墙的厚度：
Fig. 52，有着扶壁的砖墙厚度。
fig. 2.10 Handbook, plate 8:
Fig. 47-Fig. 50, thickness of brick walls,
Fig. 52, brick wall with buttresses.
(© Université Laval, Bibliothèque)

图2.11　手册图纸9：
Fig. 51，两砖厚的墙；
Fig. 53，钟楼角上的扶壁；
Fig. 55，仰瓦屋顶。
fig. 2.11 Handbook, plate 9:
Fig. 51, wall of two bricks,
Fig. 53, buttresses at the corner of a bell tower,
Fig. 55, *yang wa* roof.
(© Université Laval, Bibliothèque)

舶来与本土：1926年法国传教士所撰中国北方教堂营造手册的翻译和研究
Building Churches in Northern China. A 1926 Handbook in Context

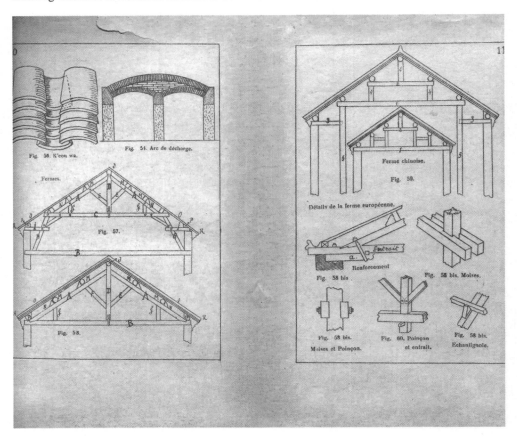

图2.12 手册图纸10：
Fig. 54，缓解负荷的辅助拱；
Fig. 56，扣瓦屋顶；
Fig. 57~Fig. 58，西式屋架。
fig. 2.12 Handbook, plate 10:
Fig. 54, relieving arch,
Fig. 56, kou wa roof,
Fig. 57-Fig. 58, Western trusses.
(© Université Laval, Bibliothèque)

图2.13 手册图纸11：
Fig. 59，中式屋架；
Fig. 58bis和Fig. 60，西式屋架的节点。
fig. 2.13 Handbook, plate 11:
Fig. 59, Chinese truss,
Fig. 58 bis and Fig. 60, joining
of Western trusses.
(© Université Laval, Bibliothèque)

第二部分　手册的内容和翻译
Part Two: The Handbook Translated

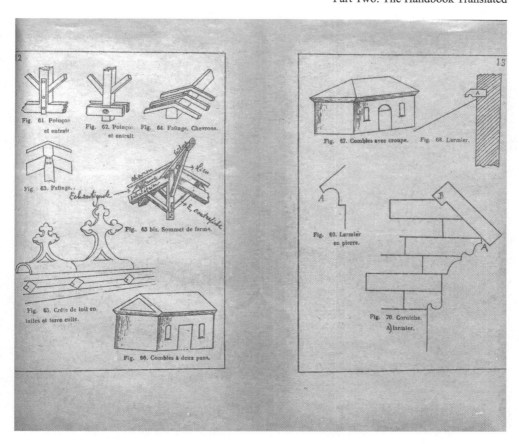

图2.14　手册图纸12：
Fig. 61~ Fig. 64，西式屋架的节点；
Fig. 65，用在屋脊处的特殊瓦；
Fig. 66，双坡屋顶。
fig. 2.14 Handbook, plate 12:
Fig. 61-Fig. 64, joining of Western trusses,
Fig. 65, special tiles for the roof's ridge,
Fig. 66, saddle roof.
(© Université Laval, Bibliothèque)

图2.15　手册图纸13：
Fig. 67，四坡屋顶；
Fig. 68~Fig. 77，滴水石。
fig. 2.15 Handbook, plate 13:
Fig. 67, 'hipped' roof,
Fig. 68-Fig. 70, dripstones.
(© Université Laval, Bibliothèque)

~ 241 ~

舶来与本土：1926年法国传教士所撰中国北方教堂营造手册的翻译和研究
Building Churches in Northern China. A 1926 Handbook in Context

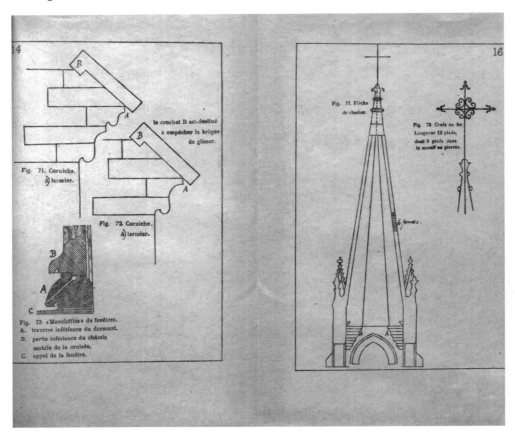

图2.16 手册图纸14：
Fig. 71~ Fig. 72，滴水石和檐口；
Fig. 73，窗框处的"喉咙"。
fig. 2.16 Handbook, plate 14:
Fig. 71-Fig. 72, dripstones and cornices,
Fig. 73, 'throat' of a window frame.
(© Université Laval, Bibliothèque)

图2.17 手册图纸16：
Fig. 77，钟楼尖顶；
Fig. 78，钟楼顶部的十字架。
fig. 2.17 Handbook, plate 16:
Fig. 77, spire of the bell tower,
Fig. 78, iron cross on top of the tower.
(© Université Laval, Bibliothèque)

第二部分 手册的内容和翻译
Part Two: The Handbook Translated

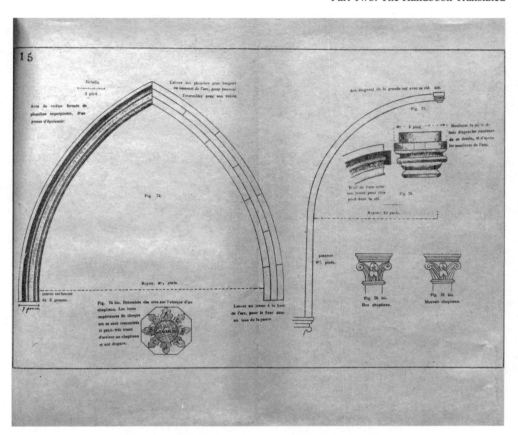

图2.18 手册图纸15：Fig. 74，用板条做成的拱；
Fig. 74bis，拱顶肋骨交接处和柱头顶部；
Fig. 75~ Fig. 76，木制拱顶石；
Fig. 76bis，做得好的柱头和做得差的柱头。
fig. 2.18 Handbook, plate 15:Fig. 74, arch made with planks,
Fig. 74bis, junction of the vault's ribs and top of the capital,
Fig. 75 -Fig. 76, wooden key stone,
Fig. 76bis, good and wrong capitals.
(© Université Laval, Bibliothèque)

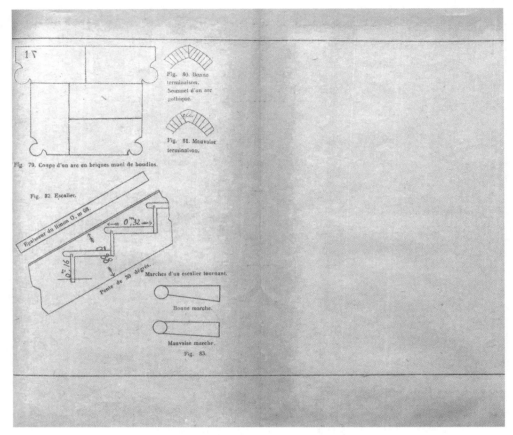

图2.19 手册图纸17：Fig. 79，馒形式拱券砖的剖面；
Fig. 80~ Fig. 81，好的拱顶做法和差的拱顶做法；
Fig. 82，木楼梯的踏步；
Fig. 83，旋转楼梯踏步好的做法和差的做法。
fig. 2.19 Handbook, plate 17:Fig. 79, section of an arch with torus,
Fig. 80-Fig. 81, good and bad top of arch,
Fig. 82, steps of a wooden stair,
Fig. 83, Good and wrong steps of a cork screw staircase.
(© Université Laval, Bibliothèque)

第二部分　手册的内容和翻译
Part Two: The Handbook Translated

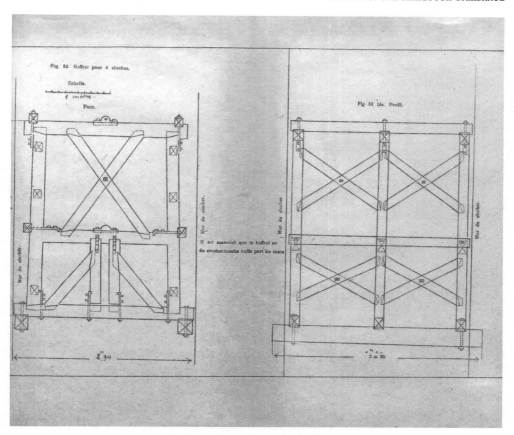

图2.20　手册图纸18：用来挂四口钟的钟架子。
fig. 2.20 Handbook, plate 18: wooden belfry for four bells.
(© Université Laval, Bibliothèque)

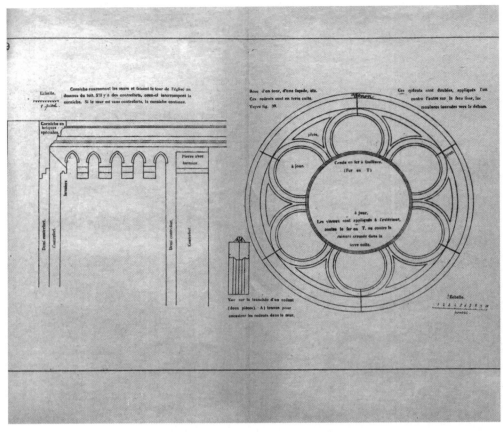

图2.21 手册原图纸19：墙上端的檐口和扶壁，
砖制玫瑰花窗和花窗中间用来装玻璃的圆形铁构件。
fig. 2.21 Handbook, plate 19: cornice on top of the wall and buttresses,
brick rose window and iron circle for glazing.
(© Université Laval, Bibliothèque)

第二部分　手册的内容和翻译
Part Two: The Handbook Translated

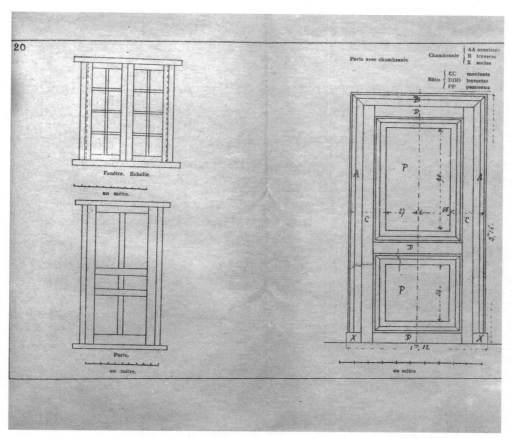

图2.22　手册图纸20：窗户和门。
fig. 2.22 Handbook, plate 20: window and door.
(© Université Laval, Bibliothèque)

舶来与本土：1926 年法国传教士所撰中国北方教堂营造手册的翻译和研究
Building Churches in Northern China. A 1926 Handbook in Context

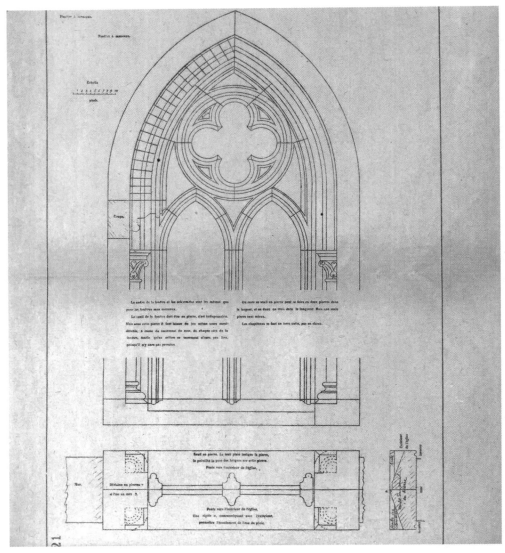

图2.23　手册图纸21：有着砖制花饰窗格、砖制柱头和石头窗沿的哥特式窗户。
fig. 2.23 Handbook, plate 21: Gothic window with brick tracery,
brick capitals, and stone threshold.
(© Université Laval, Bibliothèque)

第二部分　手册的内容和翻译
Part Two: The Handbook Translated

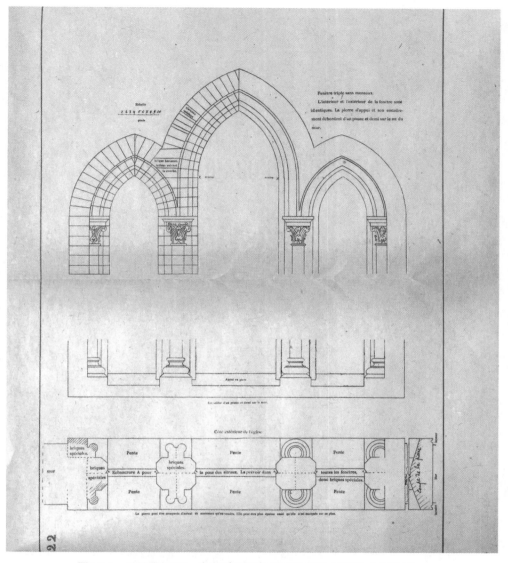

图2.24　手册图纸22：有着砖制竖框和石头窗沿的哥特式三连窗。
fig. 2.24 Handbook, plate 22: Gothic triplet window with brick mullions and stone threshold.
(© Université Laval, Bibliothèque)

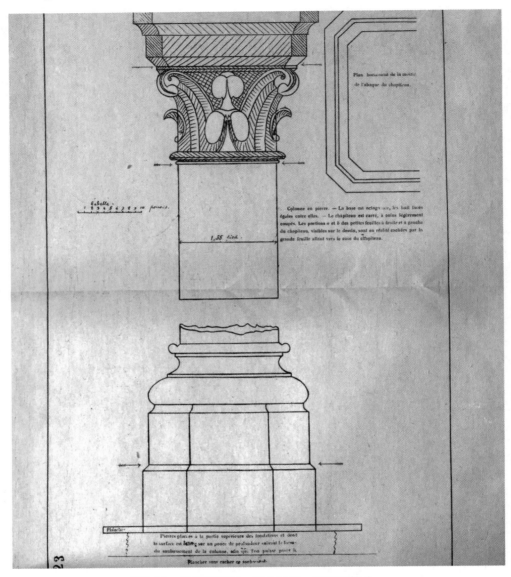

图2.25 手册图纸23：石头柱子、柱基和柱头。
fig. 2.25 Handbook, plate 23: stone column, basis, and capital.
(© Université Laval, Bibliothèque)

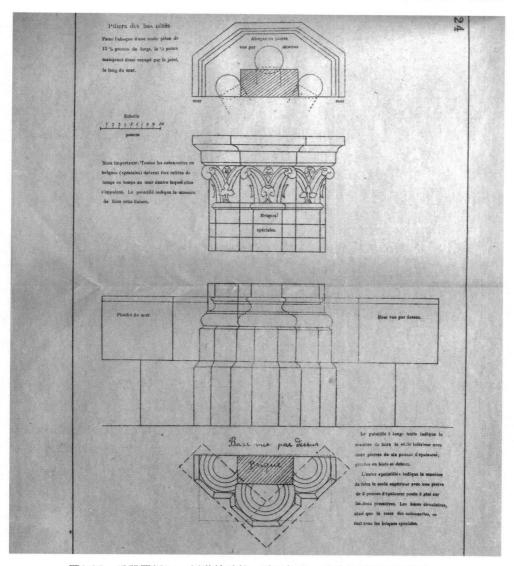

图2.26 手册图纸24：侧道的壁柱，砖制柱身，石头柱基和柱子底座。
fig. 2.26 Handbook, plate 24: embedded shaft used in aisles,
brick shaft, stone basis and plinth.
(© Université Laval, Bibliothèque)

图2.27 手册原图纸25:木制告解室。
fig. 2.27 Handbook, plate 25: wooden confessional.
(© Université Laval, Bibliothèque)

第二部分　手册的内容和翻译
Part Two: The Handbook Translated

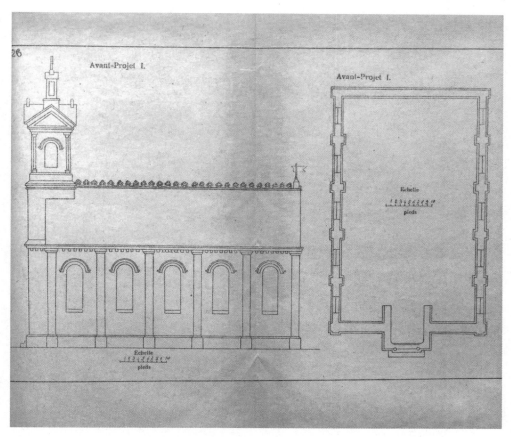

图2.28　手册图纸26~27：项目方案1，古典式教堂，单一的长方形中殿和小钟楼。
fig. 2.28 Handbook, plate 26-27: Classic style church's project 1,
rectangular single nave with small bell tower.
(© Université Laval, Bibliothèque)

舶来与本土：1926年法国传教士所撰中国北方教堂营造手册的翻译和研究
Building Churches in Northern China. A 1926 Handbook in Context

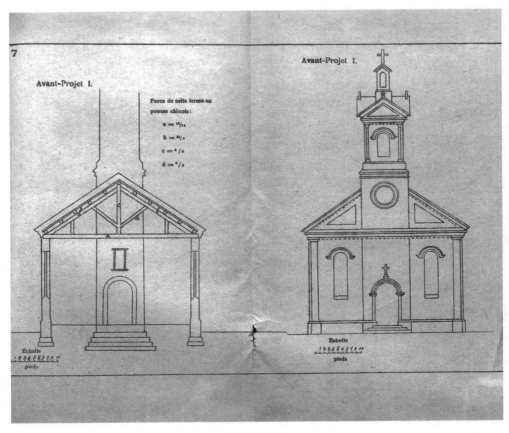

图 2.28(b)
fig. 2.28(b)

第二部分　手册的内容和翻译
Part Two: The Handbook Translated

图2.29　手册图纸28~29：项目方案2，古典式教堂，单一的长方形中殿和小钟楼。
fig. 2.29 Handbook, plates 28-29: Classic style church's project 2,
rectangular single nave with small bell tower.
(© Université Laval, Bibliothèque)

舶来与本土：1926年法国传教士所撰中国北方教堂营造手册的翻译和研究
Building Churches in Northern China. A 1926 Handbook in Context

图 2.29(b)
fig. 2.29(b)

~ 256 ~

第二部分　手册的内容和翻译
Part Two: The Handbook Translated

图2.30　手册图纸30~32：项目方案2，古典式教堂，门口和钟楼细节。
fig. 2.30 Handbook, plates 30-32: Classic style church's project 2, details of the doorway and the bell tower.
(© Université Laval, Bibliothèque universitaire)

舶来与本土：1926年法国传教士所撰中国北方教堂营造手册的翻译和研究
Building Churches in Northern China. A 1926 Handbook in Context

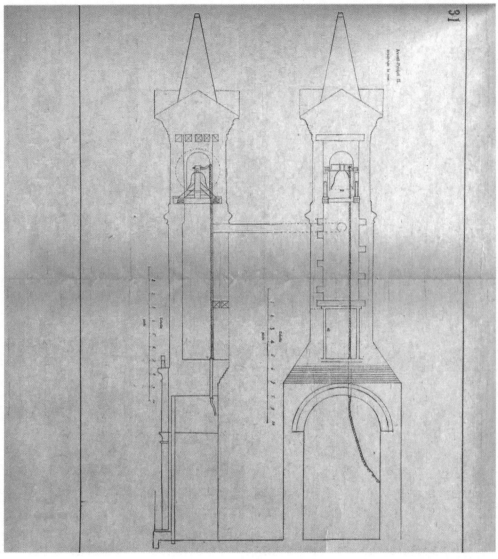

2.30(b)
fig. 2.30(b)

第二部分　手册的内容和翻译
Part Two: The Handbook Translated

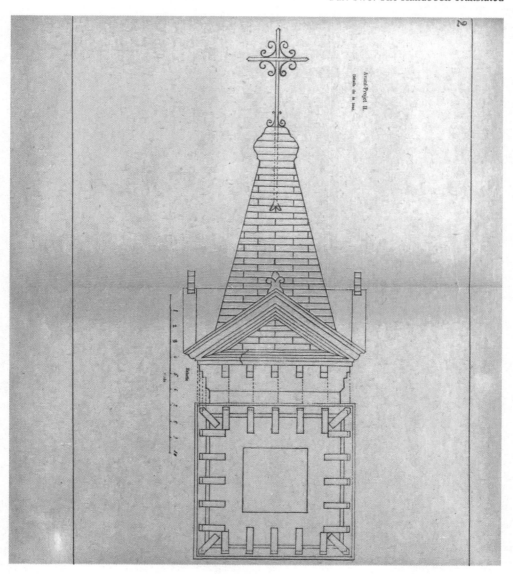

图 2.30(c)
fig. 2.30(c)

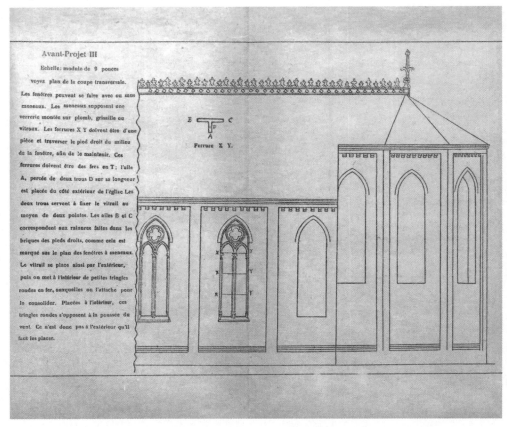

图2.31 手册图纸33：项目方案3，哥特式教堂，花窗的铁配件说明。
fig. 2.31 Handbook, plate 33: Gothic style church's project 3, indications about iron fittings in tracery windows.
(© Université Laval, Bibliothèque universitaire)

第二部分　手册的内容和翻译
Part Two: The Handbook Translated

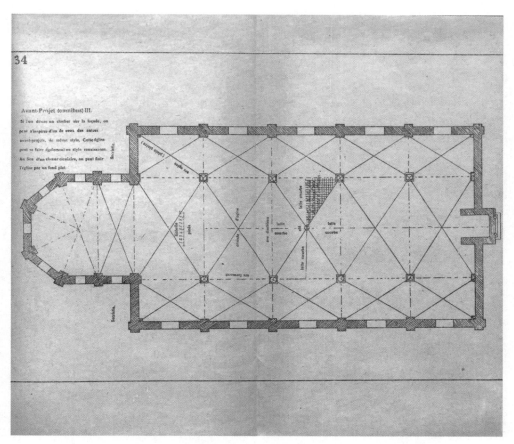

图2.32　手册图纸34~36：项目方案3，哥特式教堂，平面、里面、后殿细节。
这个设计比较灵活能用在多处，能与大钟楼组合使用或者也能用在没有后殿的古典式教堂里。
fig. 2.32 Handbook, plates 34-36: Gothic style church's project 3,
ground plan and different possible elevations, detail of the sanctuary's apse.
This flexible project, called 'omnibus', could be combined with a bigger tower
or realized in Classic style without apse.
(© Université Laval, Bibliothèque universitaire)

舶来与本土：1926年法国传教士所撰中国北方教堂营造手册的翻译和研究
Building Churches in Northern China. A 1926 Handbook in Context

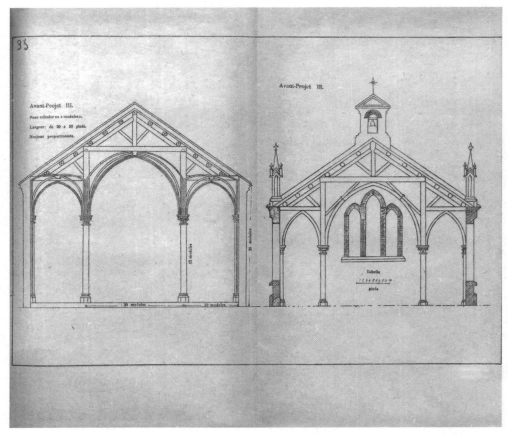

图 2.32(b)
fig. 2.32(b)

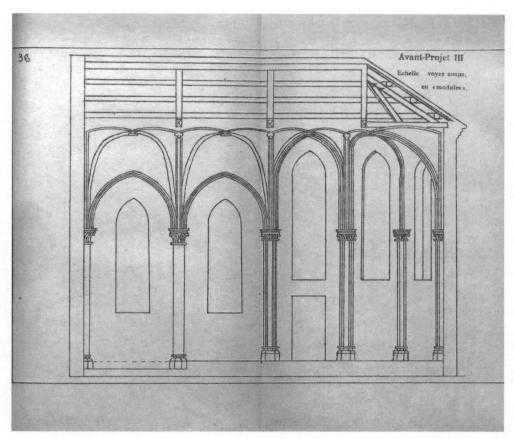

图 2.32(c)
fig. 2.32(c)

舶来与本土：1926年法国传教士所撰中国北方教堂营造手册的翻译和研究
Building Churches in Northern China. A 1926 Handbook in Context

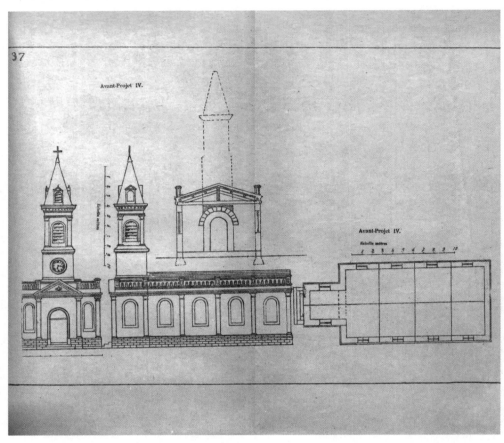

图2.33 手册图纸37：项目方案4，古典式教堂，单一的长方形中殿和方形钟楼。
fig. 2.33 Handbook, plate 37: Classic style church's project 4, rectangular single nave church with square tower.
(© Université Laval, Bibliothèque universitaire)

第二部分　手册的内容和翻译
Part Two: The Handbook Translated

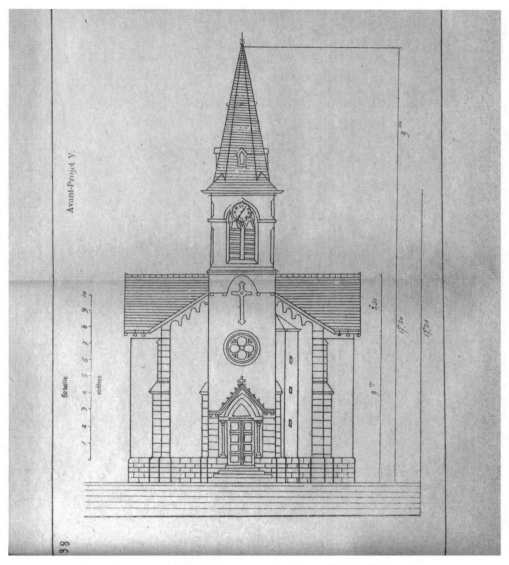

图2.34　手册原图纸38~41：项目方案5，哥特式教堂，
单一的中殿和小耳堂，多边形后殿和钟楼。
fig. 2.34 Handbook, plates 38 to 41: Gothic style church's project 5,
single nave with small transept, polygonal apse, and bell tower.
(© Université Laval, Bibliothèque universitaire)

舶来与本土：1926年法国传教士所撰中国北方教堂营造手册的翻译和研究
Building Churches in Northern China. A 1926 Handbook in Context

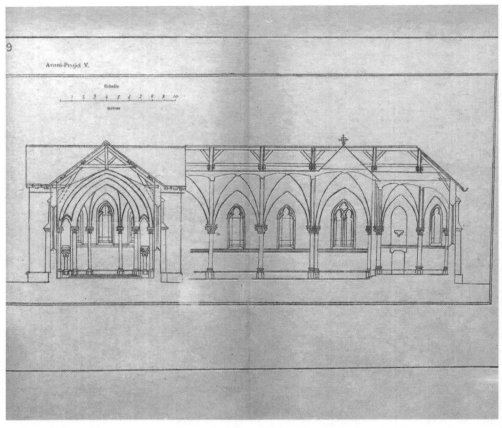

图 2.34(b)
fig. 2.34(b)

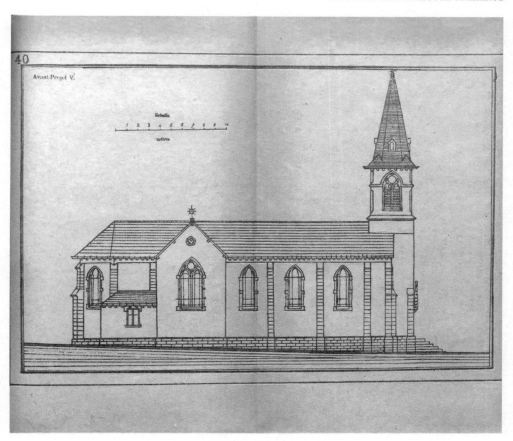

图 2.34(c)
fig. 2.34(c)

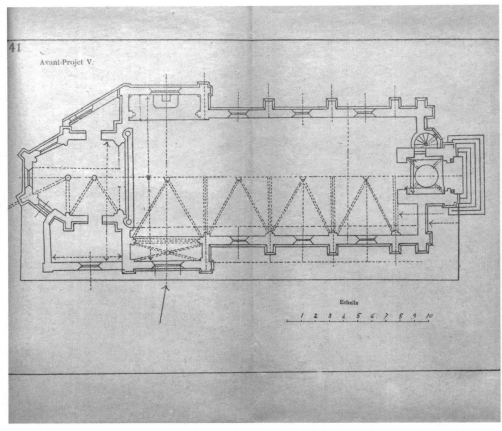

图 2.34(d)
fig. 2.34(d)

第二部分　手册的内容和翻译
Part Two: The Handbook Translated

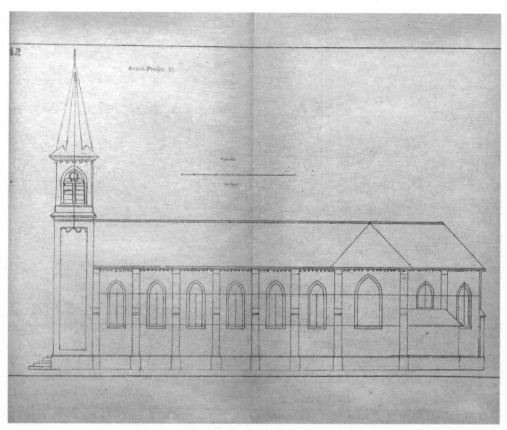

图2.35　手册原图纸42~43和45：项目方案6，哥特式教堂，
单一的中殿和小耳堂，多边形的后殿和钟楼。
fig. 2.35 Handbook, plates 42-43 and 45: Gothic style church's project 6,
single nave with small transept, polygonal apse, and bell tower.
(© Université Laval, Bibliothèque universitaire)

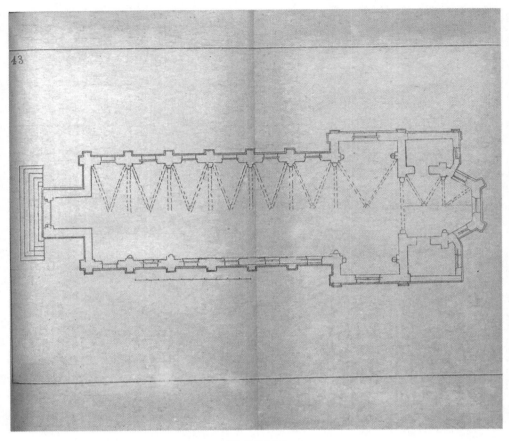

图 2.35(b)
fig. 2.35(b)

第二部分　手册的内容和翻译
Part Two: The Handbook Translated

图 2.35(c)
fig. 2.35(d)

舶来与本土：1926年法国传教士所撰中国北方教堂营造手册的翻译和研究
Building Churches in Northern China. A 1926 Handbook in Context

图2.36 手册图纸44和47~49：项目方案7，哥特式教堂，
中殿和耳堂，侧道，多边形的后殿，圣器室和钟楼。
fig. 2.36 Handbook, plates 44 and 47-49: Gothic style church's project 7,
nave with transept, small aisles, polygonal apse, sacristies, and bell tower.
(© Université Laval, Bibliothèque universitaire)

第二部分 手册的内容和翻译
Part Two: The Handbook Translated

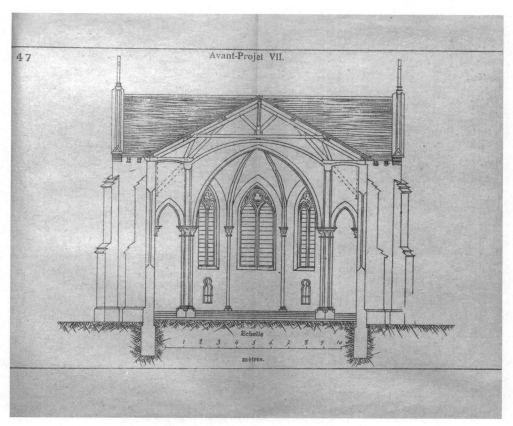

图 2.36(b)
fig. 2.36(b)

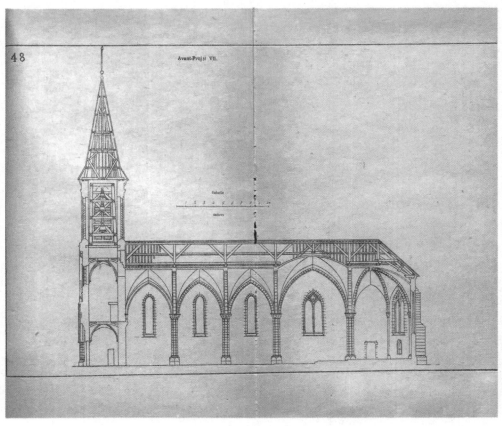

图 2.36(c)
fig. 2.36(c)

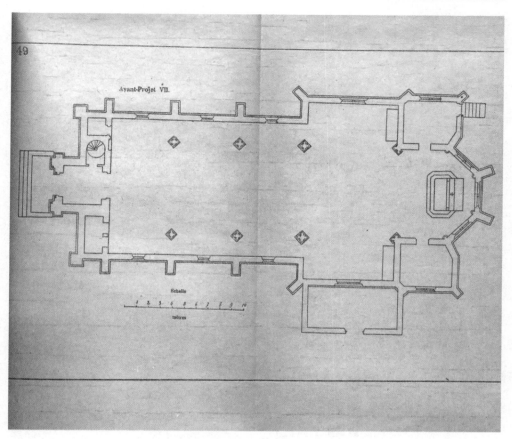

图 2.36(d)
fig. 2.36(d)

舶来与本土：1926 年法国传教士所撰中国北方教堂营造手册的翻译和研究
Building Churches in Northern China. A 1926 Handbook in Context

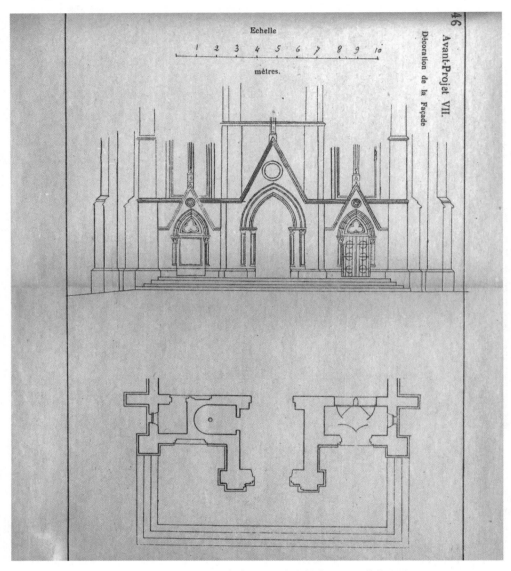

图2.37 手册图纸46：项目方案7，哥特式教堂，立面装饰和边门。
fig. 2.37 Handbook, plate 46: Gothic style church's project 7, decoration of the façade and side doorway.
(© Université Laval, Bibliothèque universitaire)

第二部分　手册的内容和翻译
Part Two: The Handbook Translated

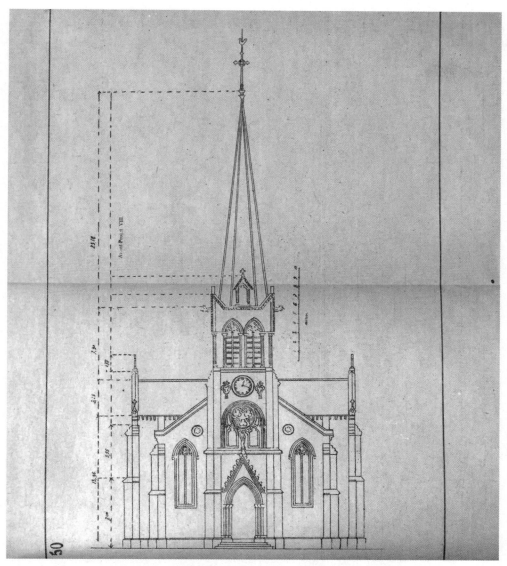

图2.38　手册图纸50~54：项目方案8，哥特式教堂，
有着大后殿的中殿，耳堂，多边形的后殿，边上的礼拜堂，圣器室和钟楼。
　　fig. 2.38 Handbook, plate 50-54: Gothic style church's project 8,
nave with large aisles, transept, polygonal apse, side chapels, sacristies, and bell tower.
(© Université Laval, Bibliothèque universitaire)

图 2.38(b)
fig. 2.38(b)

第二部分 手册的内容和翻译
Part Two: The Handbook Translated

图 2.38(c)
fig. 2.38(c)

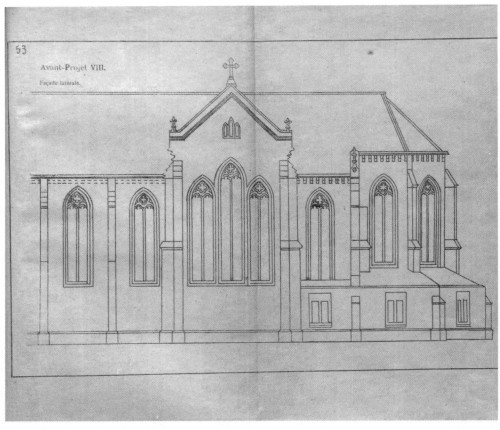

图 2.38(d)
fig. 2.38(d)

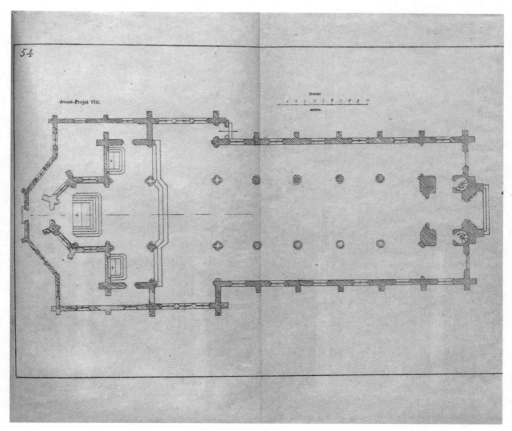

图 2.38(e)
fig. 2.38(e)

舶来与本土：1926年法国传教士所撰中国北方教堂营造手册的翻译和研究
Building Churches in Northern China. A 1926 Handbook in Context

图2.39 手册扉页和最后一页的小插图。
fig. 2.39 Two vignettes: one on the title page and the other on the last page.

第三部分
教堂建造案例分析——大名天主堂
Part Three:
The Handbook as Built

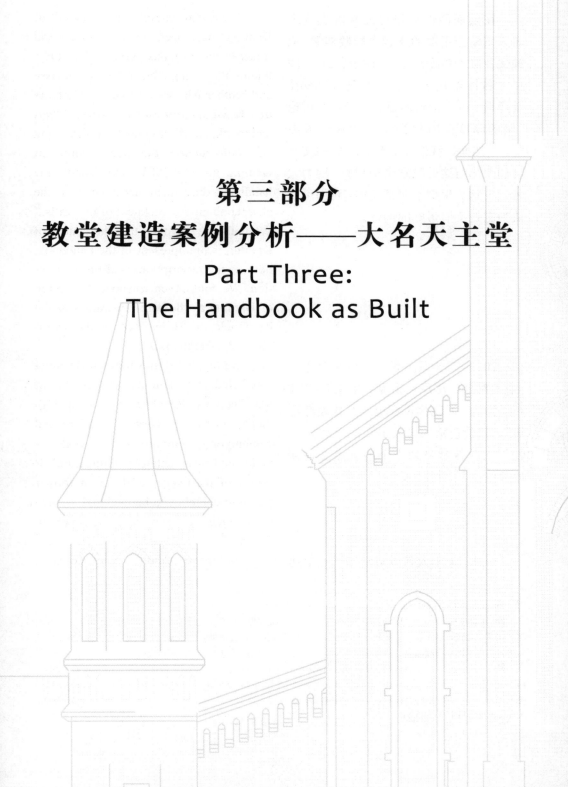

舶来与本土：1926 年法国传教士所撰中国北方教堂营造手册的翻译和研究
Building Churches in Northern China. A 1926 Handbook in Context

在前面两部分里已经数次提到大名天主堂。手册的作者雍居敬神父、梁神父及修士雷振声于 1918—1921 年间在大名传教并参与了这处教堂的设计及建造。令人惊讶的是，手册只在讨论"墙的厚度"时提到了一次大名天主堂。❶然而，我们也看到，手册的文字与插图都与这处教堂密切相关。因为大名天主堂是根据手册营建的，我们可以认为它是手册最好的例证。

2013 年 5 月大名天主堂被公布为"第七批全国重点文物保护单位"，其公布类型为"近现代重要史迹及代表性建筑"。目前包括大名天主堂在内共有 20 处天主教教堂被列为全国重点文物保护单位。❷

因此，关于大名天主堂这一杰出建

The great church or cathedral of Daming has been mentioned several times in the previous parts of this study. Father Paul Jung, father Henri Lamasse and brother Alphonse Litzler, the authors of *The Missionary-Builder: Advice-Plans* handbook, were involved in designing and constructing the new church of Daming in 1918-1921. Surprisingly, the handbook only mentions it once, in the chapter about the thickness of the walls.❶ We will see, however, that the church of Daming is omnipresent in the handbook, both in the descriptions and the illustration. We could even consider the church of Daming as the best illustration of the handbook, as if it had been built according to the handbook.

In May 2013, the church of Daming was listed as a 'national heritage site' of the People's Republic of China (7th batch), in the category of 'modern and contemporary important historical remains and representative architecture'. It is one of the twenty Catholic or former Catholic churches to be declared as such a heritage, of the highest level in China.❸

For these reasons, more research

---

❶ Le Missionnaire Constructeur [M]. Xianxian, 1926, 35.

❷ 目前共有 40 处外国教堂和有教堂在内的建筑群被列为全国重点文物保护单位，其中 20 处为天主教教堂，还有 8 处建筑群里有天主教教堂。其中最早被列为全国文物保护单位的三处天主教教堂是：第三批（1988 年）入选的天津望海楼教堂，第四批（1996 年）入选的北京南堂和广州圣心大教堂。

❸ Presently 40 Christian churches or Christian building complexes including churches are listed as national heritage sites, 20 of which being Catholic churches and 8 Catholic building complexes. The earliest national heritagized churches are Wanghailou church in Tianjin in 1988 (3rd batch); Beijing South church, Sacred Heart cathedral in Guangzhou in 1996 (4th batch).

筑的研究得以进一步展开。本部分将详述这些研究的成果，这些成果主要包括：近两年作者在法国巴黎的耶稣会档案中收集到的相关历史资料，❶ 以及2014 年北京大学考古文博学院对大名天主堂进行勘察测绘的成果。

was carried out on this remarkable building, the results of which are published in this chapter. Our historical research is based on documents collected in the archives of the Jesuits in Paris.❶ Our building-archaeological analysis is based on a complete measurement of the church realised by the School of Archaeology and Museology of Peking University during the fieldwork of 2014.

天主教传教区大名
The Catholic Mission of Daming

1858 年 6 月签订的《天津条约》中的第 13 条规定，中国境内所有基督徒信教自由必须得以保障，外籍传教士也可持官方护照在境内游历。在这之前的几十年中，传教活动仅限于通商口岸和殖民地澳门与香港。非法进入内地的传教士遭受迫害。自 19 世纪 60 年代以后，情况完全改观，传教士开始在全中国境内传布基督宗教。教廷组织了天主教传教活动，并发展了一套地域性的策略，将称为"宗座代牧区"的明确界定区域交给不同的传教修会负责。清末时，有 3 个法国天主教传教修会在中国最具影响力：遣使会、巴黎外方传教会

Article 13 of the *Treaty of Tientsin*, signed in Tianjin in June 1858, secured freedom of religious practice for all Christians in China and the right for foreign missionaries to travel everywhere in China with official passports. In the previous decades, missionary activity was limited to the treaty ports and to the colonies of Macao and Hong Kong. The missionaries who entered the mainland where illegal and were prosecuted. From the 1860s, the situation completely changed and missionaries began spreading the Christian religion throughout the whole of China. The Holy See organised

---

❶ Vanves, ASJ France, GMC series.

和耶稣会。❶

the Catholic mission and developed a territorial strategy that entrusted well defined ecclesiastical provinces called 'vicariates apostolic' to different missionary societies. During the late Qing dynasty, three French Catholic missionary societies were the most influential in China: the Congregation of the Mission (Lazarists or Vincentians), the French Foreign Missions of Paris (Missions Étrangères de Paris), and the Company of Jesus (Jesuits).❶

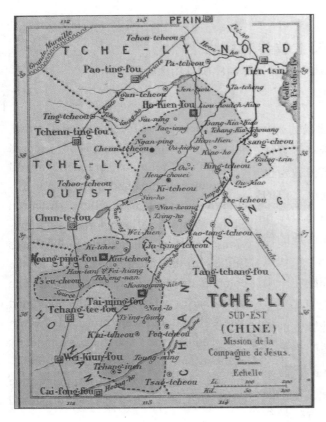

图3.1 直隶东南部宗教代牧区地图，1929年前。
fig. 3.1 Map of the vicariate apostolic of Southeastern Zhili, before 1929.
(© ASJ France, GMC)

---

❶ R. Gary Tiedemann. Reference Guide to Christian Missionary Societies in China from the Sixteenth to the Twentieth Century [M]. Armonk-London, 2009, 13-15, 19, 40-42.

第三部分　教堂建造案例分析——大名天主堂
Part Three: The Handbook as Built

这些天主教区域其中一个是直隶东南部宗教代牧区，其所辖范围包含现河北省的东南部分。17世纪与18世纪之际，耶稣会传教士已造访此地区，包括大名，并教化了几位中国人。❶ 19世纪时，教廷将直隶东南部的传教工作交给法籍耶稣会香槟省，并把献县设立为宗座代牧区的中心所在地❷（图3.1）。从此地开始，耶稣会士们向外地拓展，到达河间、深州、冀州、广平和大名。1900年义和团运动发生时，许多教徒和四名法籍耶稣会士在此代牧区被杀。❸

法籍耶稣会士于1867年从献县前

One of these ecclesiastical provinces, the vicariate apostolic of Southeastern Zhili (Tché-Li sud-est) covered the south-eastern part of the present province of Hebei. In the seventeenth and eighteenth centuries, Jesuits missionaries had travelled in this area, including in Taming, and had Christianised some Chinese.❹ In the nineteenth century, the Holy See entrusted Southeastern Zhili to Jesuits from the French Jesuit province of Champagne and established the seat of the vicariate apostolic in the town of Xianxian (Sien-Hsien)❷ (fig. 3.1). From there, missionaries went further to Hejian (Ho-kien), Shenzhou (Chen-Tcheou), Jizhou (Ki-Tcheou), Guangping (Koang-p'ing) and Daming (Tai-ming). During the Boxer Upraising in 1900, many Chinese Christians and four French Jesuits were massacred in the vicariate apostolic of Southeastern Zhili.❺

French Jesuits from Xianxian

---

❶ Paul Bornet. Notes sur l'évangélisation du Tcheli et de la Tartarie aux XVIIe et XVIIIe siècles. Notes sur l'origine de quelques chrétientés du Tchéli et de la Tartarie aux XVIIe et XVIIIe siècles [J]. Bulletin catholique de Paris, 1937 (offprint with personal notes of the author: Vanves, ASJ France, GMC 5).

❷ On the beginnings of the misson: François Xavier Leboucq. Monseigneur Edouard Dubar de la Compagnie de Jésus: évêque de Canathe et la mission catholique du Tche-Ly-Sud-Est, en Chine [M]. Paris, 1880; Henri-Joseph Leroy. En Chine au Tché-ly S.-E. Une mission d'après les missionnaires [M]. Lille-Tournai, 1900, 187-237.

❸ 在河北省的武邑县和朱家河地带，(Vanves, ASJ France, GMC, 27-61). Also: Henri Bernard. La Compagnie de Jésus en Chine. L'ancien vicariat apostolique du Tchéli sud-est, ses filiales, ses annexes [M]. Tianjin, 1940.

❹ Paul Bornet. Notes sur l'évangélisation du Tcheli et de la Tartarie aux XVIIe et XVIIIe siècles. Notes sur l'origine de quelques chrétientés du Tchéli et de la Tartarie aux XVIIe et XVIIIe siècles [J]. Bulletin catholique de Paris, 1937 (offprint with personal notes of the author: Vanves, ASJ France, GMC 5).

❺ In the districts of Wuyi and Zhujiahe (Hebei Province), (Vanves, ASJ France, GMC, 27-61). Also: Henri Bernard. La Compagnie de Jésus en Chine. L'ancien vicariat apostolique du Tchéli sud-est, ses filiales, ses annexes [M]. Tianjin, 1940.

往大名，并在该地传播福音。❶ 当时大名已经不是地方上的政治中心，但仍然是地方驻军、教育和商业中心。历史上，大名是一个重要的县城，为军队驻防地、教育和贸易中心。虽然17与18世纪时耶稣会士曾在大名传教，但是后来等传教士再回到当地时已无教徒，除了一个从北方移居至此、成员有6人的家庭以外。最开始在城市展开传教活动并不容易，因为当地人对外国人常抱有敌视的眼光，但传教士在小城镇所获得的成果倒是较为丰硕。大名的传教活动持续有所进展，在1873年时，有71名信徒，1880年约400名信徒，1900年时已达3843名信徒。❷ 义和团在大名一地活动，但之后大名的复苏也颇为显著。因为传教士所创办的学校声誉好，教徒人数发展快速，在城市里也是如此。逐渐地，大名成为信徒的区域信仰中心，后来成为了教区主教的行政中心。

自20世纪20年代中期开始，教廷准备在中国成立新的教省。这是一段很长的过程，具体事务由传信部协调，传信部是教廷传教活动行政中枢，管辖包括宗座代表刚恒毅和来自各国不同的

reached Daming in 1867 and began to evangelise the area. ❸ At that time, Daming had lost it status of regional capital, but still was an important prefecture, garrison place, educational and trade centre. Despite the ancient mission of the Jesuits in Daming in the seventeenth and eighteenth centuries, there were no more Christians in the region, except one Chinese Catholic family of six people that had migrated from the north. The beginnings were difficult in the city because people were hostile to foreigners, while in the villages missionaries achieved some results. The mission of Daming developed gradually from 71 Christians in 1873, to about 400 in 1880, and 3843 in 1900, etc.❷ The Boxers ravaged the mission of Daming, but its revival was remarkable. The Christian population increased fast, including in the city, thanks to the good reputation of the school. So, Daming became a regional centre for the Catholics and later the seat of a diocese.

From the mid-1920s, the Holy See developed projects for creating new church provinces in China. This was a long process that was coordinated from the Propaganda Fide — the central missionary administration of the Holy

---

❶ Paul Jung . Une grande ville chinoise qui s'ouvre à la foi [J]. Les missions catholiques, 1919, 198-200; Gerbes chinoises. Les Jésuites de la Mission de Sien-Hsien depuis 1856 [M]. Lille, 1934, 61-70.

❷ La Préfecture Apostolique de Ta-Ming Fu [J]. Trait d'Union. À nos Chrétiens, 1936, 20 :1.

❸ On the Daming mission: Paul Jung . Une grande ville chinoise qui s'ouvre à la foi [J]. Les missions catholiques, 1919, 198-200; Gerbes chinoises. Les Jésuites de la Mission de Sien-Hsien depuis 1856 [M]. Lille, 1934, 61-70.

第三部分　教堂建造案例分析——大名天主堂
Part Three: The Handbook as Built

传教修会。献县代牧区分为四部分，且成立了三个新宗座监牧区：永年县于 1929 年交由中国神父负责，1935 年匈牙利耶稣会士主持大名教务，1937 年奥地利耶稣会士则管理景县教务（图 3.2）。自此，法籍耶稣会士将他们在中国北方的势力集中于献县和天津。❶

1935 年 3 月 11 日教宗庇护十一世将南乐和濮阳从大名区域分割出来，并从献县分出清丰，另成立大名监牧区，仍由匈牙利耶稣会士负责教务。❸ 自此，大名一跃成为一个区域广达 8640 平方千米、人口 200 万、教徒 37903 人的新成立的教会区域中心。❹ 1947 年 7 月 10 日大名监牧区升格为代牧区，匈牙利耶稣会士隆其化被任命为宗座署理。❺ 1949

See — and involved the delegate apostolic Celso Costantini as well as several missionary societies from different countries. The vicariate apostolic of Xianxian was divided in four parts and three new apostolic prefectures were created: Yongnian (Hebei province) for Chinese secular priests in 1929, Daming for Hungarian Jesuits in 1935, and Jingxian (Hebei province) for Austrian Jesuits in 1937 (fig. 3.2). From than, the French Jesuits concentrated their Northern-China forces in Xianxian and Tianjin. ❷

On 11 March 1935, Pope Pius XI separated the territories of Daming, Nanle (Henan province), Puyang (Henan province), and Qingfeng (Henan province) from the vicariate apostolic of Xianxian, and erected the prefecture apostolic of Daming, that was entrusted to the Hungarian Jesuits.❸ So Daming became the centre of a new ecclesiastical province of 8640 km² and 2 million in-

---

❶ 类似的事情也发生在江南耶稣会传教会（江苏省和安徽省）：法国耶稣会继续留在南京和上海的宗座代牧区（1933 年），西班牙耶稣会于 1922 年在芜湖和安庆建立了新的教省，在蚌埠是意大利耶稣会（1922 年），在海门是中国牧师会（1926 年），在徐州是说法语的加拿大耶稣会（1931 年），在连云港和苏州是中国牧师会（1949 年）。

❷ A similar process happened after the division of the Jesuit mission of Kiangnan (Jiangsu and Anhui provinces): the French Jesuits remained in the vicariates apostolic of Nanjing and Shanghai (1933), and new church provinces were created in Wuhu and Anqing for Spanish Jesuits in 1922, in Bengbu for Italian Jesuits in 1922, in Haimen for Chinese secular priests in 1926, in Xuzhou for French-Canadian Jesuits in 1931,and in Lianyungang and Suzhou for Chinese seculars in 1949.

❸ Péter Vámos. Hungarian Missionaries in China [A] // Stephen Uhalley, Wu Xiaoxin. China and Christianity: Burdened Past, Hopeful Future [C]. London-New York: Routledge, 2001, 218-231 (Daming mentioned on p. 222-223).

❹ Miklós (Nicolaus) Szarvas SJ (1890-1965), prefect of Daming, appointed 31 January 1936, resigned 8 July 1947. http://www.catholic-hierarchy.org/bishop/bszar.html (accessed on 15 November 2015) .

❺ Gáspár Lischerong SJ (1889-1972), appointed apostolic administrator of Daming on 10 July 1947, moved to Taiwan in 1949 where he died in 1972. http://www.catholic-hierarchy.org/bishop/blisc.html (accessed on 15 November 2015).

年大名的外籍传教士遭驱逐出境，由中国教徒接手管理教堂。

1949 年以前，其他三组外籍传教士也在大名活动。❶

南直隶福音部的美国基督新教非教派传教士于 1902—1903 年在大名开设一个驻留传教站。1921 年，他们的传教站被美国基督新教传教士从美国清洁会接收。如今，在大名老城主街的清洁会旧址上还有一个基督教会依然存在。❷

habitants including about 37903 Catholics. The Hungarian Jesuit Miklós Szarvas became the first prefect apostolic of Daming.$^{See\ P289}$❶ On 10 July 1947, the prefecture of Daming was elevated to a diocese and the Hungarian Jesuit Gáspár Lischerong was appointed apostolic administrator.$^{See\ P289}$❷ The foreign missionaries were expelled from Daming in 1949 and the church was taken over by Chinese Catholics.

Before 1949, three other groups of foreign missionaries were active in Daming.❶

American Protestant non-denominational missionaries of the *South Chihli Mission* (SCM) opened a residential station in Daming in 1902-1903. In 1921, their station was taken over by the American Protestant missionaries from the *Mennonite General Conference of Mission* (MGC). A Christian church still exists on the former site of the Mennonites along one of the main streets of the old town. ❸

---

❶ R. Gary Tiedemann. Reference Guide to Christian Missionary Societies in China from the Sixteenth to the Twentieth Century [M]. Armonk-London: ME Sharpe, 2009, 75, 149-150, 181-182, 215.

❷ 需要感谢美国贝塞尔学院（Bethel College）门诺图书馆和档案馆的档案专家 John D. Thiesen 先生。

❸ With thanks to John D. Thiesen, Archivist of the Mennonite Library and Archives, Bethel College, North Newton, USA.

第三部分　教堂建造案例分析——大名天主堂
Part Three: The Handbook as Built

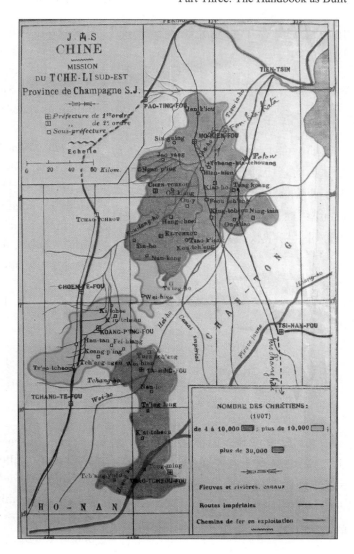

图3.2　早先直隶东南部宗教代牧区地图，1929—1935年间分四个教省，大名为南部的那个教省。
fig. 3.2 Map of the former vicariate apostolic of Southeastern Zhili, divided in four ecclesiastic provinces in 1929-1935. Daming is the southern one. (© ASJ France, GMC)

宣圣会的美国新教传教士于1919年在大名落脚，于1925年建造了一座教堂、圣经学校和医院，并在宣圣会医院发展医疗工作。这一规模大、功能多的建筑群位于大名城北的城墙外，如今这一地区已几乎全部被城镇化了，只有

The American Protestant missionaries of the *Pentecostal Church of the Nazarene* (CN) settled in Daming in 1919, built a church, a Bible college and a hospital in 1925, and developed medical activities in the Breese Memorial

几幢建物尚存于此但处于荒废状态❶（图 3.53）。

耶稣会士于 1928 年请考洛乔圣母修女会前来协助。这群匈牙利修女们创办了两座修道院，一座在大名，另一座在河南省濮阳，并主持女子学校、女性教理讲授师学校、诊所和孤儿院。

Hospital. This important complex of buildings was located outside the city wall, at the northern side, in an area that is presently totally urbanized. Several buildings remain in a ruinous state❷ (fig. 3.53).

The School Sisters of Our Lady of Kalocsa (CSSK) were called by the Jesuits in 1928. These Hungarian Catholic sisters founded two convents in China, in Daming and Puyang and operated girls' schools, female catechists' schools, dispensaries and orphanages.

大名法文学校
The French College of Daming

大名法文学校，通常也被叫做"法文"，是耶稣会士在全世界（包括西方国家，也包括传教地国家）创办的众多高中中的一座（图 3.3）。耶稣会所办的高水平教育体系包含小学、中学及大学。❸ 法籍耶稣会士在中国创办了两所大学：上海的震旦大学（于 1903—1952 年运作）和天津工商学院，后来改名为

The French College of Daming, commonly called the *Fawenn* (法文），was one of the many high schools that the Jesuits built throughout the world, both in western countries and in mission countries (fig. 3.3). The Jesuit high level educational system included primary and secondary schools as well as universities.❸ In China, French Jesuits founded

---

❶ Hoyd Cunningham. Holiness Abroad: Nazarene Missions in Asia (Pietist and Wesleyan Studies 16) [M]. Lanham: Scarecrow Press, 2003. 需要感谢宣圣会(Church of Nazarene)的档案专家 Stan Ingersol 先生，以及 Floyd Cunningham 先生，他于 1989 年、1999 年和 2003 年去看过这些建筑。

❷ With thanks to Stan Ingersol, Archivist of the Church of Nazarene, and Floyd Cunningham who visited these buildings in 1989, 1999 and 2003.

❸ Alexandre Brou. Cent ans de missions, 1815-1934: les Jésuites missionnaires aux XIXe et XXe siècles [M]. Paris, 1934; Marianne Monestier. Les Jésuites et l'Extrême-Orient [M]. Paris, 1956.

第三部分　教堂建造案例分析——大名天主堂
Part Three: The Handbook as Built

"津沽大学"（1923— 1951 年）。❶ 在天主教教育体制中，学院是当时最好的高中，为学生上大学做准备。在大名有这样的学校对该城市和地区而言是极佳的机会。最优秀的学生被送到学院就读，之后可以到天津、上海或其他地方的大学继续深造。

many schools and several colleges as well as two universities: Aurora University in Shanghai, active from 1903 to 1952, and the College of Industry and Commerce, later Tianjin Commercial University, active from 1923 to 1951.❶ In the Catholic educational system, their colleges were considered the best high schools for preparing to go to university. Having such a school in Daming was a great opportunity for the city and its region. The best students were sent to the college and could study further in Tianjin, Shanghai or in other universities.

图3.3　大名，从东门看法国耶稣会建造的宗教建筑群，1913年7月。
fig. 3.3 Daming, building complex of the French Jesuits seen from the East Gate, picture dated July 1913. (© ASJ France, GMC)

大名的第一所小学初创之时较为简陋，所吸引的大多是来自乡村的男学童。这个县城是个知识分子集聚的中

The first elementary school of Daming was small and attracted primarily boys from the countryside. The city

---

❶ George S oulié De Morant. L'épopée des Jésuites en Chine, 1534-1928. [M]. Paris, 1928.

心，有许多初级学校、多座中等学校和一所著名的公立学校（Tchoung hiao tang），在物理、化学、英文、艺术与手工艺方面的教育在当地很有名。

耶稣会创办的学院因教育品质好而声名鹊起，社会精英分子，甚至官员，都将其子弟送往这所法文学校就读。❶ 传教士的目的是要培养出一名中国精英，他必须是天主教徒且文化上是法国或西方化的（图 3.4）。在耶稣会的地盘上还有几所学校：一所专收天主教男学童的初级小学，一所非天主教男学童的初级小学，以及男学童求学的法文学校。由于皈依天主教是通往高质量教育的开门钥匙，所有的课程均包含教理讲授在内。1903 年法文学校成为寄宿学校，共有 20 名学生，1911 年则向非寄宿生开放，1917 年学生人数增至 100 名。同年，该校学生总人数为 240 名。

20 世纪 20 年代和 30 年代，共和体制尚处早期，排外情绪日益高涨，许多不同的社会团体开始时常批评耶稣会的教育体制。❷

was an intellectual centre and counted many lower primary schools, several upper primary schools and one famous state school, the *Tchoung hiao tang*, that was famous for its education in physics, chemistry, English, arts and crafts.

When the quality of the education of the Jesuit college became obvious to the urban elite, even mandarins sent their children to the French School.❶ The missionaries aimed to educate a Chinese elite that should be Catholic and culturally French-Westernized (fig. 3.4). There were several schools on the Jesuit compound: one elementary school for Catholic boys, one other elementary school for non-Catholic boys, and the French college for boys. Because conversion to Catholicism was the key to access quality education, all the programs included classes of catechism. The French college began as a boarding school with twenty students in 1903, opened to non-boarders in 1911, and reached hundred students in 1917. In that year, the total number of students in the school was 240.

In the context of the young Republic of China and growing xenophobia, different social groups increasingly criticized the Jesuit educational system in the 1920s and 1930s.❷

---

❶ Paul Jung. Une grande ville chinoise qui s'ouvre à la foi [J]. Les missions catholiques, 1919, 198-200, 211-212, 223-225, 233-235; Paul Jung. Fleurs du céleste empire. Touchante histoire de Stanislas Wang [J]. Les missions catholiques, 1918, 317-319.

❷ R. Gary Tiedemann. Educational Apostolate [A] // R. Gary Tiedemann. Handbook of Christianity in China. Volume Two: 1800-Present [C]. Leiden-Boston: Brill, 2010, 678-691.

第三部分　教堂建造案例分析——大名天主堂
Part Three: The Handbook as Built

图3.4　大名，耶稣会会士在教中国小孩弹脚踏式风琴。
fig. 3.4 Daming, Jesuit education and cultural transfer: Chinese student playing on the chapel's harmonium. (© ASJ France, GMC)

第一，对于那些不接受其他宗教信仰的学生的学校和大学，民国政府警告不予补助；第二，天主教自20世纪20年代中期开始推广本地化的新运动，并且欲切断法国殖民地和天主教之间的互动关系；第三，中国学生拒绝皈依，并要求学习更多有关中国文化与英文等方面的知识；第四，来自基督新教所创办的大学和高等学府的竞争愈加激烈，这些学校大部分是用英文授课，经费由美国赞助；第五，耶稣会专门提倡针对男学生的高等教育，一般而言，提供给女子的天主教教育并非高等教

First, the Chinese government threatened not to subsidize schools and universities that did not accept students of all religions. Second, the Catholic hierarchy itself promoted from the mid-1920s the new movement of indigenization and wanted to interrupt the interactions between French colonialism and the Catholic religion. Third, Chinese students refused to convert and wanted to learn more Chinese culture and English language too. Fourth, there was an increasing competition from Protestant universities and higher education, most of which were English-speaking and

育。大名学院也没能逃脱当时的紧张局势。❶ 为了促进该校毕业生之间的联系，耶稣会自 1922 年到 1936 年间出版了《大名法文学校同学季刊》❷（图 3.5）。

American-sponsored. And finally, Jesuits were particularly promoting higher education for boys, and in general, Catholic education for girls was not of a high level. The college of Daming did not escape these tensions.❶ Aiming to promote a network of alumni, the Jesuits published from 1922 to 1936 the journal *Le Trait-d'Union. Bulletin des anciens élèves du Collège Français de Taming*❷ (fig. 3.5).

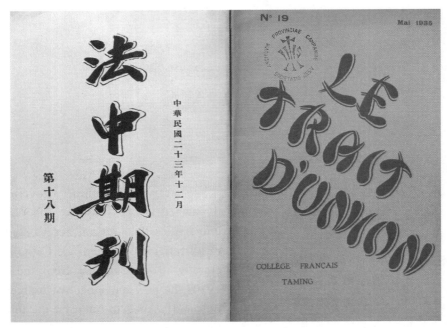

图3.5 《法中期刊》，1935年出版。
fig. 3.5 *Le Trait d'Union*, the journal of the French College's alumni, issues of 1935. (© ASJ France, GMC)

---

❶ Vanves, ASJ France, GMC 22/1, 1-2 (about nationalist students). R. Gary Tiedemann. Educational Apostolate [A] // R. Gary Tiedemann. Handbook of Christianity in China. Volume Two: 1800-Present [C]. Leiden-Boston: Brill, 2010, 678-691.

❷ Le Trait-d'Union. Bulletin des anciens élèves du Collège Français de Taming / 法中期刊。大名法文学校同学季刊, 1-20, July 1922 to January 1936. Vanves, ASJ France, GMC 87, 4.

第三部分　教堂建造案例分析——大名天主堂
Part Three: The Handbook as Built

耶稣会在城墙内东城门附近建造了一处重要建筑群。除了一幅画以外，几乎无从知晓 1900 年被义和团破坏之前的最早的建筑群的情况：一座正立面朝南的带哥特式窗户的进深五间的小教堂，供学校及神父们使用的两处院落，其他围合成小院落的附属建筑以及一处花园（图 3.6）。

The Jesuits developed an important building complex within the city walls near the East Gate. Except for one drawing, little is known about the earliest settlement and the buildings that the Boxers ravaged in 1900: a chapel of five bays with Gothic style windows and its main façade turned to the south, two courtyards for the school and the fathers, ancillary buildings organized around smaller courtyards, and a garden (fig. 3.6).

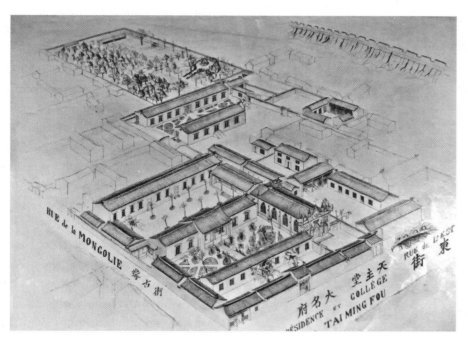

图3.6　大名，1900年前法国耶稣会在大名建造的宗教建筑群。
fig. 3.6 Daming, building complex of the French Jesuits before 1900.
(© ASJ France, GMC)

1900 年之后，建筑群沿着东大街两侧逐渐被重建。这得益于教会的快速扩张和连续地获取土地。建筑群空间分

After 1900, the whole complex was gradually rebuilt at both sides of the eastern main street (*dong dajie*). This

~ 297 ~

布的基本原则是性别分区：耶稣会神父们的住宅、男修院及男校分布在街北（图 3.7），女校、女修院、中国少女及年长妇女的住所分布在街南。❶ 男女各自有自己的礼拜堂。1917 年，教会在街南买下一处已经被废弃的曾为兵营所在的大地块，1918—1921 年在此营建了这处大教堂。除了教堂以外，还有其他几处值得一提的建筑也保留至今（图 3.8）。整个建筑群是由一代人重建的整体。这一代人就是教堂建造时担任法国学院主任的雍居敬神父及在 1902 年来到大名监督建筑工程的雷振声修士。❷

目前，这处校舍建筑群为邯郸教区圣心修院使用。修院沿主街兴建了学校的新翼，但内院仍保留了耶稣会时期留下来的三处主要建筑（图 3.9）。

resulted from fast growth and the successive acquisition of plots. The basic spatial principle was the separation of genders: the residence of the Jesuit fathers, the college and the boys' schools were on the northern side of the street (fig. 3.7). The girls' school, the convent of the Chinese Virgins and the house for elderly women were on the southern side.❶ Each group had its own chapel. In 1917, a large plot of an abandoned *yamen* was bought on the southern side of the street, and there the great church was built from 1918 to 1921. Besides the church, several other buildings are still conserved and deserve a brief description (fig. 3.8). The whole complex was thus a coherent ensemble rebuilt in the span of one generation. That is the generation of father Paul Jung, who was the director of the French College when the church was constructed, and of brother Alphonse Litzler, who arrived in Daming in 1902 to supervise the construction works. ❸

Today, this cluster of school buildings is being used by the Handan Parish Sacred Heart Christian College. A new school wing was recently erected along the main street, but four significant buildings from the time of the Jesuits are still standing around the inner courtyards (fig. 3.9).

---

❶ Paul Jung. Une grande ville chinoise qui s'ouvre à la foi [J]. Les missions catholiques, 1919, 235, 249.
❷ 相关内容见第一部
❸ See Part 1, p. 35-38.

第三部分　教堂建造案例分析——大名天主堂
Part Three: The Handbook as Built

图3.7　大名，从东门看大名东街两边的天主教建筑，1920年和2014年。
fig. 3.7 Daming, Catholic compounds at both sides of East Street, view from the East Gate, 1920 and 2014. (© ASJ France, GMC, and © THOC 2014)

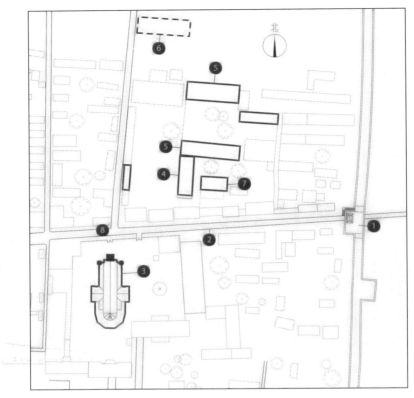

图3.8 大名，东门附近的耶稣会建筑。
1.东城门；2.东街；3.教堂；4.学校的礼拜堂；5.以前神父的住宅；
6.法文学校所在地；7.报告厅；8.牌楼所在地.
fig. 3.8 Daming, neighborhood of the East Gate and former Jesuits' building
(© School of Archaeology and Museology, Peking University, 北京大学考古文博学院).
1.East Gate; 2.East Street; 3.church; 4.schools' chapel; 5.former residences of the father;
6.location of the French College; 7.lecture hall; 8.location of the pailou.

在第一进院的东侧为最主要的一处建筑——原男修院的礼拜堂（图3.10）。这座红砖礼拜堂曾经历多次改建并很有可能还保留了部分1900年以前的遗存。1900年以后，礼拜堂扩建了两间。目前，这处建筑被用作修习室及其他多种功能。南侧的外立面及主入口为新修，但侧立面的哥特式双尖角窗、扶壁及精细的带状砖饰得以保留

The main one, along the eastern side of the first courtyard, is the former chapel of the college (fig. 3.10). This red brick chapel has been transformed several times and probably includes some parts from before 1900. After 1900, the chapel was enlarged with two more bays to the south. Presently the building is used as an all-purpose and study room. Outside, the southern façade and the main entrance are new, but the two long

第三部分 教堂建造案例分析——大名天主堂
Part Three: The Handbook as Built

（图 3.11）。室内两排支撑原有木结构屋顶的木柱被去除，并更换了原有的天花板，使中厅的空间更宽敞。原圣所位于高大的尖拱之下，包含一处主圣坛及两侧的走廊。小教堂北侧部分为一处圣器收藏室及通往侧廊的楼梯。

sides conserve their Gothic double lancet windows, buttresses and a refined brick frieze (fig. 3.11). Inside, the two rows of wooden posts that supported the original wooden roof structure have been removed and the ceiling replaced in order to have a spacious single nave space. The original sanctuary consists of a main altar under a high pointed arch, flanked by two galleries. A sacristy, and the stairs to the galleries, complete the chapel to the north.

图3.9 大名，从教堂钟楼看大名的学校建筑群，1920年和2014年。
fig. 3.9 Daming, school complex seen from the church's tower, 1920 and 2014. (© ASJ France, GMC, and THOC 2014)

~ 301 ~

图3.10 大名，扩建前的学校礼拜堂东侧，1910年前后。
fig. 3.10 Daming, eastern side of the school's chapel before enlargement, around 1910. (© ASJ France, GMC)

图3.11 大名，原学校礼拜堂的西侧。
fig. 3.11 Daming, western side of the former school's chapel. (© THOC 2014)

这座礼拜堂坐北朝南，面向主街，建筑的最北端几乎紧挨着两处原法国传教士住所中的一处（图3.8）。这两翼平行相对的建筑面阔十一开间，两层，东西向。它们的平面非常相似，是西方19世纪理性建筑适应恶劣气候的

The former chapel is north-south oriented, with its main gable turned to the south, that is to say to the main street. Its northern extremity nearly touches one of the two former residence buildings of the French missionaries (fig. 3.8). These two parallel wings of two floors and

第三部分　教堂建造案例分析——大名天主堂
Part Three: The Handbook as Built

很好的例子。北立面很拙朴，窗朝向宽敞的内廊，在冬天可以抵御严寒，同时在夏天保证空气的流通（图 3.12）。木楼梯很宽敞舒适。所有的房间都朝南并有开敞的木结构阳台，在夏天提供荫蔽的同时传教士可以享受内院的宁静（图 1.38）。两处住所均为砖结构建筑，除了装饰为模制成型砖外，其余均刷成灰白色。从二层的阳台可以看到建筑群漂亮的全景，包括围合成小院落的附属建筑、城墙的一部分、东城门、有假山的园林、男修院的主要建筑及街对面的大教堂。

大名法国男修院的主体在更往北的位置。这处 1913 年始建的西式建筑已经被拆毁。与居住部分建筑类似，学校建筑亦朝南。正立面为九开间，两层，有规律地分布着扶壁及不同大小的圆拱（图 3.13）。一层为开敞的走廊，天气太热或下雨时男孩子们可以在室内玩耍。此外，后部还有房间及通向上层教室的楼梯。

eleven bays are east-west oriented and built according to a similar plan. They are good examples of western nineteenth-century rational architecture, well adapted to the harsh climate. The northern side is severe-looking, and the windows open to large inner corridors that protect from the cold in the winter and maintain coolness in the summer (fig. 3.12). The wooden staircases are large and comfortable. All the rooms are at the southern side and open onto a wooden veranda that provided shade in the summer and allowed the missionaries to enjoy the peace of the inner garden (fig. 1.38). Both residential wings are brick buildings, which today are painted in grey and white, except for the frieze made of molded bricks. From the veranda of the residence's upper floor, a beautiful panoramic view embraced the secondary buildings around small courtyards, a section of the city wall and the East Gate, a garden with rockery, the main building of the college, and the great church at the other side of the street.

The main wing of the French College of Daming was located further to the north. This western style building, erected in 1913, has been demolished. Like the residential wings, the school faced south. The main façade, with its nine bays and two levels, was ordered by buttresses and round arches of different sizes (fig. 3.13). The first floor was an open gallery where the boys could play when it was either too hot, or raining outside. There were rooms in the back as well as a staircase leading to the classrooms at the upper floor.

图3.12 大名，原法国耶稣会神父住宅北侧。
fig. 3.12 Daming, northern side of the former residence of the French Jesuits. (© THOC 2014)

图3.13 大名，法文学校的主立面，建于1913年。
fig. 3.13 Daming, main façade of the French College, built in 1913. (© ASJ France, GMC)

第三部分　教堂建造案例分析——大名天主堂
Part Three: The Handbook as Built

一位曾经（在这里上学）的学生在1922年写的一段简短的描述中提到，校舍建筑群的其他建筑被用作寄宿初等学校、宿舍、为约 190 位非寄宿的中国男学生提供免费教育的初等高等学校及一处供戏剧演出和其他公共活动的礼堂。❶ 另外可以肯定的是这里曾经还有一处图书馆、一些储物用房、一处食堂、一处厨房建筑及养殖动物的小型农场等（图 3.3）。

还有一处有趣的建筑为位于东南侧的礼拜堂。这有可能是学生用作戏剧演出及神父们用来上公开课的礼堂。神父雍居敬当时拥有一台幻灯投影仪，用来作图文并茂的演讲，曾吸引很多大名的居民前来听讲。他的演讲包括科学和历史的题材，以及圣经的故事。❷

A short description written by a former student in 1922 mentions that other buildings of the school complex served as primary boarding school, with a dormitory for the boarders, lower and upper primary school for the free education that was offered to about 190 Chinese boys who were not boarders, and a hall used for theatre and other common activities.❶ There were certainly storage rooms, a library, a refectory, a kitchen building, a small farm with animals, etc. (fig. 3.3).

Another interesting building is located south-east of the chapel. This could have been the hall that was used for theatre performances by the students as well as for the public lectures by the fathers. Father Jung had a lanterna magica projector and gave illustrated lectures that attracted many citizens of Daming. His lectures were both on scientific and historic topics, including biblical stories.❸

---

❶ Stanislas King P'ei-yuan. Taming d'aujourd'hui [J]. Le Trait d'Unio, 1922, 1:10-13.

❷ 雍神父称之为"公共教义"。自 1917 年始耶稣会就有无线电报，可以在大名接受中国各地乃至世界的新闻。在当时吸引了不少人。见文章： Paul Jung .Une grande ville chinoise qui s'ouvre à la foi [J]. Les missions catholiques, 1919, 233-234.

❸ Father Jung called it "public catechism". Furthermore, the Jesuits had a wireless telegraph connection since 1917 and so could spread news from China and the world to Daming. This attracted a lot of people. Paul Jung . Une grande ville chinoise qui s'ouvre à la foi [J]. Les missions catholiques, 1919, 233-234.

舶来与本土：1926年法国传教士所撰中国北方教堂营造手册的翻译和研究
Building Churches in Northern China. A 1926 Handbook in Context

大名天主堂
A Marian Cathedral in Daming

为了完善这处耶稣会建筑群，需要一座大教堂。通常，在传教士的建筑工程中，教堂为最后修建的建筑，这既有物质条件的原因也有宗教精神方面的原因。传教士们的繁荣社区需要拥有足够的群众和社会基础及足够多的基督教家庭通过祈祷和捐助来支持这些工程。修建一座教堂需要长时间的准备及关于设计、材料、资金和信众参与等的讨论。原本男性及女性在分开的礼拜堂内进行宗教活动，但后来这些礼拜堂变得太过拥挤。因此，在1917年建造一处新的教堂成为首要考虑的问题。

这次建筑工程背后的三位关键人物为神父雍居敬，梁神父及雷振声修士。神父雍居敬为当时男修院的主持，同时也是建筑工程的总监及财务总管。梁神父为建筑平面的设计者。雷振声修士为教堂的建造者并负责所有施工图纸。他们的生平在关于手册作者的

To be complete, the Jesuit building complex of Daming needed a large church. Usually, in a missionary construction program, the church is the last building to be erected. The reason is both material and spiritual. Missionaries needed a flourishing parish, with enough critical mass from the social works and enough Christian families supporting the project with prayers and donations. Building a church required long preparations and discussions about design, materials, financing, involvement of parishioners, etc. From the beginning, however, separate chapels for men and women were used as places of worship. Because these chapels were overcrowded, building a new church became a priority around 1917.

The three key people behind this project were father Paul Jung, the director of the College, coordinator and financer of the works, father Henri Lamasse, the designer of the plans, and brother Alphonse Litzler, the builder of the church and author of all the technical

~ 306 ~

第三部分　教堂建造案例分析——大名天主堂
Part Three: The Handbook as Built

章节中已经有所提及。❶

这里还要提到第四位传教士：主教刘钦明，1917—1937 年任直隶东南部宗教代牧区的法国耶稣会的宗教代牧。❸

跟大多数 19 世纪、20 世纪的法国传教士一样，刘钦明主教推崇对法国 1858 年著名的卢尔德(法国西南部城市）圣母玛利亚显圣的崇拜。❺ 而主教刘钦明对位于他的家乡里尔(法国北部城市）的中世纪圣地特雷耶圣母院有着特殊的崇拜。他将直隶东南部宗教代牧区置于特雷耶圣母院的保护之下。大名的这处新教堂在他的主持之下修建，同样是为了向特雷耶圣母院致敬。这在主入口上方的玛利亚像龛及教堂的钟的铭文上均有体现❻（图 3.14）。里尔的圣母大教堂及大名天主堂均为哥特式复兴样式。

drawings. Their biographies have been provided in the chapter about the authors of the handbook. ❷

A fourth missionary must be evoked here: bishop Henri Lécroart (*Liu Qinming*), the French Jesuit vicar apostolic of Southeastern Zhili from 1917 to 1937.❹

Like most nineteenth and twentieth-century French missionaries, bishop Lécroart promoted the devotion to Our Lady Maria of Lourdes, the famous French Marian apparitions of 1858. Bishop Lécroart, however, had a particular devotion to Notre-Dame de la Treille, a medieval shrine from his hometown Lille (North of France). He placed the vicariate apostolic of Southeastern Zhili under the protection of Notre-Dame de la Treille. The new church of Daming, built under his auspices, was also dedicated to Our Lady of Grace / Notre-Dame de la Treille as carved on the Marian niche above the main entrance and cast on the

---

❶ 见第一部分相关内容。

❷ See Part 1.

❸ 主教刘钦明于 1864 年生于里尔，1879 年成为耶稣会牧师，1901 年被派往中国的山东省胶州传教区，直到 1912 年。1913—1917 年期间负责河北省献县和河间两地传教区的所有材料购置事务，1917—1937 年期间为宗教代牧，1939 年在献县过世。http://www.catholic-hierarchy.org/bishop/blecr.html (2015 年 11 月 15 日浏览)。

❹ Born in Lille in 1864, ordained Jesuit priest in 1879, sent to China in 1901, missionary in Jiaozhou (Shandong province) until 1912, than responsible for all material affairs of the missions of Xianxian and Hejian (Hebei province) from 1913 to 1917, vicar apostolic from 1917 to 1937, died in Xianxian in 1939. http://www.catholic-hierarchy.org/bishop/blecr.html (accessed on 15 November 2015).

❺ P. M. Compagnon. Le culte de Notre-Dame de Lourdes dans la Société des Missions-Étrangères [M]. Paris, 1910, 111-156.

❻ 这些铭文被刻在壁龛周围:寵愛之母保障大名（上），欲識其寵請看懷中所抱（右），要知厥能試觀掌上何持（左）。围绕像龛有三处题记：寵愛之母保障大名（上），欲識其寵請看懷中所抱（右），要知厥能試觀掌上何持（左）。

bells❶ (fig. 3.14). Both the cathedral of Our Lady in Lille and the church in Daming are in Gothic-revival style.

图3.14 圣母,"大名的守护神",教堂钟楼和钟上的一幅中国圣画。
fig. 3.14 Our Lady of Grace, patron saint of Daming, on a chinese holy picture, on the church's tower and on the bells. (© ASJ France, GMC, and © THOC 2014)

---

❶ Three inscriptions are carved around the niche: The Loved Lady protects Daming (top), If you want to know her beloved, please look at the one in her arms (right), If you want to learn about her other abilities, please look at what is in her hands (left).

第三部分　教堂建造案例分析——大名天主堂
Part Three: The Handbook as Built

立在教堂主入口两侧的两通碑刻提供了进一步的历史信息（图3.15）。西侧的碑刻为所有捐赠超过10银圆的建堂施主的名单（由崔金泽抄录：（背面））。

Two steles placed at both sides of the main entry of the church provide complementary historical data (fig. 3.15). The western stele has a long list with the names of all the benefactors who gave more than 10 Yuan for the construction of the church.

王鎔	潘兆槐	張均廷	陳鳳庭	獻縣公學	岔道口
李辛莊	崔神父	葉樹勛	張清江	天利長	獻縣女學
閆黃古莊	常兒寨	沈仲芳	宋葵	賀瑪加利大	張慶霖
獻縣西路	朱家河	濮陽西街	張殿英	馮振興	劉方溥
劉廷士	張家莊	青草河	南湖	張吟清	趙鶴清
盧老萬	張賜禎	大郭家莊	劉八莊	紙房	張啟清
吳承泰	吳保堂	張銳	西大過村	馮家莊	焦家砦
程玉田	胡際辰	高瑞峰	張煥儒	東大過村	魚台
東明縣	陳茂隆	譚慶餘	田貴元	高通海	南立車村
北土路口	任隆品	霍誠	朱得貴	薄長庚	北立車村
小李村					

張增元	顏士篯	趙鳳廷	貫鶴亭	河間縣	趙家莊
勝營	劉尚新	彭榮	趙國卿	張鳳河	河間公學
前潘村	柳林	范相如	趙錫寶	張文茂	邯青雲
交河縣	張家莊	艾束	魏邦彥	程路濟亞	蕭郭氏
武將成	郝村	魏村	大寨	吳鴻誥	程清海
蕭葉氏	趙金凱	范家圪塔	大名本城	王盡忠記	程殿甲居
葛景榮	張道生	臥佛堂	大名公學	義記	紀廣仁
隱名氏	明加辣	吳橋縣	大名女學	錦記	任興立
耿雙玉	張善仁	故城縣	來二莊	張福同	張新立位
金春溥	蕭瑪尔大	留信女學	楊善村	張墨林	張新富
王立志	七里莊	西樓底	李友仁	張新文	趙充富
石家營	漳河村				

~ 309 ~

Wang Rong, Pan Zhaohuai, Zhang Junting, Chen Fengting, Public School of Xianxian, Chadaokou, Lixinzhuang / Priest Cui, Ye Shuxun, Zhang Qingjiang, Tian Lichang, Girls' School of Xianxian, Yanhuangguzhuang, Chang'erzhai / Shen Zhongfang, Song Kui, He Majialida, Zhang Qinglin, West Road of Xianxian, Zhujiahe, West Street of Puyang / Zhang Dianying, Feng Zhenxing, Liu Fangpu, Liu Tingshi, Zhangjiazhuang, Qingcaohe, Nanhu / Zhang Yinqing, Zhao He, Lu Laowan, Zhang Cizhen, Daguojiazhuang, Liubazhuang, Zhifang / Zhang Qiqing, Wu Chengtai, Wu Baotang, Zhang Rui, West Daguo Village, Fengjiazhuang, Jiaojiazhai / Cheng Yutian, Hu Jichen, Gao Ruifeng, Zhang Huanru, East Daguo Village, Yutai, Dongming County / Chen Maolong, Tan Qingyu, Tian Guiyuan, Gao Tonghai, South Liche Village, North Tulukou / Ren Longpin, Huo Cheng, Zhu Degui, Bo Changgeng, North Liche Village, Xiaoli Village /

Zhang Zengyuan, Yan Shizheng, Zhao Fengting, Jia Heting, Hejian County, Zhaojiazhuang, Shengying / Liu Shangxin, Peng Rong, Zhao Guoqing, Zhang Fenghe, Public School of Hejian, Qianpan Village, Liulin / Fan Xiangru, Zhao Xibao, Zhang Wenmao, Gao Qingyun, Jiaohe County, Zhangjiazhuang, Aici / Wei Bangyan, Cheng Lujiya, Ms Xiao Guo, Wu Jiangcheng, Hao Village, Wei Village, Dazhai / Wu Honggao, Cheng Qinghai, Ms Xiao Ye, Zhao Jinkai, Fanjiagada, Town of Daming / Wang Jinzhong, Cheng Dianjia, Ge Jingrong, Zhang Daosheng, Wofotang, Public School of Daming / Yi's Shop, Ji Guangju, Mr Anonymous, Ming Jiala, Wuqiao County, Girls' School of Daming / Jin's Shop, Ren Xingren, Geng Shuangyu, Zhang Shanren, Gucheng County, Laierzhuang / Zhang Futong, Zhang Xinli, Jin Chunpu, Xiao Maerda, Girls' School of Liuxin, Yangshan Village / Zhang Molin, Zhang Xinwei, Wang Lizhi, Qilizhuang, Xiloudi / Li Youren, Zhang Xinwen, Zhao Chongfu, Shijiaying, Zhanghe Village.

第三部分　教堂建造案例分析——大名天主堂
Part Three: The Handbook as Built

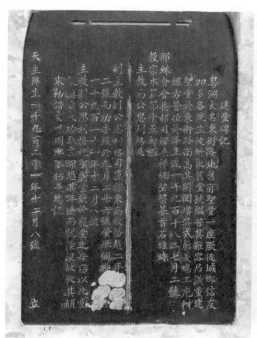 

图3.15　大名，天主堂的两块碑，左边是建造碑，
右面是捐献超过10银元的人的名字，1921年。
fig. 3.15 Daming, the two steles with the church's history (left)
and the names of the benefactors (right), 1921. (© THOC 2014)

标榜捐赠者的姓名是为了证明教堂是在中国天主教徒的资助下修建的，他们的慷慨和记忆应该被钦仰。为了筹措捐款，耶稣会展出了一个塔楼的木结构模型，以激发人们的想象力（图3.16）。

Listing the names of the donors provided proof that the church had been financed by Chinese Catholics, whose generosity and memory should be venerated. In order to raise funds, the Jesuits exhibited a wooden model of the tower, to stimulate the imagination of the people (fig. 3.16).

图3.16 大名，为募捐而做的教堂模型，1918—1919年。
fig. 3.16 Daming, a model of the church was used for fundraising, 1918-1919.
(© ASJ France, GMC)

东侧的碑刻立于 1921 年 12 月 8 日，为刘钦明主教祝圣教堂之日。碑文描述了教堂建造的事件：

【正面】
建堂碑记
粤溯大名東街路北舊有聖堂一座厥後城鄉信友加多各院生徒漸衆舊堂狹隘苦其難容乃議重建聖堂於東街路南高其閎闊增其式廊爰鳩工庀材經方營位於降生后一千九百十八年七月二

The eastern stele was erected on 8 December 1921, the day of the consecration of the church by bishop Lécroart. It tells the story of the construction:

"Originally, there was a church at the northern side of the East Street in Daming. Later the believers from urban and rural areas and the converted increased, and the old church became too small, so it was decided to reconstruct a new church at the southern side of the

~ 312 ~

第三部分　教堂建造案例分析——大名天主堂
Part Three: The Handbook as Built

號耶穌會會長郝司鐸❶嘉祿祝聖築基首石維時教宗本篤第十五御極主教馬公恩利格及副主教劉公恩利格司直隸東南教務越二年於八月二號而功告竣於九月二十六號肇舉彌撒大祭於一千九百二十一年十二月八號主教劉公恩利格祝聖新堂獻於寵愛之母窃以此堂規模宏大功烈高深恐其年遠而就湮沒故敘其顛末勒諸貞珉用垂不朽是為記

天主降生一千九百二十一年
十二月八號　立

大名天主堂并非普通的乡村教堂，而是被设计为一处位于城市环境中的大型教堂，后来成为了主教座教堂（图 3.17）。其巨大的尺寸、哥特式风格和红砖都给予这座教堂很强烈的外国色彩，与大名地方建筑形成鲜明的对比。它显示了天主教的权力和法文学校的高水平教育。一名耶稣会士在 1920 年发表了一篇关于大名的文章，题目为"大名府大教堂的落成"。❸ 在另一篇发表于 1919 年的文章中，神父雍居敬写

East street, with a higher gate and of a larger size. The craftsmen were gathered and materials were prepared and the location and direction were planned. On 2 July 1918 AD, the president of the Society of Jesus, priest Hao Jialu [Charles Héraulle]❷ consecrated the foundation stone. This was during the reign of pope Benedict XV, and bishop Ma Enlige [Henri Maquet] and vice bishop Liu Enlige [Henri Lécroart] administered the southeastern part of Zhili. Two years later, the construction was finished on 2 August. On 8 December 1921, bishop Liu Enlige consecrated the new church to Our Beloved Lady. This church is both large in scale and in merit. To avoid being forgotten and to render it eternal, the inscription of the whole story was carved in stone. Erected on 8 December 1921".

The church of Daming was not an average village church, but was designed as a large urban one that later became a cathedral (fig. 3.17). Its huge size, Gothic Style, and red brick colour gave the church a strong foreign character that was in total contrast with all other buildings in Daming. It had to express the power of the Catholic Church and the high level of education provided by the French college. An article on Daming published by a Jesuit in 1920 is entitled:

---

❶ 郝裕修（即郝司铎）（Charles Héraulle）神父在 1915—1921 年期间是直隶东南部耶稣传教会院长。
❷ Father Charles Héraulle was the superior of the Jesuit mission in Southeastern Zhili from 1915 to 1921.
❸ Pierre Mertens. Une grande ville chinoise qui s'ouvre à la foi: Achèvement de la cathédrale de Tai-ming-fu [J]. Les missions catholiques, 1920, 356-357, 367-369.

道：“梁神父无偿画了一座美丽的主教座堂的方案图纸。”❶ 这证明了这处教堂从概念上一直被认为是主教座堂。主教座堂为一个教区或代牧区的主教堂，是主教的宗座。尽管大名天主堂在1935 年才获得真正的主教座堂的身份，它的建筑形式、高耸的塔楼及巨大的体量都显然是主教座堂的标准。

"*The completion of the cathedral of Taming-fu*". See P313❸ In another article dated from 1919, father Jung wrote: "*Father Lamasse (...) drew for free the plans for a beautiful cathedral*".❶ These quotes prove that the church was called a cathedral from its conception. A cathedral is the main church of a diocese or vicariate apostolic, the seat of a bishop. Even though the church of Daming only received the canonical status of cathedral in 1935, it is obvious that its architectural form, its high tower and its huge size are that of a cathedral.

图3.17 大名，从西北看大名天主堂完工后的照片，20世纪20年代早期。
fig. 3.17 Daming, church after completion, seen from northeast, early 1920s. (© ASJ France, GMC)

---

❶ Paul Jung. Une grande ville chinoise qui s'ouvre à la foi [J]. Les missions catholiques, 1919, 234.

第三部分　教堂建造案例分析——大名天主堂
Part Three: The Handbook as Built

梁神父的参与显示了耶稣会士对大名天主堂的追求目标。他们本可以让修士雷振声来设计一处普通的教堂，但他们想要一处有很高声望的建筑。1917—1919 年，梁神父在吉林建造了法国哥特复兴式的主教座堂，直隶东南部的耶稣会士很是欣赏（图 3.18）。吉林主教座堂的设计方案也为大名天主堂所用，只有三个地方有改动。首先，吉林主教座堂比大名天主堂进深短四间。其次，吉林主教座堂的后殿屋顶的内立面与中殿类似，有拱廊及带有三重窗的天窗，而大名的后殿则没有拱廊和天窗。最后一点不同，吉林教堂的花窗、柱身和其他建筑装饰都是石头做的，而大名天主堂在这些方面则是用砖头做的。除上述区别及朝向之外，两座教堂几乎完全一样：吉林主教座堂为东西向，仍然遵循西方的传统，面向河流；而大名天主堂为南北向，面向主街。根据手册，南北朝向对于大型的教堂来说是最好的选择。❶ 教堂的设计方案在两座相隔 1400 千米的城市之间的传递，以及不同群体的传教士们之间的互相帮助，这些都让人惊叹。

The involvement of father Henri Lamasse reveals the ambitions of the Jesuits for the church of Daming. They could have asked brother Alphonse Litzler to design another average church, but they wanted a prestigious building. In 1917-1919, father Lamasse was building the cathedral of Jilin in a French Gothic Revival style that was appreciated by the Jesuits of Southeastern Zhili (fig. 3.18). The general silhouette of the cathedral, with its high spire and two corner towers looking like big pinnacles refers to the archetypal French Marian basilica of Lourdes, built in 1883-1889. The plans of Jilin served for Daming but were adapted on three points. First, the nave of Jilin is four bays shorter. Second, the inner elevation of the apse of Jilin has a triforium and a clerestory with triplet windows like in the nave, while in Daming the apse has no triforium and higher windows. Third, the traceries, shafts, and other architectural ornaments are in stone in Jilin, but in brick in Daming. For the rest, both churches are perfectly identical, except for their orientation: Jilin is west-east oriented, in conformity with the Western tradition, and faces the river, while Daming is north-south oriented and faces the main street. According to the handbook, the north-south orientation was the best one

---

❶《传教士建造者：建议-方案》手册第 4 页内容："对于两侧有窗户的大教堂，最好的朝向是南北向，这样的话窗户往东西向开，门可以放在北边或者放在南边更好。"

for large churches. ● It is fascinating to see how plans circulated between two cities, 1400 km distant from one another, and how missionaries from different societies helped each other.

图3.18 吉林，巴黎外方传教会的主教座堂，梁神父建，1917—1919年。
fig. 3.18 Jilin, cathedral of the French Foreign Missions of Paris, built by father Lamasse, 1917-1919. (© MEP archives)

大名天主堂的塔楼和正立面朝北，按照西方传统和中国传统来看，这样的朝向是不正确的。西方的传统要求正面朝西，而中国的传统则要求正面朝南。大名天主堂的朝向与当地的城市地形有关。教堂与城市的东街垂直，并正对着南北向的二级街道蒙古街与东街的路口。因此它的塔楼可以从两个方向看到（图3.19）。此外，在这两条街的

With its tower and main entrance turned to the north, the Daming church is not correctly oriented according to both the Western tradition and the Chinese tradition, which require respectively the main entrance to the west or to the south. The reason is the local urban topography. The church is perpendicular to the East street and is axed on a secondary street called Mongolia street, so that its tower

---

❶ Le Missionnaire Constructeur, Xianxian, 1926, p. 4: "For large churches, with windows on both sides, the best orientation seems to be the north-south direction, that is to say with windows opening to the east and the west, and the door or doors, placed to the north or better still, to the south".

第三部分 教堂建造案例分析——大名天主堂
Part Three: The Handbook as Built

交汇处立有一座高大的石牌楼，与红砖的哥特式教堂相映成趣❶（图 3.8）。

can be seen from two perspectives (fig. 3.19). Furthermore, there was a monumental stone archway or pailou at the junction of these two streets, which contrasted with the red brick Gothic style of the church❷ (fig. 3.8).

图3.19 大名天主堂北侧，1920年。
fig. 3.19 Daming, church from north, 1920. (© ASJ France, GMC)

---

❶ 第二个小一点的木拱门位于更西边一点，两个都被拆毁了。
❷ A second smaller wooden (?) archway was located more to the west; both have been demolished.

大名天主堂 42 米高的塔楼使其成为大名唯一的竖向建筑物（图 3.7，图 3.17，图 3.19 和图 3.20）。其塔尖占领了城市的天际线并据说从 20 里❶（11.5 千米）甚至 30 里（15 千米）外都能看见。塔楼上的钟声应该在方圆 20 里外都能听见。在大名修建这样一座砖结构建筑是前所未有的。可以想象大名那些非天主教徒的质疑——1920 年时大名约有80000人口——他们唯恐这是外来宗教的高傲表达及对该地区风水环境的破坏。❷ 1920 年的大旱期间，人们责怪教堂的钟声吓跑了雨神并认为这是导致旱灾的原因。人们曾想要攻击教堂并把钟拆毁，后被阻止。❸

With its 42 m high tower, the Catholic church was the only vertical construction in Daming. (fig. 3.7, fig. 3.17 to fig. 3.20) Its spire dominated the city's skyline and is said to have been visible from 20 li (11.5 km) or even 30 li (17 km) away. The bells hanging in the tower could be heard at more than 20 li. Erecting such a brick building was unprecedented in Daming. One can imagine that it provoked criticism from the non-Catholic population — about 80000 inhabitants in 1920 — that feared such an arrogant expression of foreign religion as well as a rupture of the fengshui harmony of the place. ❹ During the great drought of 1920, the church bell was accused of frightening the rain spirits and was considered the cause of the calamity. People planned to attack the church and destroy the bell, but that was finally avoided. ❺

---

❶ 1 里=500 米。

❷ 在大名并没有明确的迹象显示，但这是基督徒修建高楼时通常会面临的批评。比较著名的事件是广州在建两座天主教教堂时发生的冲突，相关内容见： Jean-Paul Wiest. The Building of the Cathedral of Canton: Political, Cultural and Religious Clashes [A] // Religion and Culture: Past Approaches, Present Globalisation, Futures Challenges (International Symposium on Religion and Culture, 2002: Macau), Macau, 2004, 231-252.

❸ 见 Pierre Mertens 神父写于 1920 年 7 月 24 日的一封信，见：La famine en Chine. Les missions catholiques, 1920, 458-459.

❹ We have no precise indications in the case of Daming, but this was a usual criticism when Christians built high towers. The case of the two towers of the cathedral of Guangzhou is famous: Jean-Paul Wiest The Building of the Cathedral of Canton: Political, Cultural and Religious Clashes [A] // Religion and Culture: Past Approaches, Present Globalisation, Futures Challenges (International Symposium on Religion and Culture, 2002: Macau), Macau, 2004, 231-252.

❺ Letter of father Pierre Mertens, dated 24 July 1920, in: La famine en Chine. Les missions catholiques. 1920, 458-459.

第三部分　教堂建造案例分析——大名天主堂
Part Three: The Handbook as Built

图3.20　大名天主堂西南侧，1925年前后。
fig. 3.20 Daming, church from southwest, around 1925. (© ASJ France, GMC)

大名天主堂的营建
Constructing the Church of Daming

　　由于并没有关于教堂建筑工程的档案或工程日记，对工匠及建筑材料的来源、建筑工程的组织等信息，我们所知甚少。1921年的两通碑刻上的碑文及神父雍居敬和其同事在1918—1920年为了募捐而发表的文章是目前所知

　　Because there are no building archives of the church or diary of the works, little is known about the construction, the origin of the workers or of the materials, and the works' organization. The inscriptions on the two steles of 1921 and the articles published in 1918-

仅有的关于此工程的文字资料。此外还有三幅照片记录了正在修建中的教堂的情况（图 3.21~图 3.23）。

1920 by father Jung and colleagues to solicit donations are the only written sources we know about the works. Three remarkable pictures show the church under construction (fig. 3.21 to fig. 3.23).

图3.21　建造中的大名天主堂，中殿的东北角，1918年秋天。
fig. 3.21 Daming, church under construction, northeastern corner of the nave, fall 1918. (© ASJ France, GMC)

第三部分　教堂建造案例分析——大名天主堂
Part Three: The Handbook as Built

图3.22　建造中的大名天主堂，中殿北端，1918年秋天。
fig. 3.22 Daming, church under construction,
nave to the north, fall 1918. (© ASJ France, GMC)

图3.23 建造中的大名天主堂，1919年6月。
fig. 3.23 Daming, church under construction, June 1919. (Les missions catholiques, © KADOC)

建造的主要步骤按编年顺序如下：❶

1917年：获得建造教堂的地块。

1918年7月2日：放置并赐福奠基石。

1919年8月2日：置塔楼顶端的十字架及顶石。

1920年10月2日：第一次在教堂

The main steps of the building chronology are:❶

1917: acquisition of the grounds where the church will be erected.

1918, 2 July: blessing of the foundation stone.

1919, 2 August: placing of the stone and the cross on the top of the tower.

1920, 2 October: first mass cele-

---

❶ Gerbes chinoises. Les Jésuites de la Mission de Sien-Hsien depuis 1856 [M]. Lille, 1934, 65-66.

里做弥撒。

1921年12月8日：由代牧区宗座刘钦明祝圣教堂并降福于钟。

这些日期显示建筑工程进行得相当快：一年半的时间完成从奠基到塔尖整个教堂外壳的建造，再一年的时间完成内部装饰（铺地、拱顶、窗户等）。如果考虑到工程在两个冬天会有所停滞，则营建这座巨大的"主教座堂"的实际时间甚至短于两年。

要达到这样的效率，很显然需要很强的工程组织性并由专业人员来管理。正是建造这座教堂所获得的经验后来被总结到了手册中。一方面，组织管理显示了设计方案、建筑材料、后勤组织、伙食备办及资金支持等方面都准备得很充分。另一方面，也表明了技能娴熟的人员及所有参与其中的人，从监理到承包人，从工头到工匠之间非常好的沟通。神父雍居敬作为工程的监理及主要协调人，雷振声修士作为工程建造及实际操作的协调人，他们的角色在前面已经提到数次。当时30岁的"小主"是石作及砖作的领班。❶ 而其他中国承包人和工匠的姓名均无从知晓。神父雍居敬曾提到有150人参与夯实基础，以及有40位石雕工匠，日工资均为300房

brated in the church.

1921, 8 December: consecration of the church and blessing of the bells by vicar apostolic Henri Lécroart.

These dates reveal that construction went quickly: one and a half years for building the shell of the church, from the foundations to the top of the spire, and one more year for the finishing works of the interior (pavement, vaults, windows, etc.). If we consider that construction was interrupted during two winters, the effective work time for erecting this 'cathedral' was less than two years.

To achieve such results, it is obvious that the works were well organized and run by professionals. It is precisely the experience developed during the works of Daming that is summarized in *The Missionary-Builder: Advice-Plans* handbook. This organization implied well prepared plans, building materials, logistics, catering, and financing. It also implied skilled people and good communication between all involved, from the patron and the contractor, to the masters and the workers. The role of father Jung, the patron and general coordinator, and brother Litzler, the constructor and practical coordinator, has been evoked several times. 'Little Master' (*xiaozhu*), aged 30, was the leader of the masonry works.❶ All the other Chinese contractors

---

❶ Father Mertens. Une grande ville chinoise qui s'ouvre à la foi: Achèvement de la cathédrale de Tai-ming-fu [J]. Les missions catholiques, 1920, 356-357, 367-369.

钱。❶ 手册的确建议付给工人们"日薪",而不是以工程计算的定价。❷ 有两幅照片记录了在脚手架上的石匠及正在午餐休息的工人们（图 3.21 和图 3.22）。

值得注意的是耶稣会士们还倡导工匠们进行文化宗教活动。神父雍居敬会给工匠们上晚课,用他的幻灯讲圣经里的故事。尽管工匠们白天工作强度很大已筋疲力尽,但因为他们从来没有被这样对待过,他们还是会去听雍神父的讲座。雍神父还指导工匠们祈祷、唱诗,并让他们理解他们正在建造的是上帝的房屋。根据传教士们的（记录）,"小主"和很多工匠们都皈依了天主教。❺

and workers remain anonymous. Once father Jung mentions 150 men ramming the foundation and 40 stone carvers, all payed 300 sapeque a day. ❸ The handbook indeed recommends paying workers at a 'day rate' and not at a 'flat rate' for the whole enterprise. ❹ Two pictures of the works show masons on scaffoldings and workers during a lunchbreak with their bowls (fig. 3.21 and fig. 3.22).

It is worth mentioning that the Jesuits also proposed cultural-religious activities to the workers. Father Jung gave evening lectures, using his 'magic lantern' and telling stories from the Bible. Despite being exhausted after a hard day's work, the workers, who had never been treated as such before, attended father Jung's lectures massively. He also taught the workers to pray, sing and understand that the building they were erecting was the house of God. According to the missionaries, Little Master and many workers converted to Catholicism. ❺

---

❶ 相当于当时法郎的 30 分。Paul Jung. Une grande ville chinoise qui s'ouvre à la foi [J]. Les missions catholiques, 1919, 248.

❷ 相关内容见手册《传教士建造者：建议-方案》第三章第三部分。

❸ Equaling 30 French cents of that time. Paul Jung. Une grande ville chinoise qui s'ouvre à la foi [J]. Les missions catholiques, 1919, 248.

❹ Le Missionnaire Constructeur [M]. Xianxian, 1926, 11-12.

❺ Gerbes chinoises. Les Jésuites de la Mission de Sien-Hsien depuis 1856 [M]. Lille, 1934, 65-66.

第三部分　教堂建造案例分析——大名天主堂
Part Three: The Handbook as Built

有一篇文章讲述了 1919 年 8 月 2 号置顶石和十字架于红砖塔尖的事件❶（图 3.23 和图 3.36）。在大名天际线的最高点树立基督教十字架是一个很冒险的举动。万一发生不幸的事情，人们会指责是教堂带来的厄运，天主教会会颜面尽失。而一旦成功，对他们来说则是极大的胜利。（当时有）上千人聚集在教堂周围及街道上帮忙。当雍神父在男修院的礼拜堂内祈祷时，"小主"则在脚手架的顶端，指挥着 90 人拉着四根长长的绳子，一起吆喝着把极重的石头（来自河南的 500 千克重的石材）缓缓抬升到它应在的位置。中午 11 点半巨大的铁十字架被固定在了顶石之上，正如该手册所示。❷

大名天主堂的平面和立面都与吉林主教座堂很相似，两者都依据了 19 世纪法国哥特复兴式的建筑样式。教堂总长超过了 50 米（图 3.24）。平面从北向南依次为方形塔楼，后为长长的进深相等的 11 进中殿，最后为进深一间的圣所及五面多边形后殿（图 3.57）。

One article tells how the summit stone and the cross were placed on top of the red brick spire on 2 August 1919❶ (fig. 3.23 and fig. 3.36). Erecting a Christian cross as the highest point in Daming's skyline was a very risky operation. If something should go wrong, the church would be accused of bringing bad luck to the city and the Catholics would totally lose face. In case of success, however, they would achieve a great victory. Thousands of people were around the church and in the streets to be present at the event. While father Jung was praying in the college chapel, Little Master was on the top of the scaffolding, directing 90 men pulling four long ropes, singing together and gradually lifting the heavy stone — a 500 kg stone from Henan — to its definitive position. At 11:30 am, the big iron cross was positioned in its stone base, as illustrated in the handbook.❷

The plan and the elevation of the church of Daming and the cathedral of Jilin reflect the nineteenth-century French Gothic Revival style. The total length of the church is more than 50 m (fig. 3.24). Its plan, from north to south, is composed of a square tower, a long and homogeneous nave of 11 bays ending with a sanctuary of one bay and a five sided polygonal apse (fig. 3.57).

---

❶ Father Mertens. Une grande ville chinoise qui s'ouvre à la foi: Achèvement de la cathédrale de Tai-ming-fu [J]. Les missions catholiques, 1920, 356-357, 367-369.

❷ Le Missionnaire Constructeur [M]. Xianxian, 1926, plate 16 (fig.2.18) .

图3.24 大名天主堂东侧。
fig. 3.24 Daming, church from the east. (© THOC 2014)

中殿长39.85米，宽9米，两侧有侧道，因此中央走道宽度为16.35米（图3.25和图3.26）。进深第一间没有侧道，而是两个侧入口及一个多边形小角楼，内为通往上层管风琴及塔楼上层的旋转楼梯。第七、八、九间两侧为教堂的两臂，使教堂内部平面形成十字，总宽为28.35米。然而这两臂并非真正的耳堂，因为它们并不比侧廊高且并没有与中厅形成十字（图3.27和图3.56）。两个假耳堂的北侧各开一门。圣所的第一级台阶在中厅进深第十间。两侧廊的尽

The nave-39.85 m long, 9 m wide-is flanked by aisles that give the church an inner width of 16.35 m (fig. 3.25 and fig. 3.26). The first bay has no aisles but side entrances and small polygonal corner towers containing spiral staircases leading to the organ gallery and the upper levels of the tower. The seventh, eighth and ninth bays of the nave are flanked by arms that give the church the form of a cross and a total inner width of 28.35 m. These arms, however, are not a real transept because they are not higher than the aisles and do not form a crossing with the

头为两个小礼拜堂，内有祭台。紧邻教堂的两臂前各有一个储藏室，大圣器室围绕着后殿（图 3.28）。

nave (fig. 3.27 and fig. 3.56). Each arm of this pseudo-transept has a doorway at the northern side. The first step of the sanctuary is located at the tenth bay of the nave. So the aisles end as two side chapels with altars. Two storage rooms are located along these side chapels, while a vast sacristy develops as a fan around the apse (fig. 3.28).

舶来与本土：1926年法国传教士所撰中国北方教堂营造手册的翻译和研究
Building Churches in Northern China. A 1926 Handbook in Context

图3.25　大名天主堂中殿南端，20世纪20年代和近期。
fig. 3.25 Daming, church's nave to the south,
1920s and present. (© ASJ France, GMC, and © THOC 2014)

第三部分 教堂建造案例分析——大名天主堂
Part Three: The Handbook as Built

图3.25(b)
fig. 3.25 (b)

第三部分　教堂建造案例分析——大名天主堂
Part Three: The Handbook as Built

图3.26　大名天主堂中殿北端，20世纪20年代和近期。
fig. 3.26 Daming, church's nave to the north,
1920s and present. (© ASJ France, GMC, and © THOC 2014)

舶来与本土：1926年法国传教士所撰中国北方教堂营造手册的翻译和研究
Building Churches in Northern China. A 1926 Handbook in Context

图3.27 大名天主堂西假耳堂和石柱。
fig. 3.27 Daming, church, western pseudotransept and stone columns. (© THOC 2014)

第三部分　教堂建造案例分析——大名天主堂
Part Three: The Handbook as Built

图 3.27(b)
fig. 3.27(b)

图3.28 大名天主堂西南面。
fig. 3.28 Daming, church from the southwest. (© THOC 2014)

中殿的内立面为典型的哥特式风格，分为三层：①一排柱子支撑着7米高的朝向侧道的尖拱（图3.29）；②由实心三瓣拱组成的拱廊；③13世纪哥特式三重窗式的天窗。中殿的柱头为13世纪的钩形柱头，上承柱身，延伸为拱券的主肋（图3.30）。拱廊、天窗及拱券重复的节奏使整个室内非常连贯。只有教堂中最神圣的部分，即圣所具有独特的立面。圣所内并没有柱廊，而是哥特式的高窗（图3.31）。

The inner elevation of the nave is typically Gothic style, with three levels: 1) a row of columns supporting the 7 m high pointed arches opening to the aisles (fig. 3.29), 2) a triforium consisting of a blind trefoil arcade, 3) and the clerestory with thirteenth-century Gothic triplet windows. The capitals of the nave are thirteenth-century hook capitals, on top of which a shaft rises up to the springing of the main ribs of the vaults (fig. 3.30). The repetitive rhythm of the arcades, triforium, clerestory and vaults gives its coherence to the whole interior. Only the sanctuary, the most sacred part of the

第三部分　教堂建造案例分析——大名天主堂
Part Three: The Handbook as Built

church, has a different elevation, without arcade and triforium, but with high Gothic windows (fig. 3.31).

图3.29　大名天主堂中殿的砖拱。
fig. 3.29 Daming, church, brick arches of the nave. (© THOC 2014)

图3.30　大名天主堂的砖柱身和柱头。
fig. 3.30 Daming, church, brick shaft and capitals at the triforium's level. (© THOC 2014)

图3.31 大名天主堂后殿和西南角的礼拜堂。
fig. 3.31 Daming, church, apse of the sanctuary and southwestern side chapel. (© THOC 2014)

该教堂有 4 种共 62 扇窗户：后殿的 7 扇高哥特式窗，带有花式窗格并分隔为双尖角；共 32 扇三重窗，包括 22 扇高窗上的三重窗，两臂假耳堂内的各 3 扇三重窗及圣所两侧的小礼拜堂内的各 2 扇三重窗；8 扇简单的尖角窗；4 扇在假耳堂，4 扇在正立面塔楼的两

The church has a total of 62 windows of four different types: the 7 high windows of the apse, with traceries and divided into two lancets; the 22 triplet windows of the clerestory, 3 in each arm of the pseudo-transept, and 2 in each side chapel of the sanctuary, all together 32 triplets; 8 simple lancet windows, 4 in

面；正立面上的 5 扇小玫瑰窗或花环形窗，3 扇在塔楼上，2 扇在次入口上方（图 3.32）。所有这些窗户种类都在手册中有所图示。所有中厅及侧廊的柱式、柱础、柱头也均有图示。❶ 此外还有 13 世纪哥特式柱式、柱础、正厅及侧廊的柱头的图示❷（图 3.33）。

the pseudo-transept and 4 in the main façade at both sides of the tower; 5 small rose windows or rosettes in the main façade, 3 in the tower and 2 above the side entrances (fig. 3.32). All these window types are illustrated in the handbook.❶ The thirteenth-century Gothic style columns, bases, capitals of the nave and aisles are illustrated too❷ (fig. 3.33).

图3.32 大名天主堂中殿和耳堂的东侧窗户。
fig. 3.32 Daming, church, eastern side of the nave and transept, different types of windows. (© THOC 2014)

---

❶ Le Missionnaire Constructeur [M]. Xianxian, 1926, plates 5, 19, 21, 22 (fig.2.7, fig.2.21, fig.2.23, fig.2.24).
❷ Le Missionnaire Constructeur [M]. Xianxian, 1926, plates 23, 24 (fig.2.25, fig.2.26).

图3.33 大名天主堂的13世纪哥特式风格细节：中殿的柱头、
侧道柱身的基础、拱廊、拱廊上的柱头。
fig. 3.33 Daming, church, thirteenth-century Gothic architectural sculpture: capital of nave, base of a shaft in the aisle triforium, capitals at the triforium's level. (© THOC 2014)

用得最多的建筑材料是砖，外红内青（图3.28，图3.31~图3.34）。

手册图纸中用在钟楼高处和屋顶尖塔的砖作细节及花砖都来自大名[1]

The most used building material is brick, red on the outside but blue on the inside (fig. 3.28, fig. 3.31 to fig. 3.34).

The masonries are pure brick walls (*li wai zhuan*). All the details of masonry

---

[1] Le Missionnaire Constructeur [M]. Xianxian, 1926, plates 1-4, 7-9, 13-14, 16, 17, 19 (fig.2.3-fig.2.6, fig.2.9-fig.2.11, fig.2.18-fig.2.19, 2.21).

第三部分　教堂建造案例分析——大名天主堂
Part Three: The Handbook as Built

（图 3.36）。

砖墙的砌筑方式为里外砖。所有的砖及屋顶的瓦均为当地出产，且质量都很好❶（图 3.35）。

as well as of the 'flowered bricks' or profiled bricks, the top of the tower and the high brick spire illustrated in the handbook, are from Daming$^{\text{See P338}}$❶ (fig. 3.36).

The bricks were locally produced and reveal a high quality❷ (fig. 3.35).

图3.34　大名天主堂角楼，外面红砖，里面青砖。
fig. 3.34 Daming, church, corner turret, elaborated red brickwork outside, simple blue brickwork inside. (© THOC 2014)

图3.35　大名，学校建筑上的花砖和基督十字砖、方形屋瓦（望砖）。
fig. 3.35 Daming, brick with flower and Christian cross, square roof tile (*wang zhuan*), from the school buildings. (© THOC 2014)

---

❶ 这家被废弃的砖厂，有一个大烟囱，位于教堂东南角 30 米，在耶稣会的人离开后砖厂又开业了。这表明当地的土是可以用来烧砖的。

❷ The abandoned brick factory with its high chimney, located about 30 m southeast of the church, was installed there after the Jesuits left. It proves that local clay is good for brick production.

舶来与本土：1926年法国传教士所撰中国北方教堂营造手册的翻译和研究
Building Churches in Northern China. A 1926 Handbook in Context

图3.36 大名天主堂塔楼的暗拱、小尖塔和塔楼尖顶。
fig. 3.36 Daming, church's tower, lower level with blind arches and upper level with pinnacles and high red brick spire. (© THOC 2014)

中殿的屋顶为仰瓦屋面，瓦的凹面朝上（图 1.25），而侧道及假耳堂的屋顶则为扣瓦屋面，即瓦的凸面朝上。两种屋面覆瓦方式都在手册中有描述和图示。同时该手册还注意到来自广平府的工匠是直隶东南部地区做仰瓦屋顶最出名的。❶

该手册中的所有石雕、滴水石、柱头、柱础及壁柱的细节也均来自大名天主堂❸（图 3.37）。

The roof tiles were locally produced too. The nave is covered according to the yang wa system, with the concave side of the tiles facing upwards (fig. 1.25), while the roofs of the aisles and the pseudo-transept are covered according to the kou wa system that alternates concave and convex tiles. Both are described and illustrated in the handbook that mentions that the workers from Guangpingfu are the most renowned from Southern Zhili for making yang wa roofs. ❷

All the details of carved stones, dripstones, capitals, bases of columns and shafts in the handbook are from the church of Daming too❸ (fig. 3.37).

---

❶ Le Missionnaire Constructeur [M]. Xianxian: Imprimerie de Sien-Hsien, 1926, Plates 9, 10. 广平府位于大名的西北，距离 60 千米。2014 年 7 月，教堂屋顶全都改为仰瓦屋顶，侧道上方屋顶则是扣瓦屋顶。

❷ Le Missionnaire Constructeur [M]. Xianxian, 1926, plates 9, 10. Guangping is located about 60 km northwest of Daming. In July 2014, the roof covering of the nave was totally renewed in yang wa system and the aisles in kou wa.

❸ Le Missionnaire Constructeur [M]. Xianxian, 1926, plates 13, 14, 23, 24 (fig.2.15, fig.2.16, fig.2.25, fig.2.26) .

第三部分　教堂建造案例分析——大名天主堂
Part Three: The Handbook as Built

图3.37　大名天主堂钟楼的石柱头、后殿的砖柱头和石头柱子顶板。
fig. 3.37 Daming, church, stone capital of the tower, brick capital and stone abacus in the apse. (© THOC 2014)

该教堂的木结构屋顶 2014 年时已全部更换，但屋架的形式仍然采用了原有形制，与手册中的图示一致❶（图3.39）。作者在现场找到一些原来木结构的遗存构件，上面有当初建造时的标号❷（图3.38）。

The timber roof structure of Daming has been renewed in 2014, but the trusses reproduce the original type, as illustrated in the handbook❶ (fig. 3.39). We recorded in extremis the types of joints and the assembly marks on abandoned remains of the original timber structure❸ (fig. 3.38).

图3.38　大名天主堂原梁上的标志：十三西、□西、十東、六東。
fig.3.38 Daming, church, remains of original beams with assembly marks.
(© THOC 2014)

❶ Le Missionnaire Constructeur [M]., Xianxian, 1926, plate 10, Fig.58 (fig.2.13) .
❷ 标号黑色字体，由方向和数字一起组成：表示方向的有东西南北中，数字有一、二、三、四、五等。
❸ Black painted marks combining orientation: east, west, south, north, centre, and numbers 一、二、三、四、五, etc.

图3.39 大名天主堂、2014年翻新的大名天主堂的屋顶桁架。
fig. 3.39 Daming, church, roof truss renewed in 2014. (© THOC 2014)

最后,中殿、侧廊及后殿的拱顶为木制拱顶,形制仿造哥特式肋骨拱顶(图3.40~图3.42)。拱和肋骨都涂成仿青砖砌体(图3.43)。其拱背结构是用木板、板条和石灰做成的,与手册中的描述亦一致❶(图3.44)。拱及肋条油饰成仿砖结构。以上列举的材料再一次证明了手册中理论化的经验来源于雍神父及雷振声修士建造大名天主堂的工程。

Finally, the vaults of the nave, the apse and the aisles are wooden imitations of Gothic rib vaults (fig. 3.40 to fig. 3.42). The arches and the ribs are meticulously painted with mock blue brickwork (fig. 3.43). The vaults' extrados reveal their timber structure made of planks, lath and lime, as described in the handbook❶ (fig. 3.44). This enumeration of materials shows, once more, that the theorized experience from the handbook is actually based on the construction of the church of Daming by brother Litzler and father Jung.

---

❶ Le Missionnaire Constructeur [M]. Xianxian, 1926, plate 15 (fig. 2.17).

第三部分　教堂建造案例分析——大名天主堂
Part Three: The Handbook as Built

图3.40　大名天主堂后殿的哥特式木制拱顶。
fig. 3.40 Daming, church, wooden Gothic style rib vaults of the apse. (© THOC 2014)

舶来与本土：1926年法国传教士所撰中国北方教堂营造手册的翻译和研究
Building Churches in Northern China. A 1926 Handbook in Context

图3.41 大名天主堂中殿和东侧道的哥特式木制交叉拱顶。
fig. 3.41 Daming, church, wooden Gothic style rib vaults of the nave an the eastern aisle. (© THOC 2014)

图3.42 大名天主堂西侧道的石柱和木制交叉拱顶。
fig. 3.42 Daming, church, stone columns and wooden Gothic vaults of the western aisle. (© THOC 2014)

第三部分　教堂建造案例分析——大名天主堂
Part Three: The Handbook as Built

图3.43　大名天主堂中殿的横向拱，用木板、板条和石灰做成。
fig. 3.43 Daming, church, transversal arch of the nave, made of planks, lath and lime. (© THOC 2014)

图3.44　大名天主堂中殿的木制拱顶、拱背部分。
fig. 3.44 Daming, church, wooden vaults of the nave, extrados. (© THOC 2014)

有着法国 13 世纪哥特式风格的大名天主堂用的是河南的石头，外面用的是红砖，里面用的是青砖。里外对比鲜明的用砖颜色在建造教堂的当时应该是一件很有意思的事情：外面用红砖是想在公共空间中表达一种西方身份感，里面用青砖是为了让到教堂里来的中国人在外国建筑里有"家"的感觉。如果到教堂来的中国人发现这里是一个他不了解的世界——由拱券形成的巨

The French thirteenth-century Gothic church of Daming combined Henan stone and locally produced red bricks outside and blue bricks inside. Such contrast must have been very meaningful at the time of the church's construction: using red bricks outside expressed Western-universal identity in the public space, while using blue bricks inside invited the Chinese to feel 'at home' in this foreign building. When inside, however, the

~ 345 ~

大空阔的空间，从彩色玻璃窗照进来的很虚幻的彩色光线，还有那陌生的家具陈设和雕像（图 3.25 和图 3.26），以及用中文写的不明意思的祈祷文，可以想象一下他们的感受，如果此时管风琴再奏乐的话，也许他们会被完全惊吓到，当然他们也许也会完全被吸引住。

作者还找到了关于室内装修和家具的一些信息。

在（教堂内的）不同区域还有原来的玻璃的遗存。天窗的玻璃为几何图案，由白色、蓝色、绿色和红色玻璃拼成（图 3.45）。在后殿的窗户的上层还留有哥特图案及小天使形象的原始彩色玻璃遗存（图 3.46）。根据档案中的一张图片，后殿的窗户上并没有耶稣、圣母玛丽及其他圣徒的形象❶，但有两竖行中文❷（图 3.47）。这与手册中关于彩色玻璃应为"中国制造"的建议相符合。❸

Chinese visitor discovered a world that was unknown to him, with a vast empty space rhythmed with arches, unreal colored light coming through stained-glass windows, strange furniture and statues, as well as unknown prayers written with Chinese characters on banners (fig. 3.25 and fig. 3.26). One can imagine the impression of such experience on the visitor, especially if the organist was playing. The visitor could either be deeply frightened, or totally fascinated.

Some information has been found about the furniture and the interior decoration.

Fragments of the original glass are still present at different places. The windows of the clerestory are glassed with geometric patterns including white, blue, green and red glass (fig. 3.45). In the upper part of the windows of the apse remain little fragments of colored glass with Gothic forms and little angels (fig. 3.46). According to an archive picture, there were no figures of Christ, Mary and saints on the axial windows of the apse,❹ but two Chinese Marian inscription❷ (fig. 3.47). This is in conformity

---

❶ 在中国，有人物图像的彩色玻璃窗只在上海土山湾的耶稣会工作室生产，或者就从欧洲进口，相关内容见：宋浩杰. 土山湾记忆 [M]. 上海学林出版社，2010. 关于从比利时进口彩色玻璃窗到中国的内容见文章：Thomas Coomans. Sint-Lucasneogotiek in Noord-China: Alphonse De Moerloose, missionaris en architect [J]. M&L. Monumenten, landschappen en archeologie, 2013, 32(5):6-33.

❷ 顧命傳心諄託母（德）信友 (左)，徒表意切期子慈親 (右).

❸ Le Missionnaire Constructeur [M]. Xianxian, 1926, 57.

❹ Figurative windows were only produced in the workshop of Tushanwan (T'ou-sè-wè) in Shanghai or were imported from Europe: Song Haojie. Memory of T'ou-sè-wè / 土山湾记忆 [M]. Shanghai, 2010. About the import of Belgian stained-glass windows in China, see: Thomas Coomans. Sint-Lucasneogotiek in Noord-China: Alphonse De Moerloose, missionaris en architect [J]. M&L. Monumenten, landschappen en archeologie, 2013, 32(5):6-33.

第三部分　教堂建造案例分析——大名天主堂
Part Three: The Handbook as Built

with the handbook's advice about stained glass 'made in China', which should be non-figurative. See P346❸

图3.45　大名天主堂中殿的窗户和原彩色玻璃。
fig. 3.45 Daming, church, triplet window of the nave and original stained-glass. (© THOC 2014)

图3.46　大名天主堂后殿窗户和花饰窗格的原玻璃遗存。
fig. 3.46 Daming, church, window of the apse and remains of original glass in the tracery. (© THOC 2014)

现有的祭台是新的，但主祭台的石龛被保留了下来，❶并与立于后殿的一座大各各他山融为一体。在后殿立这样一座各各他山在基督教堂中并不常见（图3.47）。

有数幢木制的忏悔室与手册中描述的几乎完全一致❸（图3.48）。

The present altars are new, but the stone tabernacle of the main altar has been conserved, ❷ and has been integrated into a great Calvary erected in the apse. Such a Calvary in an apse is unusual in Christian churches (fig. 3.47).

Several old wooden confessionals are nearly the same as the one depicted in the handbook❸ (fig. 3.48).

---

❶ 希腊大写字母可以作为基督教的符号象征：Alpha (A) 和 Omega (Ω)表示一切造物的起始和终末，Chi-rho (X P)是耶稣基督的象征。

❷ With two symbols of Christ in Greek capital letters: Alpha and Omega (A Ω) meaning that Christ is the beginning and the end, and Chi-rho (X P), the monogram of Christ.

❸ Le Missionnaire Constructeur [M]. Xianxian, 1926, plate 25 (fig. 2.27).

舶来与本土：1926年法国传教士所撰中国北方教堂营造手册的翻译和研究
Building Churches in Northern China. A 1926 Handbook in Context

图3.47 大名天主堂后殿的原祭台、各各他山、彩色玻璃窗和横幅，20世纪20年代。
fig. 3.47 Daming, church, apse with the original main altar, Calvary, stained-glass windows and banners, 1920s. (© ASJ France, GMC)

第三部分　教堂建造案例分析——大名天主堂
Part Three: The Handbook as Built

图3.48　大名天主堂的告解室。
fig. 3.48 Daming, church, confessional. (© THOC 2014)

在中殿第一进的木阳台上有一座巨大的管风琴（图 3.49）。13 世纪哥特式的外壳和几根琴管还保留着，但乐器已经无法弹奏。这样一座巨大的管风琴在中国非常特别，通常中国的教堂内以簧风琴为主。

A great organ was installed on the wooden gallery in the first bay of the nave. (fig. 3.49) The thirteenth-century Gothic-style case and several pipes have been conserved but the instrument is no longer working. Such a great organ was very exceptional in China, where the churches were usually equipped with harmoniums.

雍神父提到该乐器是在维也纳（奥地利）制作并于 1922 年运到天津。❶管风琴下方挂着的一块牌子，上面的题记显示这可能是用加拿大天主教徒的捐款购置的。❷

Father Jung mentions that the instrument was made at Vienna (Austria) and shipped to Tianjin in 1922.❸ An inscription on a board hanging under the organ suggests that it could have been bought with donations from Canadian Catholics. ❹

图3.49 大名天主堂主入口处的管风琴，产自奥地利，20世纪20年代。
fig. 3.49 Daming, church, Austrian organ on the gallery above
the main entrance, 1920s.
(© ASJ France, GMC)

❶ 见雍神父于 1921 年 8 月 25 日写给 Gagnon 修士的信 (Vanves, ASJ Fr, GMC 101, Jung)。在管风琴盒子里面，有用粉笔写的德语铭文。

❷ "A Notre Dame de Grace: Des Canadiens Catholiques / filial hommage"。这需要感谢北京中国研究中心的 Jean-Paul Wiest 博士和香港浸会大学的 David Urrows 博士。后者写了一本关于中国境内的管风琴的书，是比利时鲁汶《中国研究系列丛书》之一。

❸ Letter of father Jung to brother Gagnon, 25 August 1921 (Vanves, ASJ Fr, GMC 101, Jung). Inside the organ's case, German assembly inscriptions are written in chalk.

❹ "A Notre Dame de Grace: / Des Canadiens Catholiques / filial hommage" (to Our Lady of Grace, Catholic Canadians' filial tribute).

第三部分　教堂建造案例分析——大名天主堂
Part Three: The Handbook as Built

原来的石制领洗池是哥特式，仍然放置在靠近主入口的位置，在管风琴的下方（图 3.50）。

最后，前面已经提到的三座钟❶来自法国并于 1921 年底运到中国。它们挂在按照手册图纸所做的钟架子上❷（图 3.51）。

The stone Gothic-style baptismal font is still located near the main entrance, under the organ gallery (fig. 3.50).

And finally, the three bells, already mentioned, ❸ came from France and were delivered at the end of 1921. They are hanging in a wooden belfry the plans of which are reproduced in the handbook❷ (fig. 3.51).

图3.50　大名天主堂的领洗池。
fig. 3.50 Daming, church, stone baptismal font. (© THOC 2014)

图3.51　大名天主堂的木头钟架子和在法国铸造的钟。
fig. 3.51 Daming, church, wooden belfry and bells from France in the tower. (© THOC 2014)

---

❶ 相关内容见第一部分。
❷ Le Missionnaire Constructeur [M]. Xianxian, 1926, plate 18 (fig.2.20)．
❸ See Part 1, p. 31-32.

舶来与本土：1926 年法国传教士所撰中国北方教堂营造手册的翻译和研究
Building Churches in Northern China. A 1926 Handbook in Context

大名天主堂的测绘
Measuring the Cathedral of Daming

数千年历史的沧桑在大名的土地上留下了数以百计的各类历史文化遗产，❶ 其中，流经大名的大运河，于 2014 年成为世界文化遗产。2014 年，大名县也成为了河北省省级历史文化名城。虽然大名有悠久辉煌的历史，但由于处于军事战略的关键位置，古代建筑遭受了严重的破坏，北宋的大名府，现仅剩下夯土残垣，而明代的大名府城内也未见明代建筑遗存。

从陪都到府、道、市，最后到县，随着大名经济、政治和军事地位的不断下降，至近代，铁路未经大名，而是建在

Numerous cultural heritage sites were left on the land of Daming by thousands of years of history,❶ among which there is the recently inscribed World Heritage Site of the Grand Canal. In 2014, Daming County was also listed as a Provincial Historic and Cultural City. Despite Daming's long and glorious past, however, historic buildings suffered from severe destruction in the course of time because of the city's critical location for military strategy, at the crossroad of Hebei, Shandong and Henan provinces. The ruins of the rammed earth city wall is the only trace left of the Northern Song Dynasty (960-1127) Daming Fu. Under the Ming Dynasty (1368-1644), Daming Fu moved to its present location, but, as far as we know, there is no trace of Ming Dynasty architectural remains conserved, except parts of the city wall.

Daming's economic, political and military status has been decreasing from being the provisional capital, to being of

---

❶ 国家文物局主编. 中国文物地图集（河北分册）(上、中、下). 北京：文物出版社，2013；大名县志编纂委员会编. 大名县志 [地方志]. 北京：新华出版社，1994；程廷恒等修，洪家禄等纂.大名县志. 1934(民国 23 年)；(明)潘仲骖纂修，(明)赵慎修续修.[嘉靖]大名府志 [善本]：二十九卷；(清)周邦彬修，(清)郜焕元纂.[康熙]大名府志 [善本]：三十二卷.

第三部分　教堂建造案例分析——大名天主堂
Part Three: The Handbook as Built

了邯郸，从此大名城失去了快速发展的动力，也因此其旧城内的历史街道格局得以保存。近年来，大名政府为了重塑历史上的辉煌，重建了明代大名府的城墙、城楼。如今的大名城内，主要保留着一百多年以来的建筑，在这些建筑中，最具历史、文化和艺术价值的是欧洲传教士在大名修建的教堂、修道院等近代西式建筑。

2014年7月初，在经过先期现场调查，测绘计划获得文物主管部门同意并完成准备工作后，北大考古文博学院文物建筑专业师生一行11人，乘高铁自北京至邯郸，再转乘汽车至大名县城。大名县文物局接待了北大师生，并委托管理天主教堂的大名天主教会协助北大师生的测绘工作。数日后，鲁汶大学的高曼士教授也抵达大名，与北大师生一起工作。

现存大名天主堂，是20世纪20年代欧洲传教士组织大名本地工匠建造

only regional and in the end, it has become merely a county. The fact that the railway never reached Daming, relocated development to more favourable places such as Handan. Hence, most of the historic urban fabric and the scale of the buildings within the old town are preserved. In recent years, the local government rebuilt the Ming Dynasty city wall and gates of Daming Fu, hoping to revive its glorious history. The present Daming mainly holds historic buildings from around one century ago. Among these buildings, the Gothic-style church and the school buildings erected by the French missionaries in the years 1900-1925 are among the ones with the greatest historical, cultural and artistic values.

In July 2014, after preliminary research on the site, with the approval of the survey proposal by the cultural heritage administration and their preparation effort, a crew of eleven people including students, staff and Professor Xu Yitao from the Division of Ancient Architecture of the School of Archaeology and Museology of Peking University, arrived in Daming. Professor Thomas Coomans from the University of Leuven joined the crew from Peking University. They were cordially received by the Cultural Heritage Department of Daming County. Daming Catholic Church who manages the church building was asked to assist the crew during the survey.

The fieldwork course on building archaeology and survey aimed to con-

的具有哥特式风格的西式教堂。基于该建筑的历史背景，我们制定了工作计划，测绘调研工作由以下四部分组成。

（1）北大团队利用三维扫描仪、全站仪、遥控航拍器、相机和手工测绘等多种测绘方法和测绘工具，全面记录了教堂的周边环境、建筑整体及局部尺寸，后期绘制了教堂总平面、平面、立面、剖面和若干细部详图（图 3.55~图 3.63）。

（2）高曼士教授带北大同学辨认、分析和记录教堂砖砌体的构造细部，并利用测绘的间隙，向北大师生介绍欧洲哥特式教堂建筑的发展历史。

（3）高曼士教授、北大师生及本地建筑工匠合作翻译手册词条。基本工作

front the students with a Western Gothic style building that had been designed by foreigners and was constructed by local workers and artisans. The choice had been fixed upon the church of Daming because the handbook provided additional theoretical information about its building materials and its construction. Comparing the material source with the written source, the practice with the theory, made the case of the church of Daming particularly relevant from an educational point of view. The course's program and the survey strategy consisted in four parts:

(1) Peking University crew documented the surrounding, the architectural structure and details of the church comprehensively, using multiple survey methods including 3D laser scan, total station, aerial photography, photographic documentation and hand measurement. Accurate drawings based on these measurements were produced afterwards. They include plans, elevations, cross sections, and some details of the church (fig. 3.55 to fig. 3.63).

(2) Peking University students learned to inspect, analyse and record the construction details of the church's brickworks under the instruction of Prof. Coomans, who also introduced the evolution of Gothic churches in Europe to the students during the breaks in between work.

(3) The entries in the handbook were translated with the cooperation of

第三部分　教堂建造案例分析——大名天主堂
Part Three: The Handbook as Built

程序是：高曼士教授将手册词条的法语翻译为英文，北大师生将英文翻译为中文，与当地工匠交流（图3.52）一一确认手册中所记载的条目含义，将当年传教士用法语记录的150多个中国本地建筑用语还原为中文。具体内容见附录2。

local construction workers. For each word, first Prof. Coomans pronounced the French Romanized, thereafter Peking university crew tried to identify the Chinese term thanks to the workers' knowledge of local dialect (fig. 3.52). The actual meaning of more than 150 Chinese terms recorded in French about Chinese local architecture by the missionaries was identified. These terms are mentioned in Appendix 2.

图3.52　大名，学者和当地工人一起讨论手册中的技术用语，2014年7月9日。
fig. 3.52 Daming, academic staff and local workers identifying together the technical words of the handbook, 9 July 2014.
(© School of Archaeology and Museology, Peking University 北京大学考古文博学院)

（4）普查大名现存清、民国时期的历史建筑，包括住宅、清真寺、商业建筑和教会建筑等（图3.53），普查获得了一批可与大名天主堂及传教士手册互证的建筑实例。此外，北大师生与高曼士教授还一同与大名的文物管理部门及大名天主教会负责人进行了访谈交

(4) A general survey of all the remaining historic buildings from the Qing Dynasty (1644-1911) and the era of the Republic of China (1912- 1949) in Daming, including residences, mosques, commercial buildings and Christian buildings, was conducted (fig. 3.53). The results provide some valuable architec-

流，进一步探讨了大名文化遗产保护与利用问题。

ture examples for cross-examining with the Catholic church of Daming and the handbook. Moreover, the responsible from the cultural heritage administration and Daming Catholic Church were interviewed, with further discussion on the conservation and the use of Daming's cultural heritage.

图3.53 大名，美国宣圣会的新教徒传教士的房子，1920年前后和2014年。
fig. 3.53 Daming, house of the American Protestant missionaries of the Pentecostal Church of the Nazarene under construction and present state, around 1920 and 2014.
(© China Collection, Nazarene Archives, Lenaxa, Kansas; © THOC 2014)

以上工作在文物现场进行了一周，现场工作结束后，北大师生返校开展内业工作，于2015年4月绘制完成了教堂各项图纸。

大名天主堂坐落于大名城内东侧，距离复建的明代大名府东城门仅200米之遥，站在城墙上，教堂钟楼的尖顶似乎触手可及，在当地众多的老照片当中，不乏同样的视角（图3.7）。作者可以想象到，历史上一代代的传教士和大名人，应该都曾站在这里眺望过这

The on-site survey lasted for one week. Afterwards, the Peking University crew finished the production and finalisation of all the drawings, until April 2015.

The church is located in the eastern part of Daming city, only 200 meters from the reconstructed East Gate of the Ming Dynasty Daming Fu. Standing on top of the city wall, one has the feeling that the spire of the church is almost reachable. This similar perspective is also reflected in many of historic photos

第三部分　教堂建造案例分析——大名天主堂
Part Three: The Handbook as Built

座教堂（图 3.54）。对来自欧洲的传教士来说，教堂是他们跨越万里、历尽艰辛完成宗教使命的象征，面对教堂的身影，他们应该充满神圣的使命感。而对大名当地人来说，教堂应该曾被视为来自陌生国度的怪物，它与本土建筑大相径庭，高高的尖塔一举抢占了大名城的天际线。转瞬百年，当年的欧洲传教士早已离去，而大名人也早已将这座教堂当成了他们历史文化的一部分，不能割舍。

一年过去了，当作者回想起大名测绘时，始终难以忘记的是，在夕阳下，看着教堂的身影渐渐沉入夜幕，那一刻，感觉到这座西式教堂与历史悠久的中国古城融为了一体，宁静而安详。回想我们的工作，我们的研究，为揭示历史而存在，也必将成为历史。若干年后，大名是否还能记起，曾有我们这样一批人来过（图 3.55~图 3.63）？

taken by the locals (fig. 3.7). Generations of missionaries and Daming citizens had the similar view of the church from the same spot (fig. 3.54). For the missionaries from Europe, the church was a symbol of the fulfilment of their religious mission thousands miles from home. For the Chinese Catholics and most students of the French College, the Gothic church was a symbol of identity and pride. For the other locals in Daming, however, the church probably was considered as a grotesque monster from a foreign country, because of its difference from the local architecture and its high spire dominating the skyline. A century has flown by in the blink of an eye. French and Hungarian Jesuits have left Daming since long, whilst the church remains and has become an inseparable part of the history and culture of Daming's citizens.

A year has passed since our survey in Daming. The image of the church fading into the night under the ray of sunset is still imprinted in our memory. At such tranquil moment, one can really feel that the Western church is embraced by the ancient Chinese city. Our research, which is to reveal history, will also become history. Years from now, may the memory of us being here and what we have done for the city still live in this book (fig. 3.55 to fig. 3.63).

舶来与本土：1926 年法国传教士所撰中国北方教堂营造手册的翻译和研究
Building Churches in Northern China. A 1926 Handbook in Context

图3.54 大名，两名法国耶稣会神父和一名中国神父在大名城墙东门上给学校照照片，20世纪20年代。
fig. 3.54 Daming, two French Jesuit fathers and one Chinese father taking pictures of the school from the East Gate of the city wall, 1920s. (© ASJ France, GMC)

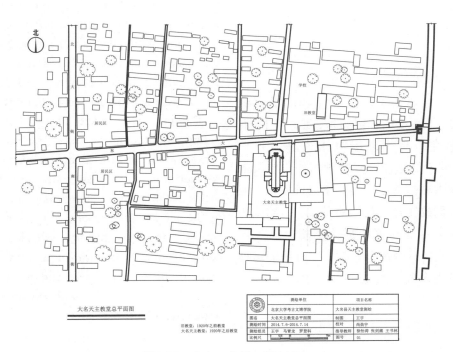

图3.55 大名，东街，2014年。
fig. 3.55 Daming, area of East Street, 2014.
(© School of Archaeology and Museology,
Peking University 北京大学考古文博学院）

~ 358 ~

第三部分 教堂建造案例分析——大名天主堂
Part Three: The Handbook as Built

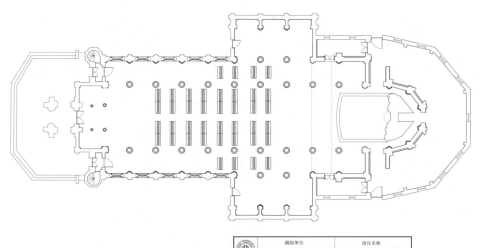

图3.56 大名，大名天主堂平面，2014年。
fig. 3.56 Daming, church, ground plan, 2014.
(© School of Archaeology and Museology,
Peking University 北京大学考古文博学院）

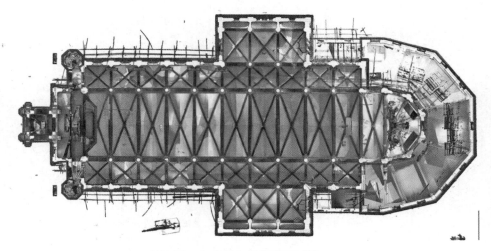

图3.57 大名天主堂仰视平面图。
fig. 3.57 Daming, church, ground plan with projection
of the rib vaults. (© School of Archaeology and Museology,
Peking University 北京大学考古文博学院）

舶来与本土：1926 年法国传教士所撰中国北方教堂营造手册的翻译和研究
Building Churches in Northern China. A 1926 Handbook in Context

大名天主教堂北立面图

第三部分　教堂建造案例分析——大名天主堂
Part Three: The Handbook as Built

图3.58　大名天主堂北立面。
fig. 3.58 Daming, church, elevation of the northern side.
(© School of Archaeology and Museology,
Peking University　北京大学考古文博学院)

图3.59 大名天主堂中殿的纵剖面。
fig. 3.59 Daming, church, longitudinal section of the nave. (© School of Archaeology Museology, Peking University 北京大学考古文博学院）

第三部分　教堂建造案例分析——大名天主堂
Part Three: The Handbook as Built

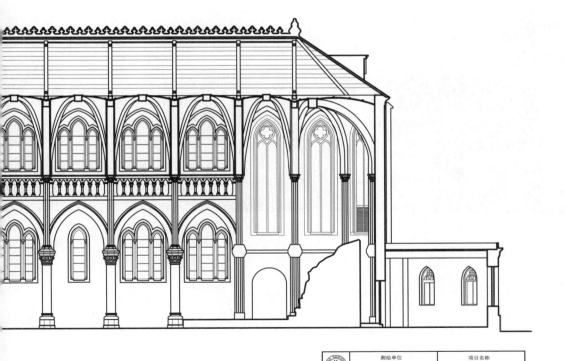

大名天主教堂III-III剖面图

舶来与本土：1926年法国传教士所撰中国北方教堂营造手册的翻译和研究
Building Churches in Northern China. A 1926 Handbook in Context

图3.59(b)
fig. 3.59(b)

第三部分　教堂建造案例分析——大名天主堂
Part Three: The Handbook as Built

图3.60　大名天主堂中殿的横剖面和假耳堂的北侧。
fig. 3.60 Daming, church, cross section of the nave and elevation of the northern side of the pseudotransept. (© School of Archaeology and Museology, Peking University 北京大学考古文博学院）

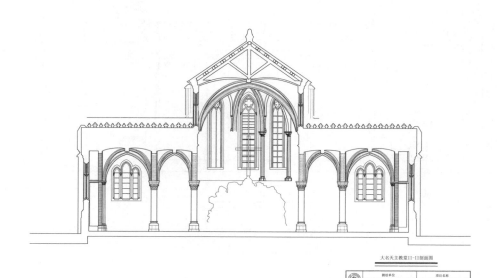

图3.61　大名天主堂中殿和假耳堂的横剖面。
fig. 3.61 Daming, church, cross section of the nave and the pseudotransep. (© School of Archaeology and Museology, Peking University 北京大学考古文博学院）

~ 365 ~

舶来与本土：1926 年法国传教士所撰中国北方教堂营造手册的翻译和研究
Building Churches in Northern China. A 1926 Handbook in Context

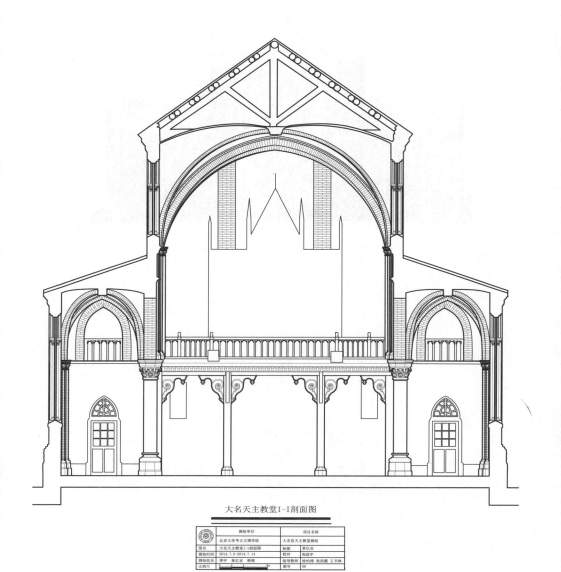

图3.62　大名天主堂中殿的横剖面。
fig. 3.62 Daming, church, cross section of the nave.
(© School of Archaeology and Museology,
Peking University 北京大学考古文博学院)

第三部分　教堂建造案例分析——大名天主堂
Part Three: The Handbook as Built

图3.63　大名天主堂中殿、假耳堂和钟楼的横剖面。
fig. 3.63 Daming, church, cross section of the nave,
the pseudotransept and the tower. (© School of Archaeology and Museology,
Peking University 北京大学考古文博学院)

附 录
Appendix

舶来与本土：1926 年法国传教士所撰中国北方教堂营造手册的翻译和研究
Building Churches in Northern China. A 1926 Handbook in Context

附录1. 手册的法语原文
Appendix 1. French Text of the Original Handbook

Le Missionnaire Constructeur. Conseils — Plans

Par des missionnaires de la Chine du Nord
Imprimerie de Sien-Hsien, 1926
Imprimi potest. J. Debeauvais S. J. Superior regularis missionis
Imprimatur. + H. Lécroart S. J. Episc [opus] Anchial [ensis] Vicar [ius] apost [olicus] de Sien-hsien

Préface 1
Chapitre premier : Choix de l'emplacement 2
Chapitre deuxième : Choix de la disposition des bâtiments 4
Chapitre troisième : plan – devis – contrat avec l'entrepreneur ... 6
    A. Plan 6
    B. Devis 6
    C. Contrat avec l'entrepreneur 10
Chapitre quatrième : Des matériaux 12
    Briques 12
    Tuiles 15
    Bois « Mouleao » 16
    Chaux « cheu hoei, ta hoei » ou seulement « hoei » 17
    Cordes « Cheng, Ma cheng » 19
    Perches pour échafaudages « Kantze » 20

Chapitre cinquième : La maçonnerie..21

    Article 1er. Les Fondations..21

    Art. 2ème. Le « ta hang » (damage)..22

    Article 3ème. La maçonnerie..23

        Le mortier..23

        Les murs..23

        Les murs chinois..24

        Appareil de briques..26

        Art 3e : Épaisseur des murs..26

    Article 4e Le salpêtre (« kien »), les isolants..28

    Article 5e. Pose des chambranles des portes et fenêtres..30

Chapitre sixième : Le toit..33

    Article 1er. Le toit chinois..33

    Article 2e. La charpente européenne..35

    Article 3e. Pente des combles..38

Chapitre septième : L'ornementation du toit..39

Chapitre huitième : Varia..53

    Varia. Comble, diverses formes..53

    Voûtes..55

    Arcs en briques..57

    Flèche d'un clocher..58

    Jointoiement de la maçonnerie..60

    Gâcher le mortier de terre..60

    Protection du dallage..61

    Polir les briques..61

    Véranda..61

Chaînage	..62
Cheminées	...63
Escalier	..63
Beffroi	.63
Goudron	..64
Ecoulement des eaux de pluie	..64
Verre	...64
Huiler	..65
Colonnes sous les poutres	..65
Conclusion	**...65**

Préface

Cette brochure n'est pas faite pour les bâtisseurs de cathédrales. Ce n'est même pas un essai de la fameuse adaptation de l'art chinois à nos églises chrétiennes.

Tous les missionnaires peuvent être obligés un jour de construire, ne serait-ce qu'une école de village de trois « *kien* » (travées). Ce sont ces modestes constructeurs, que des frères, encore plus modestes architectes, mais un peu plus anciens sur le chantier, voudraient faire profiter des expériences qu'ils ont faites.

Il s'agit avant tout de construire solidement, cela va de soi. Si l'agréable peut être mêlé à l'utile, si une petite église peut devenir jolie, tout en restant bien campée sur ses fondations, ce n'en sera que mieux. Donc, l'idéal visé, c'est d'aider à faire solide, salubre et pas trop cher, ni trop laid : une jolie église pour peu d'argent, comme les festins d'Harpagon.

Le missionnaire, nouvellement arrivé en Chine, se trouve de plus aux prises avec les conséquences de la construction de la tour de Babel : la confusion des langues. Aux conseils donnés aux constructeurs, on a donc cru bon d'ajouter, au fur et à mesure, la traduction des termes du métier.

Enfin, pour ajouter l'exemple aux préceptes, une pochette contient des figures et quelques plans d'églises, d'une exécution assez facile. On pourra du moins s'en inspirer faute de mieux.

Il va de soi que les conseils donnés et les remarques faites, s'appliquent surtout aux conditions de la Chien du Nord.

Que St. Joseph daigne bénir ces quelques pages.

Tamingfou, 19 Mars 1926.

Chapitre premier : Choix de l'emplacement[p. 1]

Il ne faut pas trop se presser pour construire dans les chrétientés nouvelles. La chrétienté peut crouler, même avec un nombre déjà assez considérable de baptisés, surtout si ce ne sont pas des familles entières, les femmes restant païennes, ou si ces néophytes sont presque tous pauvres; au contraire, la communauté chrétienne peut se développer rapidement et le local bâti trop tôt sera trop étroit.

L'idéal, quand l'occasion est bonne, est d'hypothéquer. On a un local suffisant, en attendant de nouvelles conversions, et l'argent n'est pas perdu en cas d'insuccès.

De plus, la construction est souvent une épreuve trop forte pour les tous nouveaux chrétiens. Ils ne voudront pas contribuer, ou ils contribueront d'une façon fort inégale, selon le plus ou moins grand esprit de foi. De là des jalousies, des pertes de face. Mais surtout ils seront exposés à se charger la conscience, s'ils ont mission d'acheter les matériaux, ou simplement de servir d'intermédiaire pour ces achats. Des matériaux achetés par la mission ou la communauté chrétienne seront employés pour usages privés ; on dira qu'ils sont empruntés ; mais on oubliera de les rendre ; ou bien, ils seront simplement volés. Ce qui appartient au Père, appartient aux enfants, diront volontiers certains nouveaux chrétiens, commentant à leur façon le dogme du « *Cheng eull koung hoei* » (Communion des Saints).

En achetant un terrain, si la chose est possible, que dès le début, la propriété ait au moins deux ou trois « *mou* » (arpents). Sans quoi, plus tard, quand la communauté se développera, les voisins feront des prix de « faveur ».

Autant que faire se peut, il faut éviter également les emplacements suivants :

a) Ne pas acheter hors du village. Les propriétés y sont naturellement beaucoup meilleur marché, précisément parce que personne [p. 2] n'en veut pour s'y établir. Il n'y aura pas de voisins, au moins d'un côté, souvent de trois. Les femmes et jeunes filles, surtout chez les nouveaux chrétiens, ne feront pas volontiers un si long trajet pour se rendre à l'église.

b) Ne pas se mettre non plus au beau milieu du marché (*tsi*), si c'est un village à marché. L'inconvénient serait le même, et encore plus grave, pour les femmes, si elles étaient obligées de traverser ces foules grouillantes pour se rendre à la prière. Le prêtre sera souvent troublé pendant la messe et surtout pendant le sermon, par les cris du dehors.

c) Si on a le choix, acheter de préférence un emplacement au nord de la rue (s'il s'agit d'un rue qui va de l'est à l'ouest). On pourra y réserver une bonne place, au nord, pour les constructions importantes : église, presbytère, écoles. La porte cochère, l'écurie, parfois la cuisine, pourront être construites au sud, en « *nan ou* », dont nous parlerons plus bas.

d) Enfin, plus que tout le reste, il faut éviter les terrains trop bas, sujets aux inondations. Il faut s'informer avec soin du niveau d'eau en été, surtout si l'on est près d'un fleuve. Dans ce cas, il reste la ressource de remblayer, surtout si les chrétiens s'offrent à faire le travail. Sinon, le prix, parfois considérable, de cette main d'œuvre, serait à estimer d'avance, et à faire entrer en ligne de compte dans les prévisions de la dépense.

Voici comment seront exécutés ces remblais : Ils se feront par couches de 0,20 m environ, pilonnés avec soin ; ensuite ils seront arrosés fortement ; puis on mettra une deuxième couche de même épaisseur qu'on pilonne et qu'on arrose, et ainsi de suite.

Si l'on doit ensuite charger ces remblais, il convient de les arroser copieusement, de les pilonner ensuite à nouveau, et d'employer des empattements à gradins très étendus.

Il serait souverainement imprudent d'abandonner des paysans à leur seule inspiration pour faire ces remblais, surtout s'il s'agit de la construction d'une église.

Chapitre deuxième : Choix de la disposition des bâtiments[p. 3]

Dans cette partie de la Chine, il est de la plus grande importance de donner une bonne orientation aux bâtiments, en rapport avec leur destination. Ne pas prendre modèle sur les concessions européennes, ni sur les Protestants d'Amérique, mais imiter fidèlement les gens du pays.

« *You ts'ien pou kai tong nan fang; Tong pou noan, hia pou leang* ». Un richard ne construit ni maison à l'est, ni maison au sud; en hiver, ce n'est pas chaud; en été, pas frais.

Ainsi le veut le proverbe chinois, et à bon droit. La meilleure exposition est celle des bâtiments dont portes et fenêtres sont orientées vers le sud, les « *pei ou* », ou « *t'ang ou* » (maison au nord ou maison principale). Et si l'on veut viser à la dernière perfection en ce genre, on obliquera d'environ 10 degrés vers le sud-est. De cette façon, en hiver on aura le soleil de très bonne heure le matin, et en été on sera délivré plus vite, l'après-midi, de ses rayons directs. — La direction vers le sud-ouest prodiguerait naturellement l'effet contraire.

Si on veut rendre ces maisons (*pei ou*) plus agréables en été, on peut percer des fenêtres dans le mur du nord pour obtenir un courant d'air. Peut-être ce courant d'air sera-t-il désagréable en hiver, à moins d'avoir des fenêtres très bien faites, ou même des fenêtres doubles. Et encore ! On peut aussi se contenter de simples prises d'air ou toutes petites fenêtres à hauteur d'homme ; on l'ouvre en été, on les ferme et calfeutre en hiver.

Après le « *pei ou* » vient le « *si ou* » (maison de l'ouest), dont l'exposition est à l'est. En hiver, elle est ensoleillée dès le lever du soleil. En été, dès onze heures, elle est à l'abri des rayons directs.

L'orientation la plus défavorable est celle du « *tong ou* » (maison à l'est), exposition à l'ouest. Ces maisons sont quasi inhabitables en été. Aussi ne faudrait-il jamais les choisir pour petites [p. 4] chapelles, (portes avec fenêtres d'un seul côté, c'est-à-dire ouvrant seulement vers l'ouest). On pourrait tenir dans ces « *tong ou* » le matin, pour la messe. Mais en été, après-midi, vous ne pourriez ni y confesser, ni y donner le salut.

Mieux vaudrait mettre la chapelle dans le « *nan ou* » (maison au sud) exposition au nord. En hiver, elle ne sera pas trop glaciale, si la chrétienté est… fervente, c'est-à-dire s'il y a beaucoup de monde. Mais l'habitation du missionnaire n'est pas à mettre dans le « *nan ou* », à moins d'y placer un bon fourneau. C'est dans le « *pei ou* » qu'il doit demeurer si possible.

Ne pas oublier qu'il faut, pour la maintenir sèche et saine, exhausser cette habitation de deux pieds environ, au-dessus du sol extérieur, ou au moins de trois marches à la porte d'entrée.

Par contre, ne pas trop élever le plafond ; onze pieds sous la poutre (leang) est une hauteur qu'il ne faut guère dépasser. « Pas élevé par le haut, mais haut par le bas », dit le proverbe chinois. Il ne s'agit ici évidemment que des pièces d'habitation particulière. Les salles communes de classes, les grands « *ko't'ing* » (salles de réception) doivent être plus hautes : 13 ou 14 pieds sous le « *leang* ».

Pour les grandes églises, avec fenêtres des deux côtés, la meilleure orientation semble être la direction du nord au sud, c'est-à-dire avec les fenêtres ouvrant vers l'est et l'ouest, et la porte ou les portes, placées au nord ou mieux au sud. En hiver, on aura le soleil toute la journée ; en été, grâce aux fenêtres des deux côtés, on pourra établir le courant d'air rafraîchissant. Ne pas oublier une manœuvre intelligente des rideaux quand le genre et les dimensions de la construction les admettent.

On a aussi constaté généralement que les églises orientées nord-sud coulent moins, à la saison des grandes pluies, que celles orientées est-ouest.

Pour ces grandes églises, l'orientation où les fenêtres ouvrent vers le nord et le sud, avec portes à l'est ou à l'ouest, peut être avantageuse en été, mais en hiver, l'église sera glaciale, dès que le vent soufflera du nord. Or l'hiver est le temps des missions et de bien des fêtes.

En faveur des églises orientées est-ouest, et qui ont bien leurs avantages, certains proposent les deux aménagements suivants, admissibles seulement d'ailleurs pour les constructions de petites [p. 5] dimensions et de peu de style :

a)ou bien, sur le côté nord, remplacer une fenêtre sur deux par un mur en retrait, dessinant la place de la fenêtre, et sur lequel on pourra suspendre de grands tableaux instructifs et édifiants ;

b)ou bien, en maintenant sur le côté nord le même nombre de fenêtres qu'au sud, les faires plus petites, v. g. de 2 pieds, au nord qu'au sud ; leur cintre supérieur sera à la même hauteur de part et d'autre, mais l'appui inférieur sera plus élevé, v. g. de 2 pieds, au-dessus du pavé, sur le côté Nord.

Chapitre troisième : Plan-devis-contrat avec l'entrepreneur

A. Plan

Pour toute construction, si modeste soit-elle, faites vous-mêmes ou faites faire un plan. Il va de soi que pour une grande église, il faudrait s'adresser à un architecte, diplômé ou non, mais que ce soit un homme de métier. — Les plans doivent être soumis à l'approbation du Vicaire apostolique. Voyez à ce sujet le J. C. [*Jus canonicum*] Canon 1162. Plus d'une église avec colonnes a été construite tandis que le plan existait uniquement dans la tête du Père. Hélas on ne s'en aperçoit que trop.

Ce plan ne doit pas être seulement le plan par terre, composé d'un tracé d'une seule ligne. Il faut y ajouter le plan en élévation, celui de la charpente, les plans de détail : tour, portail, fenêtres, voûtes, etc. etc.

Sur ces plans, on lira l'épaisseur des murs, l'appareil pour les maçons peu expérimentés, etc. de manière à éviter les surprises, les erreurs fâcheuses pendant la construction. Il est bon de faire un plan général de toute la propriété, en désignant l'emplacement de chaque bâtiment. La position topographique de la propriété par rapport au reste de la localité pourrait être indiquée au moins grosso modo.

B. Devis

Si les plans sont bien faits, il sera relativement facile d'établir le devis des dépenses. Et quand le devis sera bien établi, quand on sera sûr de n'avoir rien oublié, ni cordes pour les échafaudages, ni clous pour les chevrons, ni instruments de travail à mettre à la disposition des ouvriers, ni gratifications à leur accorder, majorez hardiment vos prévisions d'un tiers. Sans cette précaution, on risquera le sort de l'homme de l'Évangile qui ne put achever son édifice parce qu'il avait mal calculé les dépenses.

Une règle de prudence et même du droit canon vous décidera à ne pas compter à la légère sur les aumônes « possibles ». Du reste, s'il s'agit de la construction d'une église, il sera utile de relire le J. C., là où il parle de *Locis sacris*.

Note. Il y aurait lieu de traiter ici de la question du style chinois pour nos églises. Qu'il nous soit permis de citer la conclusion d'un article, paru dans le Bulletin des Missions Etrangères édité à Hong kong, n° 42, Juin 1925, pour ce qui regarde l'architecture. Le mot « architecture » est peut-être pompeux dans l'état actuel de la science de l'art chinois. Le mot « construction » conviendrait mieux. Ce que nous voyons en Chine actuellement, dit un ingénieur éminent, ne dépasse pas les limites de la « construction »… À l'heure actuelle, il n'est guère possible de rien augurer (de ce qu'on pourrait découvrir dans les nombreuses ruines ensablées dans les déserts du Kansou et dans les pays parcouru par le P. Huc). On ne trouvera jamais d'ailleurs de monuments grandioses, comparables à ceux de l'Inde ou d'Angkor, parce que les matériaux employés n'étaient pas la pierre, et parce que les rites chinois qui ont « créé cette nation », unis aux superstitions du Dragon et du Tigre, avaient enlevé à l'architecture l'essentielle condition de son développement : la liberté. « La liberté est l'atmosphère de l'art » dit Mgr Baudrillart ; or la liberté a été tellement réglementée en Chine, que les artistes se sont rejetés en désespoir de cause sur les arts secondaires d'ornementation, préférant orner leur logis et leurs pagodes que d'en faire une construction d'art. (loc. cit. p. 343).

Quant à cette décoration sur bois, pierre, soie, porcelaine, etc. , voici le jugement qu'en porte l'auteur de la « causerie sur l'art chinois », causerie qui mérite d'être lue en entier.

« Je connais une petite église d'un cachet particulier : l'extérieur, [p. 7] la façade, présentent un cachet chinois très bien rendu. J'y entrai une première fois, curieux d'admirer la décoration qui, pensais-je, serait la clef, l'exemple à suivre pour l'utilisation des arts décoratifs chinois. Je ne pus réprimer un sentiment d'étonnement d'abord, de malaise ensuite, en n'y voyant que des souhaits de bonheur et de longue vie, mélangés aux inscriptions les plus fantaisistes des rébus et calembours cités plus haut (chauve-souris, poisson, pêche, caractère chinois « *cheou* » longévité, etc.). Le Chinois qui entre dans ce sanctuaire se trouve indubitablement entouré de conceptions qui lui sont familières ; il s'y sent peut-être au chaud ; l'atmosphère qu'il respire l'incite à la sympathie de l'idée que représente ce monument. Il y viendra plus facilement, la sympathie des choses attire, elle aussi. Mais n'est-ce pas diminuer l'idéal du Christianisme que de présenter à nos fidèles et dans nos temples des conceptions si peu en rapport avec la grande et consolante théorie de la douleur, qu'est venu enseigner le Christ ? Bonheur d'ici-bas, vivre sa vie, jouir dans un dilettantisme si peu élevé, n'est-ce pas pratiquer une morale déprimante et trop peu élevée ? » (Loc. cit. p. 555) Voyez :

1)Bulletin catholique de Péking, trois longs articles des mois d'Avril, Mai et Juin 1924, pp. 127, 177, & 218.

2)Bulletin de la Société des Missions Étrangères de Paris, N° 42, p. 431, du mois de Juin 1925.

Voici le jugement d'un missionnaire ayant vécu longtemps en Chine. Les connaissances architecturales, son expérience de ce pays, la parfaite adaptation, comme aussi la réelle beauté de nombreuses églises qu'il y a construites, lui donnent le droit d'avoir une opinion et de la formuler :

Le style chinois consiste essentiellement, en tant que mode de construction, en un toit monumental, extrêmement coûteux et très mal compris pour remplir son office, posé sur des colonnes. Un tel système de construction est voué par avance à une durée très éphémère. Libre aux Bouddhistes de voir dans l'effondrement périodique de leurs pagodes un symbole de l'instabilité et de la fragilité des choses humaines. C'est ainsi qu'au Japon, la célèbre pagode d'Ise est, depuis des milliers d'années, régulièrement *rebâtie* tous les 20 ans. Mais ce n'est pas là ce que nous cherchons pour nos édifices religieux !

En outre ce système, extrêmement dispendieux et instable, [p. 8] est aussi peu adaptable que possible aux conditions pratiques que réclame le culte catholique. Basé sur la longueur habituelle que comportent des poutres ou des poteaux en bois, il n'est applicable qu'à des édifices de très petites dimensions : Kiosques, salles de la dimension d'une pagode, etc.

« C'est pour une raison analogue que l'architecture grecque, si admirable par ses proportions et la pureté de ses lignes, s'est montrée absolument inapte à engendrer des édifices catholiques. Le Parthénon est une merveille minuscule, qui ne peut pas s'agrandir, pour cette simple raison que les linteaux de pierre, monolithes posés d'une colonne à l'autre, et qui caractérisent la construction de l'antiquité grecque, atteignent au Parthénon leur longueur pratique maxima.

« On objectera la Madeleine de Paris. Soit ; mais ce n'est là qu'une mauvaise contrefaçon de style grec à linteaux non plus monolythes, mais obtenus de longueur suffisante par des pierres ajoutées les unes aux autres, et consolidées par des armatures en fer ; recours pour les grandes portées à des arcs romans, absolument inconnus des Grecs, etc.

« Ce qui prouve qu'un tel exemple n'est pas à suivre, c'est qu'il n'a guère été imité. Ce qui a été imité, et reproduit à un nombre énorme d'exemplaires dans la construction des églises catholiques, c'est la basilique romaine, édifice païen par son origine. Pourquoi ? Parce que ce mode de construire une vaste salle, dont, par une disposition de la Providence, l'Église a trouvé le modèle dans la Rome païenne, était merveilleusement adapté aux besoins du culte catholique : réunir dans un très vaste espace un très grand nombre de fidèles, ayant tous vue sur le chœur et l'autel, placés à l'extrémité de la salle ».

Le culte païen, lui, en Chine du moins, réunit les foules à de très rares intervalles, (une fois, deux fois par an), sans ordre, sans tenue, sans dignité, sans horaire, *comme à la foire*, non pas à l'intérieur de la pagode, mais dans de vastes cours qui la précèdent à perte de vue. Ce sont ces cours, flanquées de longs et étroits bâtiments latéraux, coupées par des ponts, des arcs de triomphe, des portes monumentales, bordées, sur l'allée du milieu, d'énormes stèles aux savantes inscriptions, qui forment la beauté… de la pagode. Ici c'est la cadre qui fait la valeur du tableau. Pour transformer cela en église catholique, il n'y aurait qu'à poser un [p. 9] toit sur la cour, à fermer les quatre côtés par un grand mur, dans lequel on laisserait les ouvertures voulues pour portes et fenêtres, et à placer l'autel là où trône le gros Buddha, après avoir enlevé la cloison en bois et en papier là où serait le banc de communion.

En attendant cette transformation de la pagode chinoise en église catholique, la foule païenne qui est venue « à la foire » continue à s'agiter, dans les cours, à crier, à acheter et à consommer ses petites provisions, sous le regard bienveillant et lointain de Koan yin poussa ou du Fouo ventru. À l'intérieur de la pagode, quelques bonzes et bonzillons sont agenouillés devant l'idole en psalmodiant leurs prières en sanscrit et en s'accompagnant sur des petites clochettes en fer (les *K'ing*). Par petits groupes, les fidèles, les femmes surtout, viennent faire leur prostration et piquent quelques bâtons d'encens. Une œillade aux bonzes, et le pèlerinage est terminé : autre culte, autre style !

Mais revenons à la « Basilica » romaine et à sa transformation en temple du vrai Dieu. Au début, il est vrai, la largeur de la grande nef était limitée par la longueur des pièces en bois servant de poutres pour le plafond. Les tâtonnements romans et enfin la trouvaille de la voûte gothique (croisée d'ogive), qui était la solution parfaite au problème, a permis d'étendre pour ainsi dire indéfiniment la largeur [sic, longueur] de l'église, spécialement celle de la grande nef ; et c'est ainsi qu'a été réalisé ce système de construction éminemment adapté au culte catholique, qui permet de bâtir, avec le maximum d'économie, de solidité et de durée, depuis la plus humble église de village, jusqu'à nos grandes cathédrales.

Le seul style, qui dans nos édifices religieux, a apporté une certaine concurrence à celui de la basilique romaine, est celui de la croix grecque, à coupole centrale, dérivé de Ste Sophie de Constantinople : style qui a sa beauté, mais s'est révélé beaucoup moins pratique.

Quant aux églises empruntées à d'autres styles, pour ce qui regarde leur mode de construction, je ne crois pas qu'on en trouverait plus de une ou deux pour cent, dans le monde entier. Espérer que le style chinois, à peine existant en tant qu'architecture, et en tout cas éminemment enfantin comme mode de construction, viendra un jour, pour la Chine, donner un démenti à un pareil fait d'expérience, me semble, pour ma part, présenter tous les caractères d'une illusion.

Toutefois, je tiens à préciser que dans tout ce que je viens de dire, j'entends par style chinois la manière de construire des Chinois. Car si par style chinois on entend, non pas un mode de construire, mais un genre d'ornementation propre à l'art chinois, la question se poserait alors tout différemment. L'existence de cet art chinois d'ornementation est indéniable ; son adaptation au mode de construire européen ne paraît pas impossible, et enfin elle est dans la logique des choses et des traditions de l'art, même religieux.

Si, en effet, dans le monde entier, on trouve peu d'églises catholiques dont le plan et la méthode de construction ne soient pas des dérivés de ceux de la basilique romaine, on en trouve peu aussi qui ne présentent pas, dans leur construction d'abord (mais ici l'art chinois est trop pauvre pour intervenir), dans leur ornementation ensuite (et ici l'art chinois peut avoir son rôle à jouer), une certaine adaptation aux conditions naturelles, et à l'atmosphère artistique du pays où elles sont construites. On a remarqué, par exemple, que les architectes gothiques, pour la confection de leurs chapiteaux, ont toujours fait appel aux éléments que leur fournissait la flore, et jamais à une flore étrangère (v. g. feuille d'acanthe des Grecs).

Rien donc de plus louable et de meilleur goût que d'imiter, en Chine, une telle manière de faire, à condition évidemment — ce qui ne sera pas toujours très facile — d'éviter en même temps tout emprunt à des motifs d'ornementation superstitieux.

Mais une telle imitation de l'art chinois, dans la partie décorative de l'édifice (peinture, boiseries, pierres sculptées, faïences émaillées, etc.) suppose que cette partie décorative existe, c'est-à-dire, à moins qu'il ne s'agisse de misérables imitations en plâtre ou en chaux, un édifice de luxe, comportant des dépenses de pure ornementation.

Or, nous ne construisons pas actuellement (et le Saint Siège nous a prescrit récemment de ne pas l'oublier), des édifices de luxe, des églises somptueuses. « Et telle cathédrale, telle grande église, déjà construites en Chines, me direz-vous, ne sont-elles pas somptueuses ? — « Non, répondrai-je : c'est là une illusion qui tient à la nature même de l'art européen, et notamment au style gothique. Il produit une impression d'art et de beauté par le seul emploi rationnel des matériaux les plus ordinaires, mis en œuvre [p. 11] dans la mesure exactement suffisante pour atteindre, de la façon la plus économique, le but pratique à remplir. Rien ou presque rien (car *parum pro nihilo reputatur*) n'est sacrifié v. g. dans la cathédrale de Moukden à l'ornementation. Celle-ci n'intervient pour ainsi dire pas (vitraux mis à part) dans la dépense.

Si donc l'art chinois ne peut être introduit dans nos églises qu'au point de vue décoratif, c'est-à-dire à titre de dépense purement somptuaire, je ne crois pas être dans l'erreur en pensant que cette introduction n'est pas à envisager pratiquement dans l'état actuel des choses et des finances de nos missions. Ce sera à nos successeurs d'y songer, lorsque le moment sera venu d'introduire un peu de somptuosité dans nos édifices religieux.

C. Contrat avec l'entrepreneur

Il serait imprudent de vouloir soi-même recruter du monde pour le travail qu'on a en vue et même de vouloir donner du travail comme manœuvre, à tel ou tel chrétien pauvre auquel on s'intéresse. Cette charité mal ordonnée pourrait devenir, pendant le cours des travaux, la source de bien de petits frottements. Votre protégé n'obéira pas à l'entrepreneur ; il se considèrera comme votre homme ; ou bien l'entrepreneur sera jaloux de votre ingérence ; il n'osera pas prélever le tant pour cent sur le salaire de cet ouvrier, etc.

Que le « *ling tsouoti* » (entrepreneur) ait la charge d'embaucher tout son monde, cuisiniers compris, et qu'il assume la responsabilité de leurs faits et gestes à tous les points de vue : travail, vol, inconduite, etc.

La construction peut se faire à l'entreprise, comme cela se pratique à Tientsin et partout ailleurs où les travaux sont dirigés par un architecte responsable. En dehors de là, étant donné les mœurs du pays, ce mode de contrat n'est guère à conseiller, même si l'on dispose de très bons surveillants « *kien koung* », aussi consciencieux qu'entendus dans la partie, et qui n'auraient pas à quitter le chantier, pendant toute la durée des travaux. Fait à l'entreprise, le bâtiment s'élèvera très rapidement puisque les ouvriers seront au travail avant le lever du soleil, qu'ils s'accorderont à peine quelques moments de repos à midi, même au mois de Juillet, et qu'ils ne descendront des échafaudages que quand on n'y verra plus clair. Vous pourrez même réaliser une économie, et votre dépense sera moindre qu'en faisant travailler à la journée. Il est à peu près certain que vous vous ferez même [p. 12] du mauvais sang ; mais le bâtiment achevé, attendez une de ces pluies dont le mois de Juillet et d'Août ont la spécialité dans la plaine du Tcheli et aux environs ; vous n'aurez rien perdu à attendre pour vous faire la quantité voulue de mauvais sang ; quant à l'économie réalisée, vous la calculerez aisément aux murs qui se fendront, se gondoleront, aux toits qui crouleront et le reste.

Mieux vaut donc faire travailler à la journée et dépenser un peu plus. Prenez stoïquement votre part de ce qu'ils appellent « *mouo koung fou* » (tenere tempus) ; quand ils veulent parler élégamment, ils disent « *mouo yang houng* » (prendre le temps aux Européens). Laissez porter les paniers vides au pas de procession et ne vous fâchez pas trop fort, ni trop fort ni trop souvent, en voyant deux hommes combiner savamment l'équilibre de leur charge de six briques ; mais pour éviter, autant que possible, les abus par trop criants en cette sorte de sport, commencez par chercher un ou deux bons surveillants ; je ne dis pas un ou deux chrétiens, mais deux malins sur lesquels on peut compter ; ils n'abondent pas, je le sais ; mais grâce à Dieu, ils ne sont pas introuvables parmi nos gens qui ont la foi. Il sera bon d'établir une certaine hiérarchie, s'il y a plusieurs surveillants, et d'avoir l'œil aux différences de caractères (*p'i k'i*).

L'un de ces surveillants sera en même temps acheteur. Ici le contrôle personnel du missionnaire s'impose plus que partout ailleurs. Le tant pour cent, les primes (*young ts'ien*) comme ils les appellent, ne peuvent être prélevées par nos gens, qui ne sont nullement courtiers, « *king ki* ». Ils sont payés au mois, comme les autres catéchistes, donc, comme ceux-ci, hommes de la Sainte Église. Promettez-leur une bonne récompense pécuniaire, pas seulement un beau crucifix. Même vingt piastres ne seront pas perdues pour une grande construction, si les surveillants ont bien rempli leur office. Il est important que les surveillants aient du tact pour traiter avec leur monde, qu'ils ne brusquent pas les ouvriers, ne leur fassent pas perdre inutilement la face, tout en étant fermes dans leurs exigences. Sans ces quelques précautions, gare le sabotage. Il est si facile d'oublier de pilonner un coin des fondations, et surtout d'avoir une distraction, tout à la fin de la construction, en posant les tuiles de la toiture.

Il est bon que l'entrepreneur soit un homme connu, ayant fait ses preuves comme bon ouvrier et comme honnête homme. [p. 13] Du moins doit-il avoir de bons répondants, s'il est inconnu. Les renseignements que pourront fournir d'autres missionnaires constructeurs ne seront pas à dédaigner.

Cet entrepreneur gagnera sur ses ouvriers : ce ne sera que justice si le gain n'est pas exagéré, puisque sa responsabilité est engagée et qu'il devra pouvoir en remontrer aux meilleurs ouvriers par un savoir-faire que les autres ne possèdent pas. Donc ne vous laissez pas trop impressionner par les accusations dont il sera l'objet quand, à la fin de votre construction, il achètera trois « mou » de terre.

On conviendra avec l'entrepreneur du nombre total d'ouvriers à employer et de la proportion des manœuvres « siao koung » par rapport aux maçons « pa cheu » ; du nombre aussi des cuisiniers, meuniers, porteurs d'eau. Spécifiez aussi : le viatique, aller-retour ou retour seulement, par qui sera nourri l'âne du moulin, le sel, le tabac, les « k'aolao » (repas mieux soignés en guise de récompense), les ustensiles de cuisine, les jours de grandes fêtes, les jours de pluie, le cas de maladie, le chauffage, le chômage le dimanche pour les Chrétiens, les payens, etc. etc. Et tout cela par écrit, suivant le principe : avec les honnêtes gens, de l'écrit ; avec les coquins, pas d'affaires.

Ordinairement on compte deux manœuvres par maçon. Les « pa cheu » (maîtres ouvriers) soit maçons, soit surtout menuisiers, ont presque toujours avec eux quelques apprentis « t'ou ti ». Ceux-ci ne reçoivent pas grand-chose de leur maître (« lao cheull » ou « cheu fou », selon les régions) ; mais vous payez sans distinction de maître ou d'apprenti, le salaire des premiers.

Il est de bonne politique de laisser entrevoir une récompense pécuniaire à l'entrepreneur comme aux surveillants. Ce sera de l'argent bien placé, s'il a été mérité : 20 piastres ne seront pas de trop pour une construction tant soit peu considérable.

Il se peut que l'entrepreneur, en calculant ses profits et pertes possibles, se trompe dans ses appréciations. Les ouvriers, quand ils « mangent le patron », (d'après leur expression « tch'eu tong tia »), ont un appétit formidable ; de plus il peut y avoir des vols de pain, de grains, des visites de parents et connaissances, une augmentation subite du prix des grains ; que sais-je ? Ne vous laissez pas impressionner trop facilement par les gémissements de ce cher homme, mais si les preuves sont faites qu'il travaille à perte [p. 14] et qu'il n'y a pas de sa faute, ce serait cruauté de s'en tenir à la lettre morte du contrat, et de placer le malheureux entre l'alternative de prendre la fuite ou de ne pas payer ses ouvriers, comme cela se pratique dans certaines grandes entreprises.

[p. 14, 15, 16, 17 and 18: Chinese model contracts, see fig. 2.2]

Chapitre quatrième: Des matériaux

Briques. Briquetiers. Les « chaoyaoti » (briquetiers) sont gens de conscience fort élastique. Ils sont surtout prodigues de belles promesses. C'est un dur métier exercé par des gens sans grand capital. Et pourtant, il leur faut de l'argent comptant pour acheter paille, houille et tiges de sorgho. Prenez vos précautions, non seulement en rédigeant le contrat, spécifiant les dimensions des briques, tuiles, etc. l'époque à laquelle elles devront être livrées, etc. , mais surtout qu'on vous présente des sérieux répondants, que vous aurez vus vous-mêmes, avant de verser une somme quelconque.

Four. Pour une grande construction, il y a avantage à avoir un ou plusieurs fours à soi, et à fournir soi-même le chauffage. De cette sorte, on sera sûr de ne pas manquer de briques ou de tuiles au beau milieu des travaux. C'est à peu près aussi le seul moyen pour avoir des briques bien cuites. Mais il sera de toute nécessité, dans ce cas, d'avoir constamment aux environs du four, un surveillant entendu et consciencieux.

Dimensions. Il y a des briques de quatre livres, il y en a de neuf. C'est à peu près le minimum et le maximum de ce qui se fait actuellement, et de ce qui semble pratique avec les moyens du pays. Vous pouvez choisir dans toute la gamme, intermédiaire entre ces deux poids.

En indiquant aux briquetiers le poids de la brique que vous désirez — et ils ne vous demanderont pas d'autre indication — vous leur aurez dit les dimensions à donner au moule, retrait de la cuisson compris. Inutile donc de vous préoccuper d'aucun autre détail, si l'ouvrier sait que vous n'accepterez aucun « *tch'a pou touo* »quant au poids. Mais pour cela : il faut parler ferme, très ferme, et très clair, dès le début, et faire rédiger le contrat avec non moins de précision. Deux largeurs de brique, ou quatre épaisseurs, équivalent à une longueur : voilà les dimensions stéréotypées. Ainsi une brique de 9 pouces de long aura quatre pouces [p. 20] et demi de large, pour une épaisseur de deux pouces et deux « *fenn* » et demi (le « *fenn* » est le dixième du pouce).

La maçonnerie d'une église se fera plus facilement et plus économiquement, si dès le début on dresse les plans d'après un « module » déterminé, et si ensuite toutes les briques sont cuites d'après ce module ; en d'autres termes, ce sera la longueur de la brique qui sera l'unité de mesure. Ce module est une mesure arbitraire, que le constructeur peut préciser à son gré, entre huit et dix pouces chinois, de façon à obtenir les dimensions voulues pour l'église à construire.

Mais dans ce cas il ne suffira pas de dire au briquetier le poids de la brique désirée ; il faudrait se contenter au contraire de lui indiquer la longueur de la brique adoptée. C'est sa longueur qui dans ce cas déterminerait le poids. Tenir compte, dans ses calculs, du retrait de la brique pendant la dessiccation et pendant la cuisson. Tenir compte aussi du joint de la brique maçonnée. Ce joint est à peu près de trois « *fenn* » (dixième de pouce).

Pour obtenir ces dernières, il faudra fournir soi-même les moules, car les gens du pays, sauf les très grands richards, n'emploient guère de si bons matériaux. Les moules s'usent rapidement, si on les racle au fil de fer, ce qui se pratique ordinairement. De ce chef, il y aura une différence notable entre les pisés moulés au début ou à la fin de la saison, parfois une livre et plus. Les maçons seront gênés dans l'emploi de ces briques d'épaisseurs inégales et la maçonnerie sera moins belle. Une petite lamelle de fer, fixée sur les rebords du moule, préviendra l'usure en question. Mais le briquetier, né malin, trouvera le moyen de faire sauter vos lamelles de fer, car une brique de 8 livres coûtera moins de combustible qu'une autre de 9 livres. Il ne faudra donc point négliger de vérifier les moules de temps en temps.

Briques rouges. Les Chinois n'aiment pas les briques rouges. Pour eux une brique rouge est une brique mal cuite. Et ils ont raison à leur point de vue. Pour avoir une belle brique bleue, il faut avant tout qu'elle soit bien cuite, sans quoi elle ne supportera pas l'arrosage « *yinn yao* » qui suit la cuisson ; elle tombera en poussière avant de sortir du four. Pour faire des briques rouges au contraire, il suffit du premier coup de feu : laissez refroidir et sortez du four. Pas d'eau à ajouter, pas de mauvaise coloration à redouter, et que de chauffage épargné. Aussi il faut [p. 21] être « *Yang jenn* » (Européen = imbécile) pour exiger des briques rouges.

Briques panachées, vitrifiées. Des briques peuvent cependant être plus ou moins rouges, panachées, sans être mal cuites. C'est l'arrivage défectueux de l'eau à la fin de la cuisson, le « *yinn yao* », qui en est la cause. Ces briques se vendent au rabais, car un mur serait vraiment disgracieux s'il était construit avec ces briques panachées. On peut les employer pour les fondations ou bien à l'intérieur des maçonneries.

Il y a encore les « *tsiao tchoan* », briques vitrifiées. Ce sont celles qui ont été les plus exposées au feu, à l'ouverture du four « *yao k'eou* ». Il faut réserver précieusement ces briques pour les premières assises extérieures des murs au sortir de terre ; elles ne se désagrégeront jamais sous l'action du salpêtre. Ce sont elles aussi qui seront employées aux endroits qui auront à supporter une grande pression, v. g. au-dessus des abaques des colonnes. Le triage de ces « *tsiao tchoan* » doit être fait dès la sortie du four, sans quoi elles seront confondues au milieu des autres. Il faudra faire surveiller aussi les maçons pour les obliger à employer ces briques aux bons endroits. Ils n'aiment pas à les manipuler, car elles leur coupent les doigts et ils y cassent leurs instruments.

Terre à brique. Il y aurait lieu de parler ici de la terre à brique, surtout si on veut avoir un four à soi. Toute terre n'est pas bonne pour faire des briques. L'argile commune choisie pour la composition des briques ne doit être ni trop grasse ni trop maigre. Quand la terre est trop grasse, les produits se gauchissent, se déforment et se fendillent à la cuisson. Dans le second cas, les briques obtenues se vitrifieraient trop facilement et fondraient au feu.

Si l'argile est trop grasse, on y ajoute du sable fin ; si elle est trop maigre, on y ajoute de l'argile plastique. Mais c'est la question laissée au jugement des gens du métier. Il vous suffit de savoir que vous ne pouvez pas construire un four dans n'importe quel champ.

Briques à fleurs. Pour une église surtout, on aura besoin d'un certain nombre, fort considérable parfois, de briques d'une forme spéciale, pour faire des moulures et autres ornementations. Tailler ces briques à la main, après cuisson, demanderait un travail considérable et entraînerait une forte dépense. Les plans de l'architecte doivent prévoir la forme de ces briques. Il faut en calculer [p. 22] la quantité, puis on confectionne les moules spéciaux pour chaque genre de moulure. Ces briques doivent être faites en quantité beaucoup plus que suffisante, si on ne veut pas être surpris, car elles sont plus fragiles que les briques ordinaires à cause des échancrures. Du reste celles qui ne serviraient pas à leur vraie place, peuvent parfaitement être employées à l'intérieur des maçonneries (voir modèles de ces briques, figures 1 à 40).

Les fleurons, chapiteaux et quelques autres pièces du même genre, ne peuvent être faits au moule. Il faut les faire modeler par un modeleur, avec une bonne terre à poterie, mêlée de chanvre haché. Après dessiccation suffisante, on fait cuire au four. À cause de la « casse », très considérable, soit avant soit après la cuisson, il faut préparer environ le triple de la quantité nécessaire. Il est assez facile de trouver de bons modeleurs dans le pays, parmi ceux qui savent faire des têtes de dragons et autres ornementations du toit chinois.

Recommandation pour modeler un chapiteau. Modeler un beau chapiteau n'est pas chose très facile sans quelques indications. Il faut se rappeler que les feuilles qui ornementent un chapiteau, sont censées [être] appliquées sur une corbeille, mais à leur base se confondent avec la corbeille, en sorte que le bas du chapiteau ne doit pas être plus gros que la colonne elle-même. Le grand défaut des modeleurs inhabiles est de faire des chapiteaux trop gros à la base, ce qui est affreux. Voir figure 76, bon chapiteau ; figure 76 bis, mauvais chapiteau.

Recommander en outre de bien « galber » la feuille qui va vers le coin, c'est-à-dire d'en creuser la courbure d'une façon suffisante. Ordinairement les modeleurs maladroits font la feuille trop droite. Un chapiteau bien galbé fait toujours bon effet, même lorsque les détails du feuillage sont très sommaires. Au contraire, un chapiteau très fouillé, mais mal proportionné, sera toujours désagréable à voir.

Triage des briques. Étant donné la conscience large des briquetiers, il est bon qu'ils sachent qu'on fera le choix des briques, et que celles qui seront rejetées leur resteront pour compte. Ce détail est à stipuler sur le contrat. Pour faire le triage, on se sert d'un gros clou ou de toute autre pièce de fer. La bonne brique se reconnaît au son métallique qu'elle rend. Les autres sonnent « fêlé ».

Quantité. Il va de soi que la quantité des briques nécessaires pour une construction se calcule au mètre cube. Si les plans sont bien faits, cette supputation est facile à établir. Pour les constructions ordinaires, les maçons chinois savent la quantité par « kien » (travée). Mais il faut tenir compte des dimensions des briques aussi bien que de celles des « kien ».

« *Pa tchoan* » (Dalles carrées). Ce sont de petites dalles qui servent surtout pour le toit chinois. Elles sont de différentes dimensions, comme les briques, mais toujours carrées. Comme les briques, elles doivent rendre un son clair, quand on les frappe avec un corps dur. L'épaisseur ordinaire de ces dalles carrées est de 1 pouce. Inutile de les faire trop épaisses, elles chargeraient inutilement la charpente. Mais qu'elles soient aussi bien cuites ; si elles ne le sont pas, au bout de deux ou trois ans, les morceaux vous en tomberont sur la tête quand vous serez à table, heureux si les mauvais « *pa tchoan* », d'un bloc, ne se détache pas d'entre les deux chevrons et ne se précipite pas sur vous de la hauteur d'une quinzaine de pieds. J'en ai vu passer à travers le plafond comme un obus d'aéroplane.

Il faut prévoir plus de « *pa tchoan* » qu'il n'en faut pour la toiture ; on en emploie beaucoup pour les bordures et les moulures.

Tuiles « *wa* ». Les tuiles sont généralement cuites en même temps que les briques, dans un coin du four. À cause de leur peu d'épaisseur, exposées qu'elles sont au même feu que les briques, elles sont ordinairement bien cuites. Elles sont tournées « *suan* » par des spécialistes « *suanwati* », qu'il faut engager en dehors des briquetiers, si on a un four à soi.

Il importe que les tuiles ne soient pas trop petites pour ne pas être dérangées trop facilement par le vent ou par les corbeaux.

Il y a deux espèces de tuiles ; les « *t'oung wa* » (tuiles creuses), en demi cylindre. On les emploie presque exclusivement pour les pagodes, et jadis pour les palais impériaux. Elles sont belles à voir, disent les Chinois. Ce qui est certain, c'est qu'elles couvrent for mal. Aussi ne les employez jamais pour une église, sous prétexte que c'est « *hao k'an* » (beau à voir). — L'autre espèce est la tuile ordinaire, le quart de cylindre ou plutôt d'un cône tronqué.

Les maçons chinois comptent 20 tuiles par pied carré. Pour un « *kien* » ordinaire, avec poutre de 16 pieds, il en faut 3000 environ. [p. 24] Avec cette donnée, on pourra calculer la quantité de tuiles nécessaires pour une autre surface connue ; v. g. si les poutres ont 20 ou 30 pieds. Dans ce calcul il faudra tenir compte aussi du genre de recouvrement qu'on choisira. Nous y reviendrons en parlant de la toiture.

Chose curieuse, on assure que les Grecs et les Romains employaient le même genre de tuiles que les Chinois aujourd'hui et d'il y a des milliers d'années. En Italie, on emploie encore le même mode de couverture qu'en Chine. On pose sur les chevrons des dalles de 0m, 03 d'épaisseur et on garnit les joints en mortier ; sur l'espèce de dallage obtenu, on range les tuiles, sans doute sur un lit de boue épaisse, comme par ici, depuis longtemps avant Confucius. (Barberot, Construction civiles, p. 455).

En dehors des tuiles ordinaires, il faut songer à celles qui termineront le toit à sa partie inférieure, nommées « *ti choei* » (gouttières) et « *keou yen* » (recouvrement). Elles sont moulées d'une façon spéciale et coûtent plus cher que les autres.

Bois « *mouleao* **».** Dans le nord de notre Mission du Tcheli S.-E. , on ne connaît guère, pour la construction, que le sapin et autres bois venant de Tientsin. Ils sont fort chers, en comparaison de ce qu'on trouve sur place, plus au Sud ; et encore faut-il faire faire ses achats par un connaisseur, faute de quoi, on vous enverra du bois à moitié pourri, des troncs noueux ne pouvant être débités en planches, etc. Quand l'acheteur aura fait son choix, il faudra qu'il veille aux échanges pendant le chargement des barques, et même en cours de route. Il est nécessaire aussi de faire ses commandes à temps ; les communications sont souvent fort lentes. Puis il faudra débiter ces pièces pour leur permettre de sécher avant d'être mises en œuvre.

Dans le Sud de cette même Mission, il y a beaucoup de peupliers « *yang chou* », ormes « *u chou* » etc. , etc. Pour la charpente, ces bois sont tout aussi bons et meilleur marché que les bois de Tientsin. Seuls, portes, fenêtres et mobilier ne pourront guère être faits avec ces bois du pays qui jouent trop, même quand ils sont complètement secs.

Il faut éviter d'employer le saule pour toute construction sérieuse. Au bout d'un certain nombre d'années, il est rongé par les vers et tombe en poussière.

On augmentera d'un sixième la force des bois de charpente, [p. 25] en les écorçant quelques mois avant de les abattre. Cette opération doit se faire au moment où l'arbre est en pleine sève, en Mai par exemple, pour couper en automne.

On conserve indéfiniment les bois de charpente, si, après les avoir écorcés et sciés, on les laisse tremper pendant quelques jours dans du lait de chaux, et si on les enduit exactement.

En tout cas, l'abattage des bois doit se faire avant le mouvement de la sève, donc en hiver, d'Octobre à Février.

La dessiccation naturelle du bois consiste à le protéger contre le soleil et la pluie et à le laisser aussi sécher, à libre circulation d'air.

Pour les grosses pièces de la charpente, le « jeu » a peu d'inconvénients, et on peut employer des bois quasi verts pour ces parties de la ferme.

Les dimensions des pannes, « *linn* », en peuplier vert, sont de 6 pouces, 0m, 20 de diamètre, sur 10 pieds, 3m, 30 de longueur. Il ne faut pas exagérer la longueur des pannes : 11 pieds est un maximum pour celles qui ne sont pas posées dans les murs des pignons, surtout si c'est du bois vert. Des « linn » en orme, pas sèches, de 11 pieds, plieraient sous la charge du toit chinois.

Les chevrons ont ordinairement 2 pouces de côté, 0m, 066 à 0m, 070, pour une longueur de 4 pieds environ. La longueur des chevrons dépend naturellement de l'écartement des pannes. Il importe de connaître exactement cet écartement, d'après le plan, avant de faire la commande des chevrons, pour ne pas payer très cher du bois inutile.

Le nombre des chevrons par travée se calcule d'après les dimensions des dalles, « *pa tchoan* ». C'est ordinairement 12 chevrons par « *kien* » et par rangée. Le nombre des rangées dépend de la longueur du bâtiment.

Chaux « *cheu hoei, ta hoei* » ou seulement « *hoei* ». On distingue chaux grasse & chaux maigre. La chaux grasse est le carbonate de calcium ou calcaire fortement chauffé, qui conserve à peu près la même apparence qu'avant la cuisson. Si on mouille ces pierres, la chaux s'échauffe, fuse et foisonne, c'est-à-dire augmente considérablement de volume et dégage une chaleur considérable. Elle devient alors chaux éteinte ou chaux hydratée. La chaux vive, c'est-à-dire avant d'être combinée avec l'eau, prend le nom de chaux anhydre, en chinois, « *cheng hoei* » chaux crue. La chaux [p. 26] éteinte et en poudre se dit en chinois « *t'ang hoei* » ou « *chou cheu hoei* ». La chaux éteinte et mélangée avec de l'eau se délaie et forme une pâte liante, propre à entrer dans la composition du mortier. « *Linn hoei* » veut dire délayer ainsi la chaux et la filtrer à travers un grillage, pour en extraire pierres et petits cailloux. On conserve cette chaux délayée et débarrassée des corps étrangers, dans des fosses appelées « *cheu hoei k'eng* ». On la recouvre d'une couche de terre sèche, dès que l'eau en trop a été absorbée par la terre. Ce recouvrement avec de la terre est important pour mettre la chaux à l'abri de l'air et l'empêcher de se combiner avec l'acide carbonique. Bien recouverte dans cette fosse, la chaux éteinte peut se conserver indéfiniment jusqu'au moment de l'employer comme mortier. Il faut la préserver aussi de la gelée, donc la recouvrir d'une couche de terre assez épaisse. La chaux fraîchement éteinte ne vaut pas celle qui l'est de longue date.

Plus étendue encore, la chaux éteinte forme le lait de chaux, « *cheu hoei chouli* » ou « *cheu hoei jou* ». Elle est alors propre au badigeon, « *choa cheu hoei* ».

Chaux maigre. Les calcaires contenant des matières étrangères inertes, fournissent des chaux qui s'hydratent plus lentement, qui foisonnent moins et qui, combinés avec l'eau, forment une pâte moins liée. On les appelle chaux maigres. Pas plus que les chaux grasses, elles ne prennent sous l'eau.

Pour avoir de la chaux hydraulique, prenant sous l'eau, il faut des calcaires contenant environ 12 à 20% d'argile.

Nos chaux du Tcheli et celles qui viennent des provinces voisines sont surtout des chaux grasses ; quelques-unes sont des chaux maigres. Je n'y connais pas de chaux hydriques.

Mortier de chaux grasse. Additionnée de deux ou trois fois son volume de sable (bon sable de rivière), cette chaux forme le mortier de chaux grasse. Il faut donc préparer du sable. Ce mortier durcit lentement, aussi faut-il se garder de construire trop rapidement, mais bien laisser aux assises inférieures le temps de durcir assez pour supporter le poids des parties supérieures. Aussi, nous le répéterons plus tard, il est bon de ne pas achever les murs d'une grande église dès la première année. L'idéal sera de s'arrêter à la hauteur des fenêtres à l'automne de la première année, puis de reprendre au printemps suivant.

Achat de la chaux. Il faut stipuler dans le contrat d'achat, qu'on [p. 27] n'acceptera pas les pierres ! En chargeant la chaux sur les chars ou sur les barques, les fournisseurs peuvent être tentés d'ajouter des pierres brutes. On ne les découvrira qu'au moment du « *linnhoei* ».

Il ne faut pas faire charrier la chaux en été. Vous n'aurez que de la poussière, et vous payerez l'eau absorbée en route au prix de la chaux. Cette quantité d'eau n'est nullement négligeable. La meilleure époque pour acheter et faire transporter la chaux est le printemps ; du reste, dans la montagne, la calcination des pierres à chaux a lieu en hiver surtout.

Dès que la chaux est livrée, on la fait éteindre, puis couler dans les fosses préparées d'avance, dont nous avons parlé. Il faut qu'au moins un des ouvriers soit au courant de cette manipulation.

Pas plus que pour la fabrication des briques, il ne faut employer de l'eau saumâtre pour délayer la chaux.

Cordes « *cheng, ma cheng* ». La dépense pour cet article est considérable. Il faut donc la prévoir dans le devis.

L'échafaudage de l'ouvrier du pays n'a rien d'artistique. C'est un assemblage assez informe, souvent même trop instable s'il s'agit par exemple d'un clocher. Il est maintenu quasi exclusivement par des cordes et des cordes encore plus informes que l'échafaudage lui-même, puisqu'elles sont tressées sur place par n'importe quel manœuvre.

Pourtant il faut distinguer des cordes de deux espèces : il y a celles qui servent à la traction sous toutes ses formes : force de bras, poulies, etc. Ces cordes sont faites avec soin par des professionnels, les cordiers « *ta chengti* ». On les fabrique avec un chanvre qui reçoit différents noms, selon les régions : « *hao ma* ».

Elles se vendent naturellement à la livre et sont d'un prix assez élevé. On ne les emploie pas pour fixer les perches des échafaudages, parce qu'elles pourriront rapidement si elles restent exposées à la pluie.

Les cordes qui maintiennent les échafaudages sont faites avec du « *penn ma* » ou « *k'iou ma* » appelé encore « *k'ing ma* ».

Ces cordes sont simplement tressées ; elles ne sont pas tordues comme font les cordiers pour les bonnes cordes. Le manœuvre les tresse par longueurs très variables, selon l'usage qui en sera fait. Il va de soi qu'elles ne peuvent pas servir pour une poulie quelconque. Elles peuvent rester exposées à la pluie et à l'humidité [p. 28] pendant toute la durée de la construction, deux ans s'il le faut, sans pourrir. On les hache ensuite finement pour les mêler à la chaux du crépissage. Sous cette forme, ce chanvre prend le nom de « *ma tai* » ou « *ma tao* » et la chaux à laquelle il est incorporé « *ma ta hoei* ».

Perches pour échafaudages « *kantze* ». Encore une forte dépense qu'on pourrait être tenté d'oublier en établissant le devis. Pour une construction de moindre importance, on pourra souvent emprunter perches et planches. Il n'y a guère avantage à les louer. La location coûte assez cher et il ne vous restera rien de l'argent ainsi dépensé. Au contraire, le moindre bout de bois que vous aurez acheté trouvera son emploi, ne serait-ce que pour être débité en lattes pour le plafonnage.

À proximité de Tientsin, on n'emploie guère que les « *cha kao* », petits sapins. On en trouve aussi à bon prix aux environs des chemins de fer et des fleuves. Ailleurs les jeunes peupliers, trop faibles pour fournir des chevrons, coûtent moins cher et rendent le même service. On les revend même plus facilement, après les travaux, que les « *cha kao* ».

Si les perches doivent rester longtemps plantées en terre, v. g. un ou deux ans, pour une grande église, il est bon d'enduire préalablement la partie enterrée d'une couche de goudron.

Chapitre cinquième: La maçonnerie

Article 1er. Les Fondations. Dans notre plaine du Tcheli et aux environs, inutile de pratiquer des sondages, pour aller chercher le terrain de roches. Du reste, vous pouvez être sans inquiétude ; les bons maçons connaissent leur pays et le sous-sol. Si vous ne les gênez pas dans leurs opérations, un peu lentes, il est vrai, ils établiront le plus beau clocher sur une base inébranlable. Ce sera au moyen du battage avec la hie ou demoiselle. Cette opération se dit « *ta hang* ». La hie se dit « *hang* ». Il y a des [p.29] « *mou hang* » en bois, « *cheu hang* » en pierre, « *t'ie hang* » en fer. Au besoin, ils sauront employer béton et pilotis. Laissez-les juger.

Après la consistance du sol, l'empattement est le plus important facteur de stabilité des constructions. On appelle empattement la disposition des murs de fondation, plus larges à la base et se rétrécissant assise par assise, jusqu'à la sortie de terre. C'est l'empattement qui répartit le poids de la construction élevée au-dessus, sur une surface plus développée, diminue ainsi la pression verticale par unité de surface et permet de construire sur un sol même peu résistant. L'ouvrier chinois connait ce principe et l'emploie d'une façon fort rationnelle. Il n'y a qu'à le laisser faire sur ce point ! Voyez figure 41.

On économisera beaucoup d'argent, si l'entrepreneur consent à n'employer que peu de « *pa-cheu* » (maîtres-maçons) pour creuser et battre les fondations. En effet, tout ce travail important peut se faire par des manœuvres, sous la direction de deux ou trois maçons seulement.

Ces manœuvres creusent d'abord les tranchées « *kiue ti kiao* », à une profondeur variable, suivant la nature du terrain. Ils ont l'habitude de les creuser pas trop « *tch'a pou touo* », d'après le tracé qui a été préalablement fait aux piquets et aux cordeaux. Après qu'ils auront damé avec force cris, pendant quinze jours, un mois, au moment de commencer la maçonnerie on constatera que les murs tombent à deux pieds en dehors du terrain battu, ou du moins n'y reposent qu'à demi. Si vous n'y veillez, ils feront alors en toute hâte un damage supplémentaire, naturellement des plus sommaires, puisqu'il s'agit de s'éviter des reproches, et en avant la construction.

De fait il est assez difficile, non pas de tracer les limites du bâtiment sur le terrain encore intact, mais de creuser ensuite les tranchées, à plusieurs pieds de profondeur, d'après ce tracé. Les piquets disparaissent sous les premiers coups de pioche, la terre extraite et rejetée modifie l'aspect du terrain. Il faut donc enfoncer des piquets témoins, assez élevés, en dehors du terrain à creuser et à une certaine distance. De temps en temps, les ouvriers pourront viser d'un piquet à l'autre et constater qu'ils restent dans l'alignement. Puis, quand les tranchées sont complètement ouvertes, déblayées et nivelées, avant de donner le premier coup de hie, on refait le tracé de la tranchée, avec toute l'exactitude [p. 30] possible, en replaçant, cette fois, les piquets au fond des tranchées, pour constater surtout à l'équerre, si les angles droits ne sont pas devenus de 80° ou de 100°.

Il va de soi que les tranchées doivent être assez larges pour les empattements et pas seulement pour les murs tels qu'ils seront à la sortie du sol.

Art. 2ème. Le « *ta hang* » (damage). Le « *ta hang* » est une opération des plus fastidieuses… pour celui qui paie, surtout s'il est pressé de voir sortir sa construction de terre. Et pourtant, il est nécessaire de ne pas céder, dès ce début, à l'énervement. Pour tout le monde, sauf pour le « *tong jia* » (patron), le « *ta hang* » est une kermesse. Les enfants viendront en prendre leur part. On y mange bien, du pain blanc « *mouo mouo* » tous les jours, des « *k'aolao* » (petits festins à la viande et à l'eau de vie) de temps en temps ; on y chante beaucoup, c'est parfois un peu… gaulois ; mais surtout on y travaille fort peu, malgré ou à cause des grand cris. Quand c'est un particulier, surtout un richard qui construit, les parents « *ts'ints'i* », à tous les degrés, arrivent de loin et s'il a bonne presse, tout le village viendra aider à manger ses bons dîners.

Pour vous, prenez-en votre parti, remerciez ces braves gens qui sont venus dépenser leur cœur et vos provisions, activez le mouvement et donnez du cran aux coups de demoiselle par quelques « *k'aolao* » supplémentaires, et vous verrez que votre église ne se fendra jamais par le milieu, comme celle de x !

Ce n'est pas à dire pourtant qu'il faille tolérer tous les paresseux à n'importe quel degré, mais, soit dit une fois pour toutes, mieux vaut ne rien dire sur place, ni personnellement. On fait signaler le monsieur fatigué à l'entrepreneur. Le paresseux se corrigera ou sera éliminé sous un autre prétexte, et toutes les faces seront sauvées.

Pour savoir si le sol est suffisamment battu, on constate qu'il rend un son sec, quasi métallique. C'est à l'entrepreneur à faire cette constatation. Pour une église, un clocher surtout, on ajoutera un bétonnage pilonné. On fait concasser des bricailles [briquaillons] de la grosseur moyenne d'une noix. Ces briques concassées sont arrosées avec de l'eau de chaux assez épaisse, dans la proportion d'un volume de chaux pour cinq de bricailles. On amalgame et on laisse la chaux pénétrer dans la brique pendant un jour ou deux.

Au moment de pilonner, on jette ce mélange de bricailles et de chaux dans les tranchées des fondations, on y ajoute de la terre qu'on amalgame, puis un peu d'eau, on égalise et on recommence à damer, d'abord par petits coups, puis à grands coups de demoiselle. Si ce travail est bien exécuté, on obtiendra un bloc d'une pièce, de la forme exacte du fond de la tranchée ; il faudra employer la barre à fer pour y forer un trou. Ce sera comme une énorme dalle qui supportera d'une façon uniforme la pression des murs, en la répartissant sur toute l'étendue de la tranchée des fondations. On pourra y poser, sans crainte aucune de tassement, de fort jolis clochers de 50m et plus.

Dans ce cas, il faudra naturellement donner une épaisseur plus considérable au bétonnage, et le pilonner avec plus de soin, par couches successives. Il ne faudrait pas non plus achever la construction d'un pareil clocher, dès la première année, mais laisser tasser la moitié pendant un hiver, en recouvrant bien les fondations pour les préserver de la gelée.

Article 3ème. La maçonnerie. Le mortier. Nous en avons parlé à la page 26, à propos des matériaux à préparer. Le sable remplit ici le rôle de matière inerte, il augmente économiquement le volume du mortier ; de plus il est utile, parce qu'il divise la chaux, la rend plus perméable à l'air et lui facilite la prise par la possibilité d'absorption de l'acide carbonique. Les maçons chinois ont donc tort d'employer la chaux pure, sous prétexte que c'est plus « *hao k'an* » (beau à voir). Que de fois, chez eux, le « *hao k'an* » prime la solidité !

Cependant ce mortier de chaux grasse ne durcit qu'à l'air. Il ne faut donc pas trop se hâter pour « enterrer » les fondations, mais autant que possible les laisser se dessécher avant de les recouvrir de terre. On peut mélanger le sable à la chaux grasse à plusieurs reprises, pour faciliter l'absorption de la plus grande quantité possible d'acide carbonique ; on ramollit ensuite par addition d'eau, au fur et à mesure des besoins. Mais toujours le mélange doit être fait avec le plus grand soin, pour assurer une parfaite homogénéité au mortier.

Les murs. Nous avons parlé de l'empattement des murs de fondations. L'empattement, sur un bon sol, pour une maison à plusieurs étages, ou pour les murs ordinaires d'une église, doit être de 0m, 70 à 0m, 80 d'épaisseur.

Avant de parler des murs en élévation, c'est-à-dire au-dessus du sol, disons un mot de la préparation à donner aux briques. Le secret d'une bonne maçonnerie est tout entier dans ce précepte : mortier ferme et matériaux mouillés. Or c'est toujours l'inverse qu'on fait ; nos maçons préfèrent noyer le mortier que de mouiller la brique ou la pierre, et les constructeurs laissent faire et ont tort. Les matériaux secs, les briques surtout, ont bien vite absorbé la faible quantité d'eau que contient le mortier, puis tous deux sèchent séparément ; le mortier appauvri d'eau, est presque en poudre, et la cohésion, l'adhérence qu'on se proposait d'obtenir ne se produit que partiellement, là où pour une cause quelconque l'eau n'a pas été entièrement prise par la brique ou la pierre » (Barberot. Constructions civiles. p. 887).

Une brique de 8 à 9 livres peut absorber près de 2 livres d'eau. Mais l'absorption se fait assez lentement. Il ne faut donc pas se contenter d'arroser les briques, ni même de les plonger un instant dans l'eau. Il faut les faire *tremper*, afin d'obtenir un « *yinn t'eou leao* » (compénétré à fond). À cet effet, pour une petite construction, on peut se contenter d'un marmite « *kouo* », ou d'une grande jarre « *wong* ». Pour un grand bâtiment, on creuse un petit bassin, dont le fond sera garni de terre glaise, « *kiao ni* » pour empêcher l'eau de filtrer en terre, puis on le revêtira de briques pour que l'eau reste propre. Dans ce bassin, on range les briques à employer, on remplit d'eau et on laisse *tremper* pendant 20 à 30 minutes. Si les briques ainsi trempées ne sont pas employées immédiatement, et qu'il faille les retirer du bassin pour en préparer d'autres, en vue du moment de presse, il faudrait les arroser de nouveau au moment de s'en servir. Puis le maçon, ou un manœuvre, arrosera de nouveau la rangée de briques qu'il vient de maçonner, avant de superposer une nouvelle assise.

Il ne faut pas employer de l'eau saumâtre pour cette opération. Les murs, il est vrai, resteront humides pendant longtemps, à la suite de la grande quantité d'eau ainsi absorbée. C'est pour cela qu'il vaut mieux construire au printemps qu'en été. C'est pour cela aussi que le proverbe chinois dit : « Dans une maison neuve, la première année, on fait habiter un ennemi ; la 2ème année, son ami, et la 3ème année, on y habite soi-même ».

Les murs chinois. Le maçon du pays connaît au moins quatre espèces de murs, sans compter les sous-divisions. Au plus bas [p. 33] de l'échelle, il y a le mur en terre battue, qui se fait encore de deux manières, avec ou sans planches. C'est le « *ta ts'iang* » et le « *touo ts'iang* » (mur frappé, entassé). Puis le mur en pisés, « *p'i ts'iang* », qui lui aussi connaît des subdivisions. Cette manière de construire remonte à la plus haute antiquité, disent les livres. Nous pouvons les croire sur parole. Inutile de vouloir en remontrer au premier paysan venu en ce genre « d'architecture ». Ces maisons sont calculées pour durer vingt ans.

Puis vient le mur du petit et souvent même grand richard : « *li cheng, wai chou* », (cru au-dedans, cuit en dehors), c'est-à-dire pisés à l'intérieur, briques à l'extérieur. En réalité, ce sont deux murs, ou plutôt c'est un vrai mur en pisés, revêtu d'une doublure en briques contre la pluie.

La « doublure » se fait de bien des manières. La plus économique est comme de juste, la moins solide, c'est le revêtement en « *piao tchoan* » (briques sur champ). Ceux qui veulent faire plus solide (mais ils sont rares, car qui peut prévoir l'avenir ?), posent toutes les briques en « *wotchoan* » (briques à plat ou en panneresses). On dit aussi « *pien tchoan* » (plat). Et si tout le mur est revêtu de cette façon, c'est « *pien tchoan tao ting* » (briques à plat jusque sous la toiture).

Le plus gros inconvénient de ce mur double, c'est le tassement qui nécessairement est double aussi. Les pisés, et plus encore le mortier de terre très épais, qui les relie, ont un coefficient de tassement bien supérieur à celui des briques cuites et à surface régulière, reliées par une mince couche de mortier. Au bout de peu de temps, le mur se bombera, et on recommencera. Pourtant, quand il est bien fait, cela peut tenir 40 ans.

Enfin notre mur en « *li wai tchoan* » (briques dedans et dehors). Nous ne parlerons que de celui-là. Malheureusement nos maçons ont très peu l'habitude de construire ce mur et il faut les surveiller de très près, surtout s'ils travaillent à l'entreprise. Ils appliqueront les principes de leurs murs doubles, d'autant que de cette façon, la maçonnerie monte bien plus rapidement. Donc, ils feront avant tout un mur très épais, aussi épais que le mur en pisés, dont ils ont l'habitude. Ce mur sera double : une cloison en briques dehors, une autre dedans. L'intervalle sera rempli avec n'importe quoi, surtout si le surveillant des travaux a le dos tourné. Ainsi, dans cette oubliette, ils empileront pêle-mêle, sans même [p. 34] prendre le temps d'égaliser, des bricailles, de la terre, de la boue, des paniers de chaux pure aussi, s'ils ne sont pas contents de vos « *k'ao lao* ». Puis, sur 6 ou 8 rangées de briques posées en long, ils mettront une rangée en large, pour cacher ce qu'il y a dans l'oubliette, et vous aurez un mur en pures briques. J'ai vu construire de cette façon, à l'entreprise naturellement, les colonnes en briques sous les poutres, dans un mur en pisés. C'était une vraie cheminée bouchée.

Il faut donc avant tout s'assurer des notions d'appareil qu'ont les maçons. On appelle « appareil » en construction, le détail de la disposition des pierres ou des briques dans un édifice. Pour les pierres de taille, appareiller, c'est préparer d'avance la forme des pierres avant de composer un ensemble. C'est la besogne des tailleurs de pierres, dirigés par un chef d'atelier qui trace les épures, les coupes, etc. On pourra en avoir besoin, même dans cette contrée, pour une partie des murs d'une grande église. Il faudra trouver aussi, dans ce cas, des spécialistes pour poser ces pierres de taille. Une bonne pierre, mise en œuvre d'une façon défectueuse, ne peut donner qu'une maçonnerie mauvaise. La manière la meilleure de poser la pierre est la suivante : on présente la pierre à la place qu'elle doit occuper et on la cale près des quatre angles au moyen de morceaux de bois réglés à l'épaisseur du mortier qu'on veut intercaler, soit de 5 à 10 millimètres. On soulève alors la pierre en lui faisant faire quartier, c'est-à-dire, d'un côté seulement, on étend une couche de mortier un peu plus épaisse que les cales, on remet la pierre en place, on frappe dessus à l'aide d'une masse jusqu'à ce que le mortier reflue et que la pierre touche les cales. Quand le mortier commence à prendre une certaine consistance, on retire les cales.

Appareil de briques. Pour les briques qui sont toutes taillées, les choses se font plus simplement. L'appareil consiste uniquement dans la disposition et l'entrecroisement qu'on leur donne sur le mur, suivant son épaisseur.

L'appareil le plus simple et le plus économique et celui en briques sur champ, les « *piao tchoan* » des Chinois. On l'emploie pour les cloisons, là où le lattis enduit de chaux serait trop faible. On serait alors obligé de consolider cette cloison par des poteaux en bois, poteaux de remplissage, écartés au maximum de 2 mètres. (Figure 42).

Pour une cloison plus solide, il faudrait poser la brique à plat [p. 35] et dans le sens de la longueur, en ayant soin, bien entendu, de croiser les joints. (Figure 43). En chinois « *wo tchoan* », briques couchées.

Le mur d'une brique d'épaisseur, soit environ 0m, 24 à 0m, 25 peut être fait au moyen de différents appareils :

1° En briques boutisses (fig. 44), la brique faisant toute l'épaisseur.

2° En boutisses et carreaux (fig. 45), c'est-à-dire chaque assise formée de boutisses et panneresses alternées.

3° Par assises de boutisses alternées d'assises de panneresses. (fig. 48).

Le temple, le « tabernacle », comme ils disent, des protestants à Tamingfou, est construit avec des murs d'une seule brique d'épaisseur. Il est vrai que ces murs n'ont à supporter qu'un toit en tôle galvanisée.

Les figures 47 à 53 renseigneront sur l'appareil qu'on peut donner à un mur d'une brique et demie, de deux briques et plus.

Le maçon du pays, à moins d'avoir travaillé pour des Européens (église, chemin de fer, etc.), devra faire son apprentissage du « *li wai tchoan* » tel qu'il vient d'être décrit. Instinctivement, il fera opposition : c'est du nouveau, on ne peut pas employer n'importe quelle bricaille, etc.

Art 3e : **Épaisseur des murs.** L'avouerai-je, c'est la question de l'épaisseur à donner aux murs qui est la première inspiratrice de ces quelques pages. C'est dans l'épaisseur des murs que beaucoup de missionnaires enterrent la plus belle partie de leur maigre budget. Je voudrais les préserver de cette erreur.

J'ai dit que le grand « tabernacle » des protestants de Tamingfou a des murs d'une brique et demie au rez-de-chaussée. L'église de Tamingfu, dont les plans ont été faits par un architecte, a également des murs d'une brique et demie, Fig. 52. Seul le clocher, de près de 50 mètres de hauteur, a deux briques au premier étage. Or je pourrais citer par dizaines des églises-granges, sans cachet, sans style aucun, comme aussi sans hauteur (tout au plus 25 pieds) sous les poutres (l'église de Tamingfou en a 40) ; en guise de clocher, elles ont un abri, plus laid que tout le reste, pour une cloche de 30 kilos, qu'on ne peut pas toujours sonner à la volée. Et pourtant les murs de la façade de ces granges-à-prière ont 4, [p. 36] même 5 pieds, 1m, 65, d'épaisseur, avec des contreforts dignes d'une cathédrale. Les murs latéraux sont à l'avenant : 3 briques au moins, 0m, 80, avec contrefort de soutien à l'intérieur et à l'extérieur, pour que la charpente ne tombe pas dans l'église. Et ceux qui ont élevé ces monuments disaient aux chrétiens leurs inquiétudes pour la solidité de la construction.

Il est vrai que de ce temps-là, les briques coûtaient moins d'une sapèque la livre, de bonnes briques, certes. C'était le bon vieux temps. Or, depuis lors, nos allocations sont allées en diminuant d'année en année, et pourtant le nombre des chrétiens n'a fait qu'augmenter, doubler, tripler. Conclusion : serrons les rangs de nos briques et faisons des poutres un peu moins grosses.

Je laisse de côté les procédés techniques par lesquels les architectes déterminent l'épaisseur des murs.

Consultez-les et employez les épaisseurs qu'ils indiqueront. Voici le tableau que donne Barberot pour une maison de 6 étages :

Aux basses fondations:	0m,75 à 1m,00,	donc 3 à 4 briques de 0m,26, ou 8 livres.	
Dans le haut des caves:	0m,65 à 0m,80,	2 ½ à 3	»
Sous sol	»		»
1er étage	0,45 à 0,55	2	»
2ème »	0,55 à 0,45	1 ½	»
3e et 4e étage :	0,25 à 0,40	1 ½	»
5e »	0,40	1 briq. à 1 ½	
6e »	mansardé		

Nous voilà loin de la façade de l'église-grange à 5 briques sous toit, et de murs latéraux de 3 briques, renforcés d'un double contrefort. Qu'on réserve ces contreforts pour une grande église, dont les murs n'auront qu'une brique et demie. Ils supporteront tout le poids de la charpente et ils ne seront pas même écrasés par la très pesante charpente chinoise en dalles, terre et tuiles. L'expérience en a été faite.

Le « koantsiang » (gaver avec de la bouillie). Est-ce une invention chinoise ? Je l'ignore. Je n'en trouve pas l'indication dans les livres que j'ai sous la main. Voici le procédé :

Le « tsiang » est un mélange très liquide de chaux, de sable et d'eau, qu'un ou deux manœuvres remuent constamment dans un bassin creusé en terre. Quand les maçons ont fini de poser une ou même deux assises de briques, ils y versent à pleins [p. 37] seaux, le « tsiang ». Ils le feront couler ensuite dans tous les interstices de leur maçonnerie, égalisent et recommencent à maçonner une nouvelle assise. Ce procédé est peut-être destiné avant tout à économiser beaucoup, même quand ce n'est pas cher. Il est certain qu'il donne de très bons résultats, et les architectes ne lui sont nullement hostiles. Pourtant, comme en toutes choses, il faut de la surveillance, surtout si la construction est faite à l'entreprise.

Article 4ᵉ Le salpêtre (« *kien* »), les isolants. Salpêtre se dit « *siao* » en chinois et non pas « *kien* ». Pourtant les Chinois disent que le mur est attaqué par le « *kien tsiang* » ou bien « *tchoan kien leao* ». Et en français on a traduit : le mur est salpêtré. Laissons de côté l'analyse chimique de tous les sels qui vont se nicher dans les briques dès qu'elles sont en contact avec le sol. Il est certain que cet ancien fond de mer qu'est le Tcheli, est très riche en chlorures, en nitrates, en sels de tous genres. Et tout cela remonte à la surface et constitue la lèpre, la plaie mortelle de toute construction, si elle n'est pas en pierres. La brique est très poreuse. Nous savons la quantité d'eau que cette éponge peut absorber. L'humidité chargée de sels, qui y monte du sol par capillarité, s'évapore sous l'action de l'air et du soleil. Mais bien entendu, l'eau seule est pompée. Les sels demeurent. Ils s'accumulent en cristaux dans les pores de la brique, et la font éclater par petites couches, à commencer par la surface extérieure. La brique tombera d'autant plus vite en poussière qu'elle est mal cuite. Les briques vitrifiées, « *tsiao tchoan* » ou « *liou li tchoan* » résisteront presque indéfiniment. Mais la pierre elle-même, si elle est tendre et poreuse, finira par succomber.

Il faut donc mettre un obstacle à l'humidité remontant du sol et envahissant par capillarité une certaine hauteur des édifices. Sinon, toute la construction s'écroulera par effritement de sa base, avant l'arrivée du salpêtre au sommet. Négliger de se défendre contre le salpêtre, c'est commettre un péché mortel en construction. Et pourtant, ce péché se commet par manque d'expérience.

Les matières isolantes contre la montée du salpêtre sont fort nombreuses.

Les Chinois n'emploient guère que les différents « *kien ts'ao* » [p. 38] (herbes contre le salpêtre) : roseaux, tiges de sorgho, pailles diverses, feuilles mortes mêmes. Le premier défaut de ces matériaux, c'est leur tassement très considérable. Les murs se fendent, surtout si le bâtiment est important. Dans une église à colonnes, vous auriez un tassement quasi nul à l'endroit des colonnes, et un mouvement des murs de plusieurs centimètres.

Leur deuxième défaut, c'est leur peu de durée. Au bout de 10 ans, mettez 30 ans, le meilleur « *kients'ao* » est pourri. Les riches emploient plus volontiers le bois comme isolant. De fait il s'oppose bien au passage du salpêtre, et il n'y a aucun tassement, mais le bois lui-aussi finira par pourrir.

Aux plus fortunés il ne reste plus que la pierre. C'est sans contredit un excellent isolant, le meilleur du pays. Malheureusement, dans beaucoup de régions, on ne trouve pas de pierres, ou c'est à des prix inabordables, à cause du transport. Du reste, nous l'avons dit, la pierre elle-même ne résiste que si elle est à grain compact. C'est pour cela qu'en Europe nos bâtisseurs de cathédrales protégeaient leurs pierres-mêmes contre le salpêtre, avec d'épaisses feuilles de plomb, rendues continues par une bonne soudure. Trop vert, et bon pour les goujats !! Mais c'est dire l'importance des isolants contre le salpêtre.

On emploie encore des plaques de feutre imprégnées d'asphalte. Le bitume est un des meilleurs isolants. Viollet-le-Duc en a trouvé dans les fondations des cathédrales aussi frais que s'il venait d'être posé. On vend à Tientsin, Changhai, etc. du carton bituminé, « Asphaltina », « Maltoïde », etc. etc. Ce n'est pas cher. Les rouleaux sont de différentes largeurs, et le carton lui-même est en différentes épaisseurs.

Du reste le missionnaire peut fabriquer lui-même son carton bituminé. C'est bien moins cher, et tout aussi bon pour l'usage dont nous parlons.

On prend du gros papier jaune, dit « *ts'ao tcheu* ». On y étend, avec un gros pinceau, une couche de goudron, et on laisse sécher au soleil. Trois couches de ce papier valent n'importe quelle « Asphaltina ».

Une autre **matière isolante** contre le salpêtre, ce sont les « *suetze* » petites nattes en bandes dont on se sert pour les « *leang cheu tounn* », tas de grains. Elles sont tressées à la largeur de beaucoup de murs. Du reste, on peut les mettre en double, pour [p. 39] donner exactement la largeur voulue.

Enduites de goudron sur les deux faces au moment de la pose, elles constituent un excellent isolant, non sujet au tassement et imputrescible au moins pendant très longtemps.

Où et comment placer le corps isolant ? Pour un isolement bien fait, il vaut mieux mettre deux lits protecteurs : un au ras du sol et l'autre un peu plus haut, placé au niveau intérieur du rez-de-chaussée. Souvent l'unique lit protecteur est placé beaucoup trop haut, à un ou deux pieds non seulement du sol, mais même du niveau des chambres. C'est sacrifier inutilement toutes les assises de briques qui sont au-dessous.

Mais le plus grand défaut serait toute solution de continuité dans ce lit isolant. On se souvient comment les grandes constructeurs du moyen âge soudaient leurs plaques de plomb. Voyez également avec quel soin les Chinois recouvrent toute l'épaisseur du mur avec leur « *kien ts'ao* » et le laissent déborder de quelques millimètres. Et pourtant beaucoup de missionnaires-constructeurs croient avoir tout fait après avoir posé une belle rangée de magnifiques pierres de taille sur la moitié extérieure du mur, ne faisant absolument rien pour isoler la moitié intérieure. « Elle n'est pas exposée à l'humidité du sol » disent-ils. L'humidité du sol serait donc comme les « *chenn* » (esprits) des payens, qui vont toujours droit devant eux, et qui ne trouvent pas la porte de côté, quand un « *ying pei ts'iang* » (mur paravent » est là pour les arrêter ?

Note. Nous croyons inutile de parler des murs creux, avec couche d'air isolante, contre la pluie et le froid. Le climat sec du nord de la Chine semble dispenser de ce luxe de précautions. Pourtant l'hôpital protestant de Tamingfou est construit de cette manière. Il est vrai que c'est un hôpital. La paroi extérieure de ces murs au rez-de-chaussée a une brique d'épaisseur (0m, 30), la paroi intérieure est d'une demi-brique, la couche d'air équivaut à un quart de brique.

Article 5e. Pose des chambranles des portes et fenêtres. On peut poser ces chambranles quand la construction est achevée. D'ordinaire on les met en place au fur et à mesure que montent les murs. De cette façon on économise du travail et la pose est plus solide. Il y a pourtant alors quelques précautions à prendre, pour éviter de graves inconvénients qui ne se rencontreraient pas si la [p. 40] pose se faisait à la fin des travaux.

Avant tout il faut prévoir le tassement des murs. Sans cette précaution, les montants seront écrasés ou du moins se gondoleront, si bien qu'on ne pourra plus placer ni portes ni fenêtres. Il faut laisser du jeu, d'abord sous les traverses du bas des chambranles des fenêtres. Sans cela, elles seront soulevées par la poussée du mur, sous la fenêtre. Pour empêcher cette poussée considérable, les ouvriers habiles construisent un cintre renversé sous la fenêtre. Mais seuls des ouvriers habiles savent faire ce genre de cintre. Aux autres, il faut recommander de laisser flotter tout le chambranle. On le fixera définitivement, au moment du crépissage. Ce ne sera nullement aux dépens de la solidité.

Il faut de bons maçons aussi pour faire les cintres au-dessus des portes et fenêtres. Les Chinois ne connaissent guère que leur vieux « *kouo mou* » (linteau), même dans les plus belles pagodes et dans les palais impériaux de Pékin. Il y aurait lieu de reparler à ce propos de la fameuse adaptation du style chinois à nos églises chrétiennes. Mais passons.

Pour cintrer un arc ordinaire, on se contente de faire une forme en plâtras, sur lequel on pose les briques ou les pierres. Il ne faut pas trop se hâter pour démolir ensuite cette forme, mais attendre que la maçonnerie ait pris son assiette.

Inutile de parler ici des différentes formes d'arcs : plein cintre, surbaissé, ogive en tiers-point, ogive à lancette, ogive surbaissée, arc Tudor, anse de panier, etc. etc. C'est l'affaire de l'architecte qui fera le plan. En toute hypothèse, il faut tenir compte de la poussée, surtout s'il s'agit d'une voûte (cave, église, etc.). Du reste, pour les caves, on n'emploie plus guère les voûtes ; elles obligent à une grande hauteur pour obtenir les passages de portes, et ont une épaisseur de maçonnerie considérable. Elles sont remplacées par les planchers en fer, en ciment armé. Nous en reparlerons. On soulage les arcs des portes et fenêtres par des arcs de décharge ; fig. 54. Ce sont des arcs en contre-haut, qui reportent la charge sur les points d'appui.

Outre l'inconvénient du tassement, que nous avons signalé, si on pose les chambranles et bâtis dormants en même temps que la maçonnerie, il y en a encore deux autres. D'abord la pluie sur ces bois non protégés par huile ou vernis. Ensuite les dégradations résultant des heurts, quand les manœuvres transportent paniers, [p. 41] pierres, poutres, etc. On peut protéger les montants par des lattes de bois contre ce dernier inconvénient.

Mais souvent on préfère ne poser ces pièces de menuiserie que quand les maçons se sont retirés. On fixe alors à demeure dans la maçonnerie, au moyen de deux ou trois pattes de chaque côté, qu'on scelle dans le mur et sur lesquelles on visse les montants, ou mieux au moyens de briques de bois, encastrées dans les murs aux bons endroits.

Les menuisiers chinois font volontiers des économies sur les linteaux, en employant du bois qui n'a pas l'épaisseur voulue pour être équarri sur toute la longueur. « C'est encore assez solide », vous diront-ils. Avouez pourtant que voilà un franc « *pou hao k'an* » (pas beau à voir). Du moins que l'arête soit complète sur toute la longueur, du côté extérieur.

Si on pose les chambranles dès le début, en même temps que les murs, les maçons, de leur côté, se permettront un « *tch'a pou touo* », qui ne sera pas seulement « *pou hao k'an* ». Sous prétexte d'éviter l'écrasement des montants de ces chambranles, ils laisseront un espace de cinq pouces au moins entre le linteau et la traverse supérieure du chambranle. Pour vous consoler, ils vous feront remarquer que la fumée pourra sortir par là quand vous chaufferez, « *k'ao k'ao* », en hiver. Sans doute, mais dans une église ? Et le vent du nord ? Et les moineaux ?

À ce point de vue, il vaudrait mieux ne poser les chambranles qu'à la fin.

Là où il faudra surveiller encore plus le « *tch'a pou touo* » des ouvriers du pays, si vous ne voulez pas geler dans votre chambre ou dans votre église, en hiver, ni être envahi par les terribles moineaux, toute l'année, c'est au moment de poser la toiture sur les murs. La partie supérieure des murs, là où elle rencontre les chevrons « *tch'oantze* » et les dalles « *pa tch'oan* » de la toiture, sera achevée « *tch'a pou touo* ». Une fois les poutres posées, on ajoutera encore quelques rangées de briques, de façon à soutenir « *tch'a pou touo* » les « *t'ou linn* » (dernière panne), qui est supportée par la maçonnerie dans l'art de construire de ce pays. Sur les « *t'ou linn* », on clouera immédiatement les chevrons « *tch'oantze* » et sur les « *tch'oantze* », on maçonnera les « *pa tch'oan* » (dalles) sans s'inquiéter le moins du monde de la solution de continuité entre le mur et la toiture. Pourtant, sur la partie extérieure du mur, cette [p. 42] solution de continuité aura exactement les dimensions de l'épaisseur des chevrons, soit deux à trois pouces (joie des moineaux !), et sur la partie intérieure, elle sera de deux pieds environ.

« On terminera plus tard, quand on fera le crépissage », diront les maçons. Malheureusement, on ne fait pas le crépissage sur la partie extérieure des murs, et sur la partie intérieure, on oubliera de le faire jusqu'en haut ; ce sera encore le « *tch'a pou touo* », et le vent du nord en hiver passera encore beaucoup plus facilement que la fumée.

Il faut donc exiger que le mur soit complètement terminé, et qu'il ferme hermétiquement, jusque sous les « *pa tch'oan* », avant de commencer à poser ces derniers. Ce travail est des plus faciles et il n'y a aucun motif de le remettre à plus tard.

Chapitre sixième: Le toit

Article 1er. Le toit chinois. Les fondations et le toit sont, plus que tout autre, les parties vitales d'un édifice. Malheureusement, la bonne toiture, du moins la toiture idéale, est encore à trouver. De plus, ceux pour qui sont écrites ces quelques pages n'ont guère à leur disposition ni les charpentes métalliques, ni les couvertures savantes. Ils ne disposent même pas de tuiles à emboîtement, appelées aussi tuiles mécaniques. Bon gré mal gré, il faut prendre son parti de la tuile ronde, que les Asiatiques employaient bien avant la civilisation grecque.

Contentons-nous de parler du toit chinois tel qu'il se fait dans la plaine du Tcheli. Il s'y fait de plusieurs façons.

1) Le toit plat en terre. Quand il est posé sur une maison en terre, cet ensemble a nom « *t'ou p'ing fang* », maison plate en terre. Et de fait, c'est bien plat. Inutile de tracer des règles d'art. Demandez au premier paysan venu. Pour nous, missionnaires, ces toits en terre sont mauvais, parce que personne ne les entretiendra. Et pourtant, il faudrait ajouter une couche de terre, « *ni fangtze* », tous les deux ou trois ans. [p. 43]

2) Le toit plat, recouvert de dalles ou de « *tchoan man ting* », briques bien jointoyées. Le bon jointoiement est difficile à faire. Mais surtout, à cause de la très grande porosité de la brique, ces toits coulent infailliblement, quand la pluie est persistante.

3) Le toit en macadam « *cheuhoei tch'oeiting* » ; on dit aussi plus simplement « *hoei ting* » et encore « *tch'enn ting* ». Ce toit est très bon, quand il est bien fait.

Beaucoup de maçons ne savent pas le faire. Il faut donc s'assurer de leurs connaissances en ce genre, avant de les mettre à l'ouvrage. Son grand ennemi, c'est la neige. L'eau de la neige fondante pénètre dans les porosités de la surface, et pendant la nuit, quand le froid est revenu, la gelée fait éclater la chaux, comme le salpêtre fait éclater les briques. Chez les particuliers, on ira balayer la neige, dès qu'elle aura cessé de tomber. Mais qui donc se dévouera dans une chrétienté, pour monter sur un toit qui appartient à la communauté ?

4)Le toit en tuiles « *wa fang* ». C'est l'unique toit pratique pour les missionnaires pour lesquels nous écrivons. Il a le grand inconvénient d'être très lourd et de nécessiter une ferme en rapport avec ce poids. Mais cet inconvénient est contrebalancé par de réelles qualités : la solidité d'abord, quand le travail est bien fait ; l'imperméabilité ensuite, à condition toujours d'être bien fait ; en tous cas, à l'encontre des toits en tôle, ardoise, etc. , la grande fraîcheur en été, et en hiver la conservation de la chaleur. Pour des maisons comme les nôtres, sans étage, ce dernier avantage a sa valeur.

Le toit en tuiles se fait de plusieurs façons :

a)La façon la moins recommandée, mais qui est réputée « *haok'an* », c'est le toit en tuiles demi-cylindriques, « *t'oung wa* ». Nous en avons parlé. C'et le toit des pagodes et des palais impériaux. Avant la république, du temps des « *Ts'ing* », les simples mortels n'avaient pas le droit de mettre de si belles tuiles sur leurs maisons. Il ne faut jamais céder aux instances des chrétiens pour en mettre sur nos églises. La cathédrale de Yen tcheou fou en était couverte. C'étaient même de magnifiques « *t'oung wa* » vitrifiées, de couleur impériale (jaune), comme les Palais de Pékin. À chaque forte pluie, ça coulait lamentablement. Je crois qu'on a changé toute cette brillante couverture. Du moins, il y a près de vingt ans, le changement était décidé. [p. 44]

b)Le toit « *yang wa* », toutes les tuiles ayant leur partie concave tournée en haut, s'appelle en d'autres régions le toit « *fan mao ki* », la poule qui a les plumes retournées. Figure 55. Au sud du Tcheli, ce sont les ouvriers du Kouangpingfou qui sont réputés pour ce travail. Et de fait, on se demande pourquoi ces tuiles aussi simplement juxtaposées, sans recouvrement, sans jointoiement, ne laissent pas passer l'eau. Mais le fait est là ; qu'il pleuve à verse, pas une goutte de pluie ne traversera une toiture bien faite. Le vent n'a pas de prise non plus : toute la surface fait bloc. Mais pour que ce soit un petit chef d'œuvre, il faut qu'on puisse marcher dessus, sans casser ni même déranger une tuile. Et les bons ouvriers y arrivent. Il est vrai qu'ils y mettent le temps : trois rangées de tuiles par jour, pour une demi-douzaine d'hommes. Et il faut de bons « *k'ao lao* », et nombreux, sans quoi ça coulera.

c) La dernière façon est le toit en « *k'eou wa* », tuiles en recouvrement, fig. 56. On dit aussi « *ta loung* », grandes rigoles. Ce toit, comme le précédent est fait avec des tuiles quart de cônes tronqués. Elles sont de dimensions variées. La tendance est de les faire de plus en plus petites ; question d'économie et de vie chère. Naturellement, c'est moins solide que les grandes tuiles du bon vieux temps. Du reste, on peut imposer son modèle au modeleur et au chauffeur. Dans le nord de la mission, ce sont les « *k'eou wa* » qui ont la vogue. On le voit, il y a ici double rangée de tuiles. Les unes sont placées de manière à présenter le côté concave et les autres, présentant le côté convexe, font recouvrement des joints. Le tout est soigneusement jointoyé, avec du très bon mortier, après que les tuiles ont été complètement imprégnées d'eau par une « trempette » prolongée comme pour les briques.

Au contraire, les tuiles en « *yang wa* » sont posées à sec, sans chaux. Pour ces trois façons de tuiler, on prépare le « lit » de la même manière : sur les chevrons, on pose les petites dalles, appelées « *patchoan* ». Elles n'y sont pas clouées, mais maçonnées sur les quatre rebords avec un peu de bon mortier. Il faut donc préalablement les tremper dans l'eau, pour que le mortier adhère. Sur ce toit en dalles, qui suffit déjà à lui seul pour arrêter une petite pluie, on étale une couche de boue, mêlée de paille. On laisse sécher. Le bâtiment est à l'abri maintenant, contre une [p. 45] pluie de « deux doigts ». Ce n'est que quand tout cela est sec que commence le placement des tuiles. On remet d'abord une forte couche de boue, mêlée de paille et de chaux, et c'est dans cette boue consistante que les tuiles sont pour ainsi dire coulées, pressées.

Il y a une façon économique pour faire le « lit » d'un toit en tuiles. Elle ne serait pas bonne pour une église, mais elle peut servir pour une école, dortoir, etc. La durée d'un pareil toit est d'une trentaine d'années au moins.

On n'emploie ni chevrons « *tch'oantze* », ni dalles « *patchoan* ». Directement sur les pannes « *linn* », on pose des tiges de sorgho ou des roseaux, mais non pas étalées librement, comme pour un toit en terre. Les tiges sont liées préalablement en petites bottes. Au fur et à mesure de la pose, on réunit les bottes avec des chevilles en bois assez longues, qui les traversent de part en part dans le sens horizontal. On alterne l'emplacement de ces chevilles : si les deux premières bottes sont traversées par les chevilles aux deux extrémités, la suivante sera chevillée par le milieu, et ainsi de suite, toujours en alternant. Sur la panne de faîte, on laisse dépasser un peu les tiges, puis, quand elles sont bien chevillées, on les rabat sur l'autre pente du toit. Elles sont ainsi accrochées à cette panne.

On étale ensuite la boue, comme sur les dalles, puis on dispose les tuiles à l'ordinaire.

On ne laisse pas dépasser ces tiges de sorgho en bas, au bord du toit, comme pour les toits plats, mais on continue le mur jusqu'à la corniche « *fang yen* ». De l'extérieur, c'est tout aussi propre qu'un toit en tuiles ordinaires. Sans la corniche, les tuiles du bas glisseraient.

Le « lit » du toit peut être fait aussi en « *wei pa* », grande natte en roseaux. Tout chinois vous renseignera sur sa confection. Dans ce cas on économise les dalles « *patchoan* », mais non les chevrons « *tch'oantze* ».

Article 2ᵉ. La charpente européenne. On donne le nom de comble à l'ensemble de la charpente qui supporte la toiture. Une ferme est l'ensemble des pièces de bois qui, de distance en distance, 3 à 4 mètres, fractionne les combles. Les fermes sont reliées entre elles par les pannes. En chinois, comble se dit « *fang kia* » ou « *mouleao kia* » (support de maison) (bois-support). [p. 46] Il n'y a pas de mot spécial pour traduite « ferme ». On se contente de donner un nom à chacune des pièces de bois qui composent la ferme. Une panne se dit « *linn* ».

Il y a deux espèces de fermes : la ferme chinoise et la ferme européenne. Voyez les figures 57 à 59.

La ferme chinoise, fig. 59, composée de « *ta leang* » (grande poutre), « *eull* (2) *leang* », parfois « *san* (3) *leang* », reliés par des « *kao tchoutze* » (colonnettes), est lourde et coûteuse. Elle ne permet pas de donner une grande largeur aux bâtiments, de franchir de grandes distances, comme disent les constructeurs. De là l'emploi si fréquent de colonnes dans les bâtiments chinois et dans les églises qui ont cette ferme.

La ferme européenne, fig. 57 et 58, est beaucoup plus légère, donc moins coûteuse. Mais surtout, grâce à la combinaison des points de résistance, elle peut franchir des distances bien plus considérables que les fermes chinoises.

C'est le poinçon surtout qui fait « agent de liaison ». En effet, le poinçon, assemblé dans le tirant ou entrait, comme le montrent les figures 60, 61 et 62, et retenu par les arbalétriers, figure 63, empêche l'entrait de fléchir en son milieu s'il venait à être trop chargé par les jambettes. Il va donc de soi que le poinçon ne doit en aucune façon charger l'entrait qu'il est destiné au contraire à soutenir.

Avec cette ferme, on pourra se passer de colonnes, même pour une église de 10 m de largeur, sans donner des dimensions exagérées ni à l'entrait ni aux arbalétriers.

C'est à l'architecte à fournir les plans des fermes. À lui aussi de calculer la résistance des matériaux, et de déterminer le force de chaque pièce de la charpente.

Pour la charpente chinoise on peut s'en remettre à l'expérience de bons ouvriers du pays.

Les plans annexés pourront renseigner pour un certain nombre de cas.

Appendix

Ferme genre européen. Les principales pièces sont :

1) Arbalétrier (A). Il domine la pente du toit et supporte les pannes. Par son extrémité inférieure, l'arbalétrier s'assemble avec l'entrait (B) à embrèvement. Dans la partie supérieure, il s'assemble avec le poinçon (D), à tenon et mortaise.

2) Entrait ou tirant (B). Sa fonction est surtout d'empêcher l'écartement [p. 47] des arbalétriers, et non pas de supporter quoi que ce soit, tandis que la poutre « *leang* » chinoise est chargée à elle toute seule du poids écrasant de la toiture. Donc inutile de donner à cet entrait à l'européenne la force de la poutre chinoise. Il peut même être constitué par deux pièces parallèles, appelées moises, entre lesquelles se placent les arbalétriers et les poinçons, fig. 58 bis. L'entrait retroussé (C), parallèle à B, soulage les arbalétriers, fig. 57.

3) Poinçon (D). Il travaille à la traction comme il est dit à la page précédente. Il ne doit pas charger l'entrait, mais y être relié, soit par un boulon qui traverse l'entrait et le tenon du poinçon, soit par un ancrage en bois (on fait alors dépasser le poinçon sous l'entrait), soit par un étrier en fer qui passe sous l'entrait et qu'on boulonne dans ses deux branches supérieures sur le poinçon. Fig. 60, 61, 62. L'assemblage du haut du poinçon avec les arbalétriers se fait par tenon et mortaise chevillés. Fig. 63.

4) Contre-fiche (E). Elles soulagent les arbalétriers en s'appuyant sur le poinçon. La contre-fiche doit être placée perpendiculairement à l'arbalétrier, le plus près possible des pannes. Par son autre extrémité elle s'embrève avec le poinçon. Dans une ferme, le poinçon reçoit 4 contre-fiches ; deux soulagent les arbalétriers et deux autres, appelées « lien », soutiennent le faîtage. Fig. 63.

5) Jambette (G) est une petite pièce qui soulage l'arbalétrier en s'appuyant sur l'entrait.

6) Jambe de force (H), elle relie l'entrait à l'entrait retroussé.

7) Blochet (I).

8) Faîte ou Faîtage (J). C'est l'arête supérieure du comble, et qui reçoit les chevrons, comme les autre pannes. Il est délardé suivant la pente du toit pour qu'on puisse y clouer les chevrons. Fig. 64.

9) Pannes (M) « *linn* », réunissent les fermes entre elles, en s'appuyant sur les arbalétriers, et aux extrémités du bâtiment, sur les deux pignons. Les pannes des extrémités doivent être plus longues d'au moins un pied, 0,33 m, que les autres, pour pouvoir être posées sur le mur. Il est bon de les chaîner avec les pignons. Les charpentiers chinois connaissent et pratiquent toujours l'assemblage [p. 48] des pannes entre elles par « queues d'aronde » ou « queues d'hirondelles ». Il n'y a qu'à laisser faire.

10) Échantignole ou Chantignole (N), tasseau qui empêche le glissement de la

panne. Dans la pratique, on se contente de la clouer sur l'arbalétrier, sans embrèvement.

11)Chevrons (O) « *tch'oantze* », sont cloués sur les pannes et soutiennent les dalles « *patchoantze* » du toit chinois. Leurs dimensions dépendent de l'écartement entre elles des pannes. Ordinairement pour des chevrons de 5 pieds de long, on donne 2 pouces au moins de côté. Pour une toiture en tôle galvanisée, il faudrait naturellement beaucoup moins de force.

12)Sablière (P) « *t'ou linn* ». C'est la dernière panne, posée à plat sur le mur. Elle reçoit l'extrémité inférieure des chevrons. Elle peut être en plusieurs pièces, puisqu'elle ne porte rien.

13)Coyau (R) « *fei tch'oan* », moitié d'un chevron, taillé en biseau en haut. Il est placé sur la saillie de l'entablement pour l'adoucir la pente de la couverture au point où elle se repose sur la corniche, et aussi pour projeter le plus loin de la façade les eaux pluviales. C'est surtout au moyen de ces « *fei tch'oan* » que les architectes chinois donnent à leur toiture la forme en nez retroussé.

Combles à la manière chinoise, fig. 59.

1) Poutre principale « *ta leang* », 1.
2) Poutre secondaire « *eull leang* », 2.
3) Poutre secondaire « *tch'a leang* », 3.
4) Colonnettes « *kao tchoutze* », 4.
5) Colonnes « *mingtchou* », 5.

Le reste a même disposition et même noms que pour la première charpente dite « européenne ».

Article 3e. **Pente des combles.** La première chose qui doit préoccuper le constructeur dans l'étude d'un comble, c'est la pente à donner aux versants. Cette pente sera surtout déterminée par les matériaux de couverture dont on dispose. La raison d'économie fera toujours adopter la pente minima, à condition qu'elle reste pourtant suffisante.

Ainsi pour ne parler que des toitures de la région qui nous occupe, un toit plat, soit recouvert de briques, soit en macadam, [p. 49] peut n'avoir que cinq degrés de pente. Un toit en tôle, 26° ; un toit en « *k'eou wa* » autant, soit 26° ; un toit en « *yang wa* », 35°. Les ouvriers chinois ont des formules stéréotypées qui leur tiennent lieu de degrés auxquelles ils n'entendent pas grand-chose. Ainsi ils disent : « *tcheng ou, sie ts'i* », ce qui équivaut à une pente de 45°. La moitié de la poutre, soit a-b est « *tcheng* », mettez cinq pieds ; la pente du toit a-c, est « *sie* » ; les arbalétriers devraient donc avoir sept pieds (« *tcheng* » = droit, horizontal ; « *sie* » = oblique, en pente). de ce pays. De fait, avec cette inclinaison, on peut être sûr que l'eau s'écoulera. Mais les tuiles glisseront assez facilement (surtout si les corbeaux s'en mêlent), sans compter les tempêtes.

Cette pente : « *tcheng ou, sie ts'i* », est peut être calculée pour les pluies parfois diluviennes.

L'inclinaison de 35 degrés a donné de bons résultats, mais c'est certainement la mesure maximum, surtout pour le versant du toit orienté vers le nord. On pourrait, pour un « *pei ou* », donner 40 degrés au versant nord et 35 au sud. Pour plus de symétrie, dans nos régions, 40° me paraît la meilleure pente pour une toiture en « *yang wa* ». Qu'on fasse faire alors des tuiles un peu plus grandes que celles qui se vendent actuellement.

Chapitre septième: L'ornementation du toit

Voilà un titre qui serait déplacé en Europe, mais nous ne sommes ni en Europe, ni en Amérique, où le titre en question serait peut-être encore plus mal sonnant : orner un toit, et consacrer à cette ornementation un chapitre entier. Rassurez-vous : c'est pour vous empêcher d'orner le vôtre comme le voudraient peut-être ceux qui vous entourent.

L'architecte chinois orne son toit, parce qu'il lui faut bien orner quelque chose. À toute construction, on tâche au moins de donner du cachet, quand on ne peut lui donner de l'architecture.

La construction chinoise a été définie : un toit très lourd et [p. 50] très orné, posé sur de grosses colonnes en bois. Cette définition s'applique au Temple du Ciel, à Pékin, comme aux Palais impériaux.

J'ai là, sous les yeux, la photographie du palais de l'Impératrice Ts'eu-hi, de lugubre et méchante mémoire. Certes elle avait de quoi pour élever une grandiose demeure, digne de la souveraine incontestée de 400 millions d'hommes. Et pourtant là, comme ailleurs, c'est le toit très lourd et très orné, posé sur de grosses colonnes en bois. Inutile de faire le voyage à Pékin si vous voulez en avoir une idée de visu. Allez dans n'importe quel yamen ; vous saurez comment sont construites toutes les maisons qu'on nomme les palais impériaux.

Il n'y a pas d'étage, c'est un simple rez-de-chaussée. La façade consiste en une cloison en bois sculpté, abritée par une véranda. Cette véranda elle-même est constituée par la continuation du toit. A-t-on substitué le verre au papier qui est ordinairement collé sur la cloison ? La photographie ne le dit pas. Cela devait dépendre de la persuasion plus ou moins ferme de Ts'eu-hi, que le verre laisse passer les courants d'air.

L'architecte n'avait pas plus à se préoccuper des portes et fenêtres pour le mur du fond et les deux pignons que pour la façade. Tout l'air et toute la lumière arrive par la cloison du sud. (Ces bâtiments sont toujours rigoureusement orientés vers le sud).

À l'intérieur, pas de voûte, mais une accumulation invraisemblable de bois de charpente. Je vous fais grâce de la description. Allez voir dans la première grande pagode que vous rencontrerez. Le Temple du Ciel et la Pagode de Confucius ne sont pas faits autrement.

En désespoir de cause, l'architecte ne trouvant rien à orner, du moins avec ses briques, s'est appliqué à orner le toit avec les tuiles et leurs « dérivés ». Il commence par le faîte, « *tsi* », à l'encontre de ses collègues de l'étranger. Sur le « *tsi linn* », panne du faîtage, il a élevé un vrai mur de 4, 5, parfois 6 pieds de haut. Qu'on juge de cette seule charge sur toute la ferme ! Ce mur, le plus en vue de tout le bâtiment, est aussi le plus orné. Ce sont d'abord plusieurs rangées de briques, formant moulures ; puis d'énormes pièces moulées, creuses à l'intérieur, avec des fleurs de lotus, etc., sur les deux faces ; puis d'autres briques à moulures ; enfin des tuiles de différentes formes, disposées savamment pour [p. 51] compléter l'impression de « *hao k'an* ». Les deux extrémités du « *tsi* » sont terminées par deux « *tsi choei* » têtes de crête. Ce sont d'énormes têtes de dragons qui n'ont absolument rien d'esthétique. Il y a des dragons à la gueule béante pour les grands lettrés ; ça absorbe le « *fong choei* » de tout le voisinage. Pour les simples mortels, les dragons n'ont pas le droit d'ouvrir la gueule.

La crête bien campée, bien ornée, parfois même bien peinturlurée, on procède à l'ornementation des deux versants de chaque pignon. Là aussi, il y a des fleurs de lotus, puis toute une ménagerie de singes, petits chevaux, poissons plongeants « *kinn t'eou u* », chauve-souris, bonzes en prière.

Le plus beau, c'est quand on a le droit de piquer là haut des cornes de cerfs en fer « *kangtch'a koayinn* ». Mais tout le monde n'a pas ce droit-là. Il faut un grade très élevé dans la hiérarchie mandarinale, à moins qu'il ne s'agisse d'une pagode: les « *chenn* » (esprits) ont tous les droits.

Quand tous les « *wan yeull* » (jouets), sont en place, on pose les tuiles. Pour le genre de monuments qui nous occupe, ce sont des tuiles vernissées et vitrifiées au four, comme la faïence et la porcelaine, de couleur jaune. Quelques-unes de couleur verte ou bleue serviront à dessiner au beau milieu du toit d'énormes caractères : « *fou* » (bonheur) ou « *wan* » (dix-mille (ans). La pose commence par le bas. La première rangée est protégée contre le glissement par d'énormes clous qui traversent chacune des premières tuiles, moulées d'une façon spéciale. Ces clous eux-mêmes sont coiffés de bonshommes creux, en terre cuite, toute une rangée de petits bonzes en prière.

Puis la pose continue en arc concave. Le comble du laid serait de faire une pente en ligne droite.

En ce temps-là, il y avait une chrétienté dont les administrateurs n'obéissaient pas à leur curé. Il avait dit aux charpentiers et aux maçons de faire le toit de l'église en pente droite. Quand il fut parti pour aller donner ses missions, eux dirent aux ouvriers de faire le toit concave. Et les ouvriers le firent concave, même très concave. À la première grande pluie, le toit neuf coula, et au premier grand vent qui suivit, les tuiles du haut glissèrent. Cela coula souvent, comme qui dirait à chaque forte pluie. Si bien que les chevrons et les pannes firent mine de pourrir. On enleva alors la moitié des tuiles, là où c'était [p. 52] concave ; on y mit environ deux pieds de terre et on replaça les tuiles ; ça ne coula plus ; mais on avait trop présumé de la bonne volonté des pannes. À la première grande neige, elles cassèrent, du moins plusieurs firent ainsi ; on mit alors des pannes de secours, en travers. C'était bien laid, pas « *hao k'an* » du tout. Tellement, qu'au bout de vingt ans, on enleva toute la toiture, ferme comprise. Tout fut remis à neuf, et en ligne droite. Et ça ne coula plus. Coût : sept mille ligatures de ce temps-là.

Les clochetons [pinacles]. Ne voulant pas du Paradis bouddhique, ni de ses saints sur les toits de leurs églises, et pourtant les voulant beaux, un peu dans le goût du pays, les missionnaires ont fait une invention chrétienne, ou plutôt ils ont appliqué une invention chrétienne. L'extérieur des cathédrales est une forêt de clochetons, et tout le monde est d'avis que c'est beau. Mettons des clochetons sur les toits. Et on a mis beaucoup de clochetons partout, un peu trop même. Il y en a d'abord aux quatre coins de ces granges à prières. Puis aux quatre coins de ce qu'on dit être un clocher. Souvent on a fait semblant de croire que le clocher était à deux étages et on y a mis huit clochetons. Au bout de chaque contrefort, on a encore placé un clocheton, comme aux cathédrales du 13e siècle.

Le clocheton eut du succès : les riches chrétiens en mirent par rangées au haut de leurs grandes tours, et les payens à leur tour, adoptèrent le style nouveau. On voit des clochetons sur les portes-cochères, sur les murs d'enceinte, sur les « *hiaot'ang* » (écoles) payens. Il doit y avoir un microbe du clocheton.

Or cette profusion de bonnets de nuit, d'un mauvais goût parfait, ne rimant à rien, pas plus que les pigeons, les raisins, etc. sculptés à grands frais dans la façade, non seulement coûte cher, non seulement fait couler le toit, mais vienne un typhon ; le clocheton, mal campé, mal calculé, vous arrivera sur le pavé de l'église, à travers la toiture. Fassent les bons Anges que ce ne soit pas un jour de fête, pendant la grand' messe. Je sais une chrétienté où cet accident est arrivé pendant le chemin de croix. Heureusement, c'était à la quatorzième station. Le clocheton arrivait comme un bolide du haut du clocher au-dessus de la façade ; l'assistance était entièrement groupée près de l'autel.

Tout ornement, en architecture du moins, doit être motivé ; il [p. 53] faut qu'il ait son côté utile ; jamais l'ornement pour l'ornement. Il s'agit, v. g. de dissimuler un escalier tournant, donnant accès sur le toit des bas-côtés : on le cache dans une jolie petite tourelle. Il faut peser lourdement sur les contreforts des arcs-boutants, pour contrebalancer la poussée des voutes en pierres et briques. L'architecte fouillera les blocs en granit qu'il va jucher là-haut. « Miscuit utile dulci » ; la poésie des pierres. Mais à quoi bon vouloir donner des allures de cathédrale à une grange de prières ? N'imitons pas la grenouille de la fable !

J'oubliais un autre ornement inutile et coûteux : les créneaux « *touok'eou* » ; ils chargent les murs et font couler le toit. Les créneaux ont leur raison d'être pour les riches, obligés de se défendre contre les brigands. Mais pour un presbytère ou une toute petite église ? On n'a pas facilement le dernier mot avec les administrateurs pour ce détail des créneaux. Volontiers ils en mettraient sur la cuisine, voire même ailleurs, surtout quand c'est avec l'argent du Père. Il faut donc donner des ordres précis et surveiller.

Chapitre huitième: Varia

Comble, diverses formes. La principale, quasi l'unique forme de toiture généralement employée dans cette région de la Chine, est le toit à deux pans égaux, avec pignons montant jusqu'au faîte, et soutenant la panne du faîte « *tsilinn* ». C'est la manière de construire la plus économique, comme aussi la meilleure au point de vue de la stabilité du toit. fig. 66.

La forme européenne, avec « croupes » fig. 67, n'est pas du goût des Chinois. Je n'en vois pas les avantages. Elle coûte certainement beaucoup plus cher que la première, surtout dans un pays où le bois est rare. Le vent a plus de prise aussi sur cette toiture.

Larmier, fig. 68. Le larmier est un détail extrêmement important dans toute la construction, à la rencontre d'un toit avec un [p. 54] mur plus élevé. Le cas se présentera pour le toit des bas-côtés d'une église, à la jonction avec le mur de la nef principale. Il se présentera encore pour le toit de la sacristie, pour celui de l'abside, si celle-ci est plus basse que la nef.

Le larmier forme corniche. Il se compose de pierres ou de briques, taillées comme le montre la figure 68. C'est-à-dire que ces pierres ou ces briques portent, en dessous, un petit canal, ou mouchette, d'un pouce de profondeur, et d'un pouce de largeur, creusé dans le sens de la largeur. Le but de cette petite rigole est d'empêcher l'eau de pluie de couler le long du mur, et de la forcer au contraire à tomber verticalement en gouttes ou en « larmes ».

Pour que cette corniche remplisse sa fonction, il faut en calculer exactement l'emplacement. C'est un petit calcul assez délicat à faire, car il doit être fait pendant la construction du mur, dans lequel le larmier est naturellement encastré, donc avant la pose du toit. Si le calcul est mal fait, le toit viendra recouvrir la corniche, ou bien cette dernière sera trop élevée au-dessus du toit : dans les deux cas le larmier sera inutile. Il y a donc à tenir compte de tous les matériaux qui constitueront la toiture des bas-côtés, de la sacristie, etc., après avoir fixé très exactement la place où viendront se poser les premières pièces, les « *linn* », poutrelles [pannes], de cette toiture. Le larmier devra surmonter le toit d'un pouce au plus, et ressortir du mur de cinq pouces environ.

S'il y a un transept moins élevé que la nef, la corniche-larmier devra se conformer à la disposition du toit de ce transept. Elle devra donc remonter graduellement avec lui, de façon à rester toujours à un pouce au-dessus des tuiles qui touchent le mur de la grande nef.

Par extension on donne le nom de larmier à toute saillie pratiquée hors de l'aplomb d'un mur pour arrêter les eaux pluviales. Il est destiné dans ce cas à faire tomber l'eau de pluie à une distance convenable du pied du mur, à la manière des gargouilles, si curieusement sculptées, de nos grandes cathédrales.

Ce genre de larmier peut se faire en pierre ou en briques. S'il est en pierre on lui donne la forme fig. 69, la pointe A faisant larmier. En briques, la corniche sera construite comme le montrent les fig. 70, 71, 72.

[p. 55] Il est important aussi de munir de larmiers, à l'intérieur et à l'extérieur, les appuis des fenêtres, comme il est marqué sur les plans des grandes fenêtres doubles et triples d'une église. On y pratique également la petite rigole, marquée sur les mêmes plans.

En menuiserie une fenêtre bien faite ne manque pas de « mouchettes » pratiquées dans la partie inférieure B du châssis mobile, et du dormant A, ni de rigole, fig. 79.

Voûtes. Faut-il voûter celles de nos églises dont les dimensions sont plus considérables, et qui offrent un peu de cachet ?

« Non, répondent avec un grain d'indignation les puristes, puisque, dans l'état de vos finances, ce serait folie de songer à une vraie voûte en pierres et en briques, avec ses clefs de voûtes sculptées, avec sa forêt de contreforts, avec les colonnes en pierre et leurs riches chapiteaux qui la supporteraient. Alors qui prétendez-vous tromper avec votre voûte pastiche, en planches et en chaux ? Ne faites rien, si vous n'êtes pas capable de faire quelque chose de vrai ».

Et pourtant cette pauvre voûte pastiche, ce simulacre de voûte, présente bien des avantages. Sans vouloir tromper personne, on donnera plus de cachet à l'église. Au moins, question de cachet mise à part, ce sera plus propre ; on ne remarquera pas tous les petits défauts des bois de la charpente. Ce sera plus propre aussi (pardon pour ce détail) parce que les moineaux, ces terribles concurrents des meilleurs prédicateurs en ces régions, ne seront plus chez eux sous cette voûte… européenne ; où poser leur nid ? Puis ce sera plus frais en été, et plus chaud en hiver. Mais surtout si la voûte en vulgaire lattis est bien faite, l'acoustique de votre église sera complètement modifiée en bien.

Manière de construire une voûte en planches, lattis et chaux : Figures 74 à 76.

On trace sur le sol la courbe des arcs qu'on veut obtenir. Puis on découpe des planches selon cette courbe. Ces planches sont ensuite clouées ensemble, en alternant les joints. Ces joints doivent être coupés non pas carrément, mais dans la direction du rayon de la courbe. Moulurer les planches, comme à gauche du dessin, avant de les clouer.

Les faux arcs gothiques, en bois, sont de deux modèles : grand modèle pour les arcs parallèles ou perpendiculaires à l'axe de [p. 56] l'église (arcs formerets et doubleaux) ; petit modèle pour les arcs obliques (croisée d'ogives). Ces arcs obliques s'assemblent, à leur croisement, au milieu du « *kien* », dans une clé en bois. Cette clé est faite d'une seule pièce de bois ronde, moulurée avec les mêmes moulures que les arcs qu'elle doit recevoir (voir pointillé du dessin). Sur le dessous de la clé on peut clouer une planchette en bois, dans laquelle on aura sculpté préalablement une fleur ornementale, un chiffre etc.

À l'endroit de la clé où aboutit l'arc, on fait une mortaise pour recevoir le tenon de l'arc, et on creuse celui-ci pour que ses moulures semblent la continuation de celles de la clé. Pour les clés des bas-côtés, les dimensions seront moindres (v. g. 8 pouces si, dans la nef, la clé est d'un pied). Mais il faut leur conserver la même hauteur et les mêmes moulures. Tourner les moulures des clés, si l'on a un tour assez grand.

Les arcs réunis par la clé doivent former un appui parfaitement solide pour les voûtains en lattes et chaux, inutile de rien rattacher à la charpente.

En clouant les lattes, on leur donne une légère courbure vers le haut, afin que les voûtains eux-mêmes forment voûte plus solide. Cette précaution n'est à prendre que dans la partie supérieure des voûtains, celle qui devient horizontale et forme plafond.

Pour enduire les voûtains de chaux, espacer les lattes suffisamment pour que la chaux passe à travers. Un ouvrier, se tenant au-dessus de la voûte, rabat cette chaux par-dessus les lattes.

Le centre des arcs est ordinairement marqué sur le plan « coupe verticale » de l'église. En principe, la clé de voûte doit être placée au moins aussi haut que la pointe des arcs doubleaux et formerets qui l'encadrent. En principe aussi, les arcs formant la croisée d'ogive (ceux qui se croisent à la clé) doivent être des pleins cintres (demi-cercles) et non pas ce qu'on appelle vulgairement des ogives (arcs brisés). Ce sont les arcs formerets et doubleaux qui sont brisés dans le style gothique. La cathédrale de Reims est une des rares exceptions à cette règle : les arcs ogives y sont brisés.

Si, en employant le plein cintre pour la croisée d'ogive, la clé de voute se trouve trop bas (voir la 1ère règle ci-dessus), on y remédie en prenant le centre de ces pleins cintres plus haut que [p. 57] la ligne du niveau des chapiteaux. C'est seulement si ce surhaussement devenait trop exagéré qu'il faudrait faire comme à la cathédrale de Reims et briser le sommet des arcs ogives.

Si l'on veut un plafond (voûtains) très solide, il fait mettre deux couches de lattes croisées l'une sur l'autre, la 1ère perpendiculaire aux doubleaux et formerets, la 2e en croix sur la 1ère. Voyez aussi Avant-Projet III, plan par terre.

Arcs en briques. Pour construire ces arcs, il faut préparer des claveaux appropriés aux différents arcs. Des claveaux seront ornés avec avantage d'un boudin, pour faire moulure, fig. 79.

Pour trouver la forme des claveaux convenables pour chaque arc, il faut tracer l'arc sur une planche, en grandeur naturelle. Des arcs de rayon peu différent peuvent utiliser les mêmes claveaux.

Pour construire les arcs, planter un clou à la place du centre, y attacher une ficelle munie d'un nœud, indiquant le rayon de l'arc. Chaque fois que l'ouvrier pose une brique, il doit tendre la ficelle pour s'assurer que la brique est à la bonne distance du centre, et que son axe est bien dans la direction du rayon. Voyez fig. 79 la coupe complète d'un arc en briques. Le sommet d'un arc gothique doit toujours présenter un joint, fig. 80, et non pas une clé, fig. 81.

Vitraux. Des vitraux en pays de mission! C'est ici surtout qu'il faudrait répéter ce qui a été dit à propos des voûtes en pierres et briques: « trop vert »! De fait, même dans une cathédrale, des vitraux à scènes, ou simplement à figures isolées, soulèveraient probablement des critiques: « non erat hic locus ».

Mais si l'état actuel des finances des missions de Chine obligent à renoncer bon gré mal gré à cette ornementation grandiose de nos églises, et à ce bel enseignement par l'image, on peut envisager sans témérité la possibilité d'une belle grisaille. Pourtant, même dans ce cas, il faudrait renoncer à un chef d'œuvre composé en Europe. Il s'agit de s'en tirer avec les « moyens du bord », qui sont les ressources du pays. Évidemment votre composition portera la marque « made in China ». Où est le mal, puisque ce n'est pas un article d'exportation que voulez produire !

Voici, entre autres sans doute, une manière de s'y prendre. On trouve facilement à Tientsin, et ailleurs, des saumons de [p. 58] plomb. On y trouve également du verre de couleur. Il sera peut-être plus avantageux, pour une grande église, de faire venir du verre d'Europe. Mais demandez des feuilles de faibles dimensions, et recommandez un emballage tout à fait soigné, pas « made in China » du tout, car avec les miettes que vous recevriez alors, vous pourriez tout au plus faire du beau papier de rocher, pour une crèche de Noël.

Puis il vous faut un petit laminoir spécial pour les plombs à rainures. Je connais une mission qui en possède un fort bon. Elle le prête même très aimablement aux voisins moins fortunés. La Résidence épiscopale des missions plus anciennes et plus étendues feraient facilement l'acquisition de ce laminoir avec tous ces accessoires. La dépense serait peu considérable. — Reste à trouver — ou à former, au moins un bon ouvrier. Ce genre de « malins types », habiles à se servir de leurs dix doigts, n'est nullement « rara avis » dans ce pays.

Et vous voilà en possession de tous les éléments du succès, le nerf de la guerre présupposé. Pourtant je tiens à le répéter, réduite à ces proportions, l'entreprise n'a rien de hasardeux ni d'exagéré. Si vous avez pu faire la dépense pour une belle grande église, vous pouvez, sans prodigalité ajouter celle-ci.

Une remarque pour finir : une grisaille fera meilleur effet si elle est bien détaillée, c'est-à-dire si les morceaux de verre qui la composent sont de petites dimensions. L'ouvrier y mettra un peu plus de temps pour souder les plombs et découper les morceaux de verre, mais l'effet produit sera en harmonie avec le travail et la dépense.

Flèche d'un clocher (manière de la construire) ; fig. 77. La flèche octogonale d'un clocher carré repose sur quatre rangées de pierres (A A, fig. 77) mises en encorbellement sur chaque coin de la tour, de façon à transformer le carré en octogone régulier. À des ouvriers inexpérimentés (maçons et tailleurs de pierres) il est bon de montrer l'effet produit, avec des morceaux de bois taillés d'après la forme voulue. Ces pierres, à l'intérieur, forment comme un escalier renversé. Elles doivent être taillées à l'extérieur de manière à former parement. Elles doivent être placées de façon à ce que leurs extrémités, enchâssées dans le mur, ne coïncident pas, mais forment « harpe », de façon à s'entremêler avec les briques. La partie intérieure, n'étant [p. 59] pas vue, peut être taillée très grossièrement. Il faudra environ quatre pierres, de neuf pouces, à chaque coin. Le nombre exact des pierres est déterminé par les dimensions de la tour et par l'épaisseur des pierres, qui peuvent être plus minces que neuf pouces.

Une fois l'octogone obtenu, continuer la flèche avec des briques spéciales (fig. 6 à 10) présentant sur l'une de leurs faces la pente de la flèche. Avant de construire le moule de ces briques spéciales, pour déterminer exactement la pente de la flèche, on trace cette dernière par terre, en grandeur naturelle. L'intérieur de la flèche reste creux, les parois ayant environ 6 pouces d'épaisseur. En maçonnant les briques de ces parois, il importe que l'ouvrier ne ménage pas la chaux, mais en mette sur toute la face de la brique, et non pas seulement sur les bords, à la manière des maçons chinois. De plus il faut donner une très légère inclinaison, vers le dehors, à chaque brique. Sans ces deux précautions, l'eau de pluie s'infiltrera entre les interstices des briques, et le clocher sera inondé.

Surtout ne pas mettre de charpente dans ce genre de flèche, sous prétexte de la consolider !

La flèche se construit ainsi très facilement, si l'on a soin de rentrer chaque rangée des briques de la quantité nécessaire, ni plus, ni moins. Pour se diriger d'une façon infaillible, les ouvriers expérimentés dressent une haute perche, bien droite, exactement au milieu de la tour, à la naissance de la flèche. Au haut de cette perche ils fixent une planche octogonale, de la même forme que la flèche. Des huit coins de ce petit plateau octogonal partent huit cordons qui viennent aboutir aux huit coins, à la naissance de la flèche. Les arêtes suivent ces cordons.

Le sommet de la flèche doit être terminé par une ou plusieurs pierres de 5 à 6 pieds de hauteur en tout dont la dernière forme fleuron. Ces pierres sont percées d'un trou de part en part, et ce trou sert à fixer la croix en fer fig. 78. On fixe définitivement celle-ci après l'avoir mise d'aplomb, avec des cales en fer et du plomb fondu, ou du soufre fondu, pas avec du bois.

Les arêtes de la flèche sont faites, soit avec des demi-briques rondes, entremêlées avec des briques ayant même « tête » ronde, mais plus longues, de façon à couvrir le joint, soit avec d'autres briques spéciales, fig. 12 [p. 60] Si l'on met des « crochets », on peut se servir des briques, fig. 11.

Jointoiement de la maçonnerie. Il consiste à remplir avec du mortier les joints des matériaux, pierres ou briques, composant les maçonneries ; en chinois « *keou ts'iang* » (jointoyer le mur). Ce travail peut se faire de deux façons. La première et la plus économique, comme aussi la plus sérieuse comme solidité, consiste à finir la maçonnerie du premier coup. L'ouvrier, avec sa truelle, « *wa tao* », enlève toutes les bavures de chaux une fois que la brique est maçonnée, au besoin, il l'arrose avec un peu d'eau. Puis il trace une ligne de démarcation sur la chaux, entre les briques.

L'autre manière : on achève tout le bâtiment. Puis, au fur et à mesure qu'on enlève les échafaudages, on refouille et on gratte les joints à 0,02 m ou 0,03 m de profondeur, on nettoie à vif, on brosse, on arrose à grande eau. Puis on remplit les joints de mortier avec une petite truelle pointue.

Crépi ou enduit « *mouo ni* ». Il faut attendre que le tassement soit fait avant de crépir un mur en pisés.

Un mur en pures briques « *li wai tchoan* », n'est crépi qu'à l'intérieur mais uniquement avec de la chaux, pas avec de la boue. Pourtant l'ouvrier chinois insistera pour appliquer d'abord une couche de boue, qu'il recouvrira ensuite d'une mince couche de chaux. C'est plus économique, dira-t-il. Sans doute il faut rechercher l'économie, mais pas aux dépens de la solidité. Il y a toujours plus ou moins de salpêtre dans l'eau et la terre de ce pays. Au bout d'un certain temps, l'enduit en terre se détachera du mur et tombera par plaques.

À la chaux du crépi on mêle du chanvre finement haché « *matao* ». Cet enduit est excessivement solide.

L'ouvrier « *tch'a pou touo* » n'égalisera que très imparfaitement son enduit. La poussière qui viendra se déposer sur ces aspérités les rendra encore plus apparentes, et donnera aux murailles un air de grande malpropreté.

Le plafond se fait de la même manière, après avoir cloué les lattes.

Gâcher le mortier de terre « *houo ni* ». Pour leurs constructions ordinaires, avec mortier de terre, les maçons chinois ont la très mauvaise habitude de gâcher la boue au beau milieu de la [p. 61] chambre. Ils creusent un trou, y jettent terre et eau, et gâchent à même. On devine l'effet produit pour la future salubrité de l'appartement. On remplira le trou avec de la terre sèche, disent-ils. Mais toute l'eau qui a pénétré en terre, pendant des semaines, et à pleins seaux ? Du reste, on continuera dans la suite, de jeter l'eau et n'importe quelle eau, sur le dallage de la chambre. On le fera même à dessein, en été, pour motif de fraîcheur. Les motifs de rhumatisme n'entrent pas en ligne de compte. Ceux de propreté, encore beaucoup moins. Qu'ai-je à apprendre aux missionnaires, pour lesquels j'écris ?

Protection du dallage des chambres contre le salpêtre. Tout le monde n'est pas assez riche pour se payer un beau parquet, ni même un vulgaire plancher. Quant aux églises, elles sont presque toujours dallées. Le salpêtre monte très vite dans les briques du dallage, et l'humidité rend tout le local froid et insalubre. Un moyen économique et radical contre le salpêtre, ces sont les feuilles de gros papier jaune goudronné dont il a été question à la page 38. On prépare une bonne quantité de ces feuilles. Au moment de faire le dallage, on égalise d'abord aussi soigneusement que possible l'aire de l'appartement. On y répand une légère couche de sable, sur laquelle on étale une ou deux feuilles de papier goudronné. Il faut avoir soin qu'elles fassent recouvrement. Sur le papier on étend du mortier, ou simplement de la boue assez consistante, mais faite de bonne terre non salpêtrée. Puis on pose les briques bien sèches. Ce procédé, pas cher, a fait ses preuves. Il suppose qu'on ne viendra pas ensuite vider sa théière au beau milieu de la chambre, pour s'éviter les deux pas qu'il y aurait à faire jusqu'à la porte. Ce procédé est fort à recommander pour les sacristies.

Polir les briques « *mouo tchoan* ». J'aurais cru inutile de parler de ce travail couteux et nuisible. On assure que les constructeurs l'ont fait exécuter pour donner une belle apparence à leur église. Donc on frotte vigoureusement l'une contre l'autre les briques, sur la face qui fera parement (face apparente). Par là, vous enlevez à la brique sa meilleure protection, sa petite cuirasse contre le salpêtre, je veux dire la mince couche plus ou moins vitrifiée qui a vu le feu du four.

Véranda « *langtze* ». Les uns la préconisent, les autres la honnissent. On pourrait peut-être concilier les deux opinions, [p. 62] en donnant aux vérandas peu de profondeur et beaucoup de hauteur. De cette façon, elles protégeraient en été contre les rayons du soleil, et en hiver chaleur et lumière ne seraient pas interceptés.

Il s'agit ici de vérandas le long des « *pei-ou* ». À l'est et à l'ouest, mais surtout au nord, les inconvénients dont il vient d'être question n'existent pas. Dans un collège, les vérandas sont très utiles en cas de mauvais temps, mais ici surtout, il ne faut pas qu'elles enlèvent la lumière.

Il est inutile de faire une charpente pour ces vérandas comme si elles devaient porter des milliers de tonnes. Si encore le bois était moins cher !

Chaînage. Chaîner un édifice, c'est rendre les murs solidaires les uns des autres, de manière à se soutenir réciproquement. Ce chaînage est moins nécessaire dans les maisons ordinaires. Il faudrait l'établir dans une construction plus importante, à toiture légère. Le toit chinois, en dalles, terre et tuiles, par son poids considérable et la pression exercée ainsi sur les fermes, peut tenir lieu de chaînage même pour une église.

Les chaînages se font de bien des manières. Dans nos vieilles maisons en Europe, nous avons tous vu sur les façades, aux étages, ces grandes croix de St André, en fer forgé, des S, des Y, qui relient les murs par l'intermédiaire des solives des planchers, aux différents étages et aux extrémités des entraits de fermes. L'architecte dira s'il faut employer le chaînage et de quelle façon il faudra l'établir.

J'ai souvent vu chaîner les caves, et ceux qui les avaient construites de cette façon ne voulaient pas admettre que ce fût-là de l'argent dépensé en vain et du travail inutile. Voici ce qu'en dit Barberot, p. 357 : « Pourquoi chaîner des murs qui doivent résister à la poussée des terres ? Ils devraient au contraire être contrebutés de l'intérieur vers l'extérieur ».

En dehors de toute considération de poussée des terres, pour les caves, il faut avant tout tenir compte de la pression considérable exercée par les murs qui s'élèvent au-dessus. Cette pression rend tout chaînage inutile, et la poussée des terres n'est guère à redouter. Il en est de même pour les cintres dont les piédroits [p. 63] supportent une grande charge, v. g. les cintres du premier étage d'un clocher, etc. Tout chaînage devient superflu.

Cheminées « *tsaot'oung* ». En Europe, la question des cheminées est règlementée de la façon la plus minutieuse, ce qui n'empêche pas que la moitié des incendies ont pour cause première une cheminée mal établie.

Qu'en Chine, le bon sens tienne lieu de règlement. Dans une grande construction, les plans doivent prévoir tous les détails de l'établissement des cheminées, et il faut surveiller les ouvriers avec un soin scrupuleux aux endroits où elles sont établies. Ne vous fiez jamais à un crépissage, soi-disant fait avec soin, pour appuyer directement une pièce du plancher ou des combles contre un mur de cheminée ; a fortiori dans le mur. C'est une précaution élémentaire que le mur, où passe une cheminée, soit complètement isolé à l'endroit où a lieu ce passage.

Escalier « *leou t'i* ». Ceux des maisons chinoises, quand il y en a, comme dans les tours, sont absolument impraticables. Les charpentiers n'ont pas de traditions pour ce travail, puisque, dans la vraie Chine, l'étage aussi bien que la cave sont inconnus. Il est donc important de ne pas abandonner l'ouvrier à sa propre inspiration ; on s'en repentira amèrement. Il faut lui tracer tous les éléments de l'escalier, tel qu'on le désire. Ce dessin se fait plus facilement en grandeur naturelle, contre un mur, du moins pour les cas ordinaires. Car il y a plus d'une sorte d'escalier ; la forme peut varier à l'infini. Ici encore, c'est à l'architecte à parler s'il s'agit d'une forme spéciale.

La fig. 82 donne les différentes mesures pour un escalier ordinaire en bois.

Voyez, figure 83, bonne et mauvaise marche pour escalier tournant.

Beffroi « *tchoung kiaze* ». Tel que nous l'entendons ici, le mot beffroi signifie la charpente de bois formant un assemblage capable de supporter le poids des cloches. C'est au fondeur de cloches à fournir le plan du beffroi, d'après les dimensions du clocher et le poids des cloches.

Il ne faut pas que des cloches d'un poids considérable soient soutenues directement par le mur du clocher ; elles les ébranleraient. Elles doivent reposer sur la charpente en question et cette charpente, ce beffroi, ne doit toucher en aucun de ses points le mur du clocher, si ce n'est à la base. Les pièces de [p. 64] bois qui constituent cette base reposent sur les pierres qui sont maçonnées en « encorbellement » aux quatre coins de la tour.

Voyez le plan d'un beffroi pour quatre cloches, fig. 84, 84bis. Ces cloches sont à battant, « rétro-lancé », ce qui permet de donner beaucoup moins de développement au beffroi. Renseignez-vous auprès des fondeurs sur les différents systèmes de battants. Les suspensions ne sont pas toutes du même genre non plus.

Goudron « *tch'eou you* ». Le goudron est facilement volé par les bateliers pour lesquels il est précieux. Il vaut mieux le faire venir en bidons à pétrole qu'en tonneaux. Pour faire couler naturellement le goudron hors des tonneaux, les bateliers n'ont qu'à desserrer légèrement les cercles. Ils pourront dire ensuite que le goudron s'est perdu et qu'il n'y a pas de leur faute. Pourtant ils peuvent voler aussi le goudron qui est en bidons. Ils le remplacent alors par de l'eau. — Il faut donc s'assurer par devant les bateliers qu'il n'y a pas eu fraude, avant de prendre livraison.

Quand le goudron est trop épais, on peut le chauffer, en prenant les précautions voulues. On peut aussi y ajouter du pétrole de mauvaise qualité, pas de l'eau évidemment.

Écoulement des eaux de pluie. Il faut surveiller avec grand soin, la première année, l'écoulement des eaux de pluie. Les caniveaux « *choei keou* », peuvent être obstrués par des restes de briques, de la terre. Si on n'y veille, les fondations peuvent être inondées.

Verre « *pouoli* ». On peut trouver, au Tcheli, des ouvriers et des instruments pour poser des « grisailles » en couleur. Il suffit de se procurer le verre de couleur, soit à Tientsin, Changhai, etc., soit directement en Europe. Il est bon de préserver ces grisailles, et surtout de vrais vitraux, contre la grêle si fréquente en cette région de la Chine. On le fait au moyen de grillages en fil de fer galvanisé qu'on fait confectionner sur place.

On peut encore protéger les fenêtres en verre de couleur avec du verre blanc dépoli. Les rayons du soleil n'arrivent plus alors directement dans l'église, mais sont diffus grâce au dépolissage du verre blanc. Il est plus économique de faire dépolir sur place du verre blanc ordinaire, plutôt que de l'acheter à Tientsin ou ailleurs. [p. 65]

Huiler « *you-ts'i* ». « *You ts'i tsiang* » se dit de l'ouvrier qui applique huile ou vernis soit sur des meubles, soit sur des portes et fenêtres et sur les bois de charpente exposés à la pluie : extrémités des chevrons, colonnes d'une véranda, etc.

L'huile chinoise « *t'oung you* » qui sert à cet usage ne vaut pas l'huile de lin. Aussi, sous l'action du soleil, de la pluie et du vent, elle disparait rapidement des bois sur lesquels elle a été appliquée. Il faut la renouveler assez souvent : dès la 3e ou la 4e année au début. Sans cette précaution, le bois s'effrite rapidement. Cette même huile chinoise, « *t'oung you* », appliquée sur les meubles, devient facilement visqueuse et collante en été… Je crois même qu'elle le devient toujours au bout d'un certain nombre d'années. Mais elle le deviendra infailliblement et rapidement si elle a été mal cuite par un ouvrier qui n'avait pas le tour de main, ou si le marchand y a mêlé de l'huile d'haricots, laquelle est beaucoup moins chère. L'huile d'haricots, mélangée au « *t'oung you* », rend la peinture plus brillante ; d'où tentation pour l'ouvrier d'être de connivence avec le marchand.

Ne cédez pas à la tentation de peindre la voûte d'une grande église, si vous voulez lui laisser tout son effet de hauteur. L'impression de lointain, en ce cas, est accentuée par la couleur blanche ; toute autre couleur « rapproche ».

Colonnes sous les poutres, dans les murs en pisé. Ces colonnes supportent tout le poids de la toiture, et soulagent ainsi les murs trop peu résistants. Mais il faut en protéger soigneusement la base contre la pourriture. Il importe donc de les préserver du contact de l'humidité montante du sol, à l'intérieur même de la base des murs. On y arrive par l'emploi de socles en pierre, au niveau du « *kien ts'ao* » (la couche isolante), soit par le carton bituminé, ou simplement par plusieurs couches de papier goudronné. Sans cette précaution toute la maison risquera de s'effondrer au bout de 10 à 20 ans.

Conclusion

Enfin, et que ce soit la conclusion de ce modeste travail, il faut entretenir votre bâtiment, une fois qu'il est terminé : entretien et longévité d'une construction, « convertuntur » auraient [p. 66] dit les anciens. Vous vous écriez que voilà un beau spécimen de truisme. Malheureusement rien ne semble répugner au tempérament chinois comme un entretien quelconque, qu'il s'agisse d'un habit, d'une machine ou d'un bâtiment. Que s'il s'agit de l'entretien d'un édifice public, vous vous heurtez à une impossibilité qui est peut-être physique ! Voyez les « *yamenn* », les pagodes, les tombeaux des empereurs, les routes, les ponts, les remparts des villes. Or nos pauvres églises sont… des édifices publics ! Pour vous en convaincre vous n'avez qu'à regarder les fenêtres, et à y compter les carreaux cassés. Sur la façade, voyez le nombre de clochetons [pinacles] que le vent a fait tomber, et que personne ne songe à remplacer. Pourtant on en était si fier jadis ! Heureux si le toit a résisté, et si toutes les briques sont restées en place. En effet, c'est par là que commence la ruine définitive.

Donc vous qui avez peiné pour élever à la gloire du divin Maître non pas un édifice digne de sa majesté, mais témoignant du moins du respect que vous avez de son culte, entretenez, faites entretenir votre église. Entretenez aussi, et faites entretenir celles que vos prédécesseurs vous ont léguées. Elles sont devenues peut-être un peu étroites. Du moins qu'elles restent propres, et que vous ne soyez pas obligé de la reconstruire tous les vingt ans, comme la pagode de Jese au Japon.

Aux missionnaires qui désireraient plus et mieux que ce « Robinson constructeur », nous nous permettons de signaler un ouvrage récent :

« **Notions élémentaires d'architecture religieuse** », par l'abbé D. Duret, professeur au Petit Séminaire de Chavagnes-en-Paillers (Vendée). Un vol., in-8° de 200 pages, avec 481 dessins à la plume, 10 francs franco. Chez l'auteur. Compte courant de chèques postaux : 143 — 81 Nantes.

Voici le compte rendu de l'Ami du Clergé, N° 33, 19 Août 1926, couverture jaune, p. 210 :

M. l'abbé Duret a voulu composer un véritable manuel d'architecture religieuse. — « Les ouvrages savants sur ces matières [p. 67] ne sont pas rares, écrit-il dans sa préface ; mais s'ils sont pour les initiés de précieux instruments de travail, ils déroutent parfois les commerçants par leur caractère technique. Ce sont les livres du Maître ; ils appellent, semble-t-il, un manuel élémentaire à la portée de tous, qui serait comme le livre de l'Élève. C'est le but de ce petit traité ».

L'auteur a pleinement réalisé son but, et son cours est vraiment le livre de l'Élève. Et c'est le plus bel éloge que nous puissions en faire. Il se recommande par la simplicité de son exposition, la clarté et la précision de son enseignement, qui embrasse toute l'histoire de l'architecture religieuse depuis les origines jusqu'à nos jours.

Il renferme près de 500 croquis à la plume, dessinés de la main experte de l'auteur et qui éclairent le texte d'une façon remarquable. Nous croyons cet ouvrage appelé à rendre de réels services au clergé et aux élèves de nos Séminaires, auxquels il est spécialement destiné. Il est bien fait, comme s'exprime l'auteur, « pour inspirer un zèle éclairé et discret pour la maison de Dieu », et pour leur faire connaître et aimer les chefs-d'œuvre de notre art national et chrétien ».

附录2. 建造技术地方用语
Appendix 2. Chinese Construction Terms Mentioned in the Handbook

原始记录 Original Record	拼音 Pinyin	可能汉字 Possible Chinese	英语 English	意思 Meaning	手册页码 Provenance Page
cha kao	sha gan	沙杆	small firs	一种小树	28
chaoyaoti	shao yao de	烧窑的	brick maker		19
cheng / ma cheng	sheng / ma sheng	麻绳	rope		27
cheng eull koung hoei	sheng er gong jiao hui	诸圣相通功	communion of the Saints		1
cheng hoei	sheng hui	生灰	uncooked lime	生石灰	25
chenn	shen	神	spirit		39, 51
cheou	shou	寿	long life		7
cheu fou	shi fu	师傅	master		13
cheu hang	shi hang	石夯	stone rammer		29
cheu hoei	shi hui	石灰	lime		25
cheu hoei chouli	shi hui shui	石灰水	water lime		26
cheu hoei jou	shi hui ru	石灰乳（当地叫石灰膏）	water lime	石灰膏	26
cheu hoei k'eng	shi hui keng	石灰坑	pits used to store lime		26

附 录
Appendix

续表 continue

原始记录 Original Record	拼音 Pinyin	可能汉字 Possible Chinese	英语 English	意思 Meaning	手册页码 Provenance Page
cheuhoei tch'oeiting / hoei ting / tch'oeiting	shi hui chui ding / hui ding	石灰锤顶，又叫灰顶	flat roof covered with asphalt		43
choa cheu hoei	shua shi hui	刷石灰	whitewashing		26
choei keou	shui gou	水沟	ditch / drain		64
chou cheu hoei	shu shi hui	熟石灰	slaked lime in powder		26
eull leang	er liang	二梁	beam of a Chinese timber roof	大梁上面的梁	46
fan mao ki	fan mao ji	反毛鸡	'hen with upside down feathers'		43
fang kia	fang jia	房架	roof truss		45
fang yen	fang yan	房檐	cornice / eaves		45
fei tch'oan	fei chuan	飞椽（飞子）	sprocket		48
fenn	fen	分	1/10th of inch	尺寸，十分之一	20
fong choei	feng shui	风水	geomancy		51
fou	fu	福	happiness		51
hang	hang	夯	rammer		28
hao k'an	hao kan	好看	nice looking		23, 31, 40, 51, 52
hiao t'ang	xiao tang	校堂	school room		52
houo ni	huo ni	和泥	mix earthen mortar		60
k'ao k'ao	kao kao	烤烤	cook, bake		41
k'ao lao	kao lao	犒劳	reward with food and drink		13, 30, 34
k'eou wa	kou wa	扣瓦	tile roof		44
k'ing	ling	铃	bell		

~ 425 ~

舶来与本土：1926年法国传教士所撰中国北方教堂营造手册的翻译和研究
Building Churches in Northern China. A 1926 Handbook in Context

续表 continue

原始记录 Original Record	拼音 Pinyin	可能汉字 Possible Chinese	英语 English	意思 Meaning	手册页码 Provenance Page
k'ing ma	jing ma	槿麻	used for making ropes	应为当地的三种麻之一	27
k'iou ma	? ma	□麻	used for making ropes	应为当地的三种麻之一	27
k'o t'ing	ke ting	客厅	living room		4
kangtch'a koayinn	gang cha gao ying	钢叉高引？	put iron deer horns on top of the roof	一种正脊两头的装饰	51
kantze	gan zi	杆子	scaffolding pole	用于搭建脚手架	28
kao tchoutze	gao zhu zi	高柱子	big post		46
keou ts'iang	gou qiang	勾墙（勾缝）	grouting		60
keou yen	kou yan	勾檐	covering		24
kiao ni	jiao ni	胶泥	clay	粘土糊住周围防止水渗出	32
kien	jian	间	bay		23
kien	jian	碱	saltpetre		37
kien	jian	间	room	房间	56
kien koung	jian gong	监工	supervisor		11
kien ts'ao	jian cao	碱草	herbs used as insulation against raising saltpetre	墙与地面之间铺的一层草防潮	37, 38, 39, 65
kien tsiang	jian qiang	碱墙	wall attached by saltpetre	白硝从砖缝渗出	37

~ 426 ~

续表 continue

原始记录 Original Record	拼音 Pinyin	可能汉字 Possible Chinese	英语 English	意思 Meaning	手册页码 Provenance Page
king ki	jing ji	经纪	middleman	相当于现在的经理	12
kinn t'eou u	qian tou yu	潜头鱼？	diving fish	一种正脊两头的装饰	51
kiue ti kiao	xiu ji cao	修基槽 修基角	foundation trench	建房前挖槽做地基	29
koantsiang	guan ji-ang	灌浆	gorging with mortar	用浆来填充间隙	36
Kouangpingfou	Guang ping fu	广平府	place in South Hebei		44
kouo	guo	锅	pan		32
kouo mou	guo mu	过木	lintel	窗上的木头，过梁	40
langtze	lang zi	廊子	veranda		61
lao cheull	lao shi	老师	master		13
leang cheu tounn	liang shi dui	粮食堆	heaps of grain		38
leou t'i	lou ti	楼梯	stair		63
li cheng, wai chou	li sheng wai shu	里生外熟	raw inside, cooked outside	墙的一种，外面为砖，里面为土	33
li wai tchoan	li wai zhuan	里外砖	a pure brick wall	室内外砖都露明	33, 60
ling tsouoti	ling tou de	领头的	contractor		11
linn	lin	檩	purlin		25, 45, 47, 50, 54
linn hoei	lin hui	拎灰	dilute lime and filter it through a wire mesh to extract stones and pebbles	把灰和水分开	26

~ 427 ~

续表 continue

原始记录 Original Record	拼音 Pinyin	可能汉字 Possible Chinese	英语 English	意思 Meaning	手册页码 Provenance Page
liou li tchoan	liu li zhuan	琉璃砖	glazed brick		37
ma ta hoei	ma dao hui	麻刀灰	lime mixed with finely chopped rope	麻刀和灰掺合后	28
ma tai	ma dai	麻袋	finely chopped rope used for coating		28
ma tao	ma dao	麻刀	finely chopped rope used for coating		28
mingtchou	ming zhu	明柱	pillar	看的见的柱子	48
mou hang	mu hang	木夯	wooden rammer		29
mouleao	mu liao	木料	timber, lumber		24
mouleao kia	da mu liao	木料架（木梁架）	timber frame		45
mouo koung fou / mouo yang koung	mo gong fu	磨工夫（磨洋工）	take the time of the foreigners		12
mouo mouo	mo mo	馍馍	white bread /steamed bread		30
mouo ni	mo ni	抹泥	coating		60
mouo tchoan	mo zhuan	磨砖	polish the bricks		61
nan ou	nan wu	南屋	house on the south side		4
nifangtze	ni fang zi	泥房子	add a layer of earth	在土平房顶浇的一层粘泥，以防水	42
p'i k'i	pi qi	脾气	temperament		12
p'i ts'iang	pi qiang	坯墙	the rammed earth wall	夯土墙	33
pa tchoan	ba zhuan	八砖	square tiles	方砖，铺在椽子上	23, 25, 41, 42, 45

附　录
Appendix

续表 continue

原始记录 Original Record	拼音 Pinyin	可能汉字 Possible Chinese	英语 English	意思 Meaning	手册页码 Provenance Page
pa-cheu	ba shi	把式	master mason	技能比别人高的人，领导	13, 29
pei ou	bei wu	北屋	house on the north side		3, 49, 62
penn ma	? ma	建麻	used for making ropes	应为当地的三种麻之一	27
piao tchoan	biao zhuan	表砖	the wall faced with bricks	外面为砖，里面为土	33, 34
pien tchoan	pian zhuan	片砖	stretchers (bricks)	砖横着放	33
pien tchoan tao ting	pian zhuan dao ding	片砖到顶	horizontal bricks under the roof	片砖砌到顶	33
pouoli	bo li	玻璃	glass		64
san leang	san liang	三梁	beam of a Chinese timber roof	二梁上面的梁	46
si ou	xi wu	西屋	house on the west side		3
siao koung	xiao gong	小工	unskilled worker		13
suan	zhuan	转	turn		23
suanwati	zhuan wa de	转瓦的	the workers who turn the tiles when the tiles are fired at the kiln		23
suetze	xizi	席子	mat		38
t'ang hoei	tang hui	堂灰	slaked lime in powder		26
t'ang ou	tang wu	堂屋	central house	正屋	3
t'ie hang	tie hang	铁夯	iron rammer		29

舶来与本土：1926 年法国传教士所撰中国北方教堂营造手册的翻译和研究
Building Churches in Northern China. A 1926 Handbook in Context

续表 continue

原始记录 Original Record	拼音 Pinyin	可能汉字 Possible Chinese	英语 English	意思 Meaning	手册页码 Provenance Page
t'ou linn	tu lin	土檩（即檐檩）	lower purlin	檐檩	41, 48
t'ou ping fang	tu ping fang	土平房	flat earthen roof		42
t'oung wa	tong wa	筒瓦	round tile		23, 43
t'oung you	tong you	桐油	oil		65
ta cheng ti	da sheng de	打绳的	rope maker	工种，绑扎绳子的人	27
ta hang	da hang	打夯	ramming		28
ta hoei	da hui	大灰	lime	生石灰是大块的所以叫大灰	25
ta leang	da liang	大梁	ridgepole	最下面的梁	46
ta loung	da long	大陇	big gutters		44
ta ts'iang	da qiang	打墙	rammed earten wall	土墙的一种，全是泥，版筑	33
tch'a liang	xia liang	下梁	secondary beam		48
tch'a pou touo	cha bu duo	差不多	about the same, more or less		19, 29, 41, 60
tch'eou you	jiao you	焦油	tar		64
tch'eu toung kia	chi dong jia	吃东家	'eat the boss'		13
tch'oantze	chuan zi	椽子	rafter		41, 45
Tcheli	Zhi li	直隶	Zhili		12, 24, 26, 28, 37, 42, 44, 64
tcheng ou sie ts'i	zheng wu xie qi	正五斜七（勾股定理）	45 degrees pitch		49

~ 430 ~

附　录
Appendix

续表 continue

原始记录 Original Record	拼音 Pinyin	可能汉字 Possible Chinese	英语 English	意思 Meaning	手册页码 Provenance Page
tchoan man ting	zhuan man ding	砖墁顶	flat roof covered with tiles	房顶上满铺砖	43
tchon kien leao	zhuan jian le	砖碱了	wall attached by saltpetre		37
tchoung kiaze	zhong jia zi	钟架子	belfry		63
ti choei	di shui	滴水	gutter		24
tong kia	dong jia	东家	boss		30
tong ou	dong wu	东屋	house on the east side		4
t'ou ti	tu di	徒弟	apprentice		13
touo ts'iang	duo qiang	垛墙	'piled wall'	土墙的一种，全是泥，版筑	33
touo k'eou	duo kou	垛口	battlement, crenel		53
ts'ao tcheu	cao zhi	草纸	thick yellow paper	上刷沥青制成油毡	38
tsao t'oung	zao cong	灶囱	chimney		63
Ts'eu-hi	Ci xi	慈禧	empress		50
ts'ints'i	qinqi	亲戚	relative		30
tsi	ji	脊	ridge		50, 51
tsi choei	ji shou	脊兽	two crest heads		51
tsi linn / tsilinn	ji lin	脊檩	ridge purlin		50,53
tsiang	jiang	浆	a very liquid mixture of lime		36
tsiao tchoan	tiao zhuan	条砖	glazed bricks	从后面对琉璃砖的解释知条砖即为琉璃砖	21, 37
u chou	yushu	榆树	elm tree		24

舶来与本土：1926年法国传教士所撰中国北方教堂营造手册的翻译和研究
Building Churches in Northern China. A 1926 Handbook in Context

续表 continue

原始记录 Original Record	拼音 Pinyin	可能汉字 Possible Chinese	英语 English	意思 Meaning	手册页码 Provenance Page
wa	wa	瓦	tile		23
wa fang	wa fang	瓦房	tiled-roofed house		43
wa tao	wa dao	瓦刀	(bricklayer's) cleaver		60
wan	wan	万	ten thousand (years)		51
wan yeull	wan yi er	玩意儿	toy		51
wei pa	wei ba	苇笆	reed mat	芦苇编的笆	45
wo tchoan	wo zhuan	卧砖	horizontal bricks, stretchers	砖横着放	33, 35
wong	weng	瓮	jar		32
xiao	xiao	硝	saltpetre		37
yamenn	yamen	衙门	government office		66
yang chou	yang shu	杨树	poplar tree	以前当地构房多用杨树	24
yang jenn	yang ren	洋人	foreigner		21
yang wa	yang wa	仰瓦	the tiles have their concave side facing upwards		44
yao k'eou	yao kou	窑口	kiln's opening		21
ying pei ts'iang	ying bi qiang	影壁墙	screen wall	当地叫法也有影门墙	39
yinn t'eou leao	yin tou le	饮透了	let the water penetrate thoroughly	砖饮好了	32
yinn yao	yin yao	饮窑（现在还有）	add water after tiles being fired	给窑顶渗水，让砖变蓝色	20

~ 432 ~

附　录
Appendix

续表 continue

原始记录 Original Record	拼音 Pinyin	可能汉字 Possible Chinese	英语 English	意思 Meaning	手册页码 Provenance Page
you ts'jien pou kai tong nan fang; tong pou noan, hia pou leang		有钱不盖东南房，冬不暖，夏不凉	a rich man does not build a house to the east, nor to the south; in winter, it is not warm; in summer, not fresh		3
young ts'ien	yong qian	佣钱	primes		12
you-ts'i	you qi	油漆	oil, paint		65
you-ts'I tsiang	you qi jiang	油漆匠	painter		65

参考文献
Bibliography

1. 手册/ The Handbook

Lyon (France), Bibliothèque Municipale, SJ H 678/42.
    Le missionnaire constructeur, conseils-plans [M]. Xianxian: Imprimerie de Sien-Hsien, 1926 (in-8°, 67 p.). The plates are missing.

Québec (Canada), Bibliothèque de l'Université Laval, Bibliothèque des sciences humaines et sociales, NA 6045 M678 1935.
    Le missionnaire constructeur, conseils-plans (2nd print) [M]. Xianxian: Imprimerie de Sien-Hsien, 1935 (in-8°, 67 p. and 54 plates).

2. 文献资料/Archival sources

Leuven, University of Leuven, Central Library, East-Asia Collections
    Van Dijk, Leo. 问答像解（Wenda xiangjie）[M]. Tianjin, 1928.

Leuven, University of Leuven, KADOC, Documentation and Research Center on Religion, Culture and Society
    Archives of the Congregation of the Immaculate Heart of Mary (CICM)

Leuven, University of Leuven, Ferdinand Verbiest Institute (FVI)
    Picture collection, series CHC (constructions) and BR (buildings and residences)

Paris, Missions Étrangères de Paris (MEP)
    Picture collection, China, 31-34 (Manchuria)

Rome, Archivum Romanum Societatis Iesu (ARSI)
    Catalogus provinciae Campaniae Societatis Iesu [M] Reims: Compagnie de Jésus, 1887-1946.
    Litterae annuae, Provincia Campaniae [M]. Reims: Compagnie de Jésus, 1926-27.

Vanves, Archives Jésuites, Compagnie de Jésus, Province de France (ASJ France)
    China, Sousteastern Zhili mission / Xianxian mission: GMC, 5, 7, 12-13, 21-22, 79, 80, 87, 101-102.

舶来与本土：1926年法国传教士所撰中国北方教堂营造手册的翻译和研究
Building Churches in Northern China. A 1926 Handbook in Context

3. 中文参考书/ Chinese printed sources and works

大名县志编纂委员会. 大名县志 [地方志]. 北京：新华出版社，1994.

国家文物局. 中国文物地图集（河北分册）（上、中、下）. 北京：文物出版社，2013.

洪家禄，等.大名县志（程廷恒等修）. 1934（民国23年）.

[明]潘仲骖纂修，(明）赵慎修续修. [嘉靖]大名府志 [善本]（ 二十九卷 ）.

[清]周邦彬修，(清）郜焕元纂. [康熙]大名府志 [善本]：（三十二卷）.

[宋]李诚撰，邹其昌点校. 营造法式 (文渊阁《钦定四库全书》). 北京：人民出版社，2011.

卢薇，高曼士. 莼鲈之思——在华圣母圣心会士以比利时教堂为参考的建筑设计[J].建筑与文化，2013, 112:88-90.

沈锦标. 造屋三知[M]. 上海：土山湾印书馆，1902.

宋浩杰. 土山湾记忆 [M]. 上海：学林出版社，2010.

Guo Qinghua, Zhongguo jianzhu Ying Han shuang jie cidian 中国建筑英汉双解 / A Visual English-Chinese Dictionary of Chinese Architecture, Mulgrave: Image Publishing, 2007

4. 其他出版物/other printed Sources and works

Aux missionnaires constructeurs [J]. Bulletin catholique de Pékin, 1927, 133-137.

Catalogus provinciae Campaniae Societas Iesu, 1987-1946. Reims: Compagnie de Jésus.

In memoriam. Le Frère Alphonse Litzler (1862-1948) [J]. Chine Madagascar, Missions des Jésuites français du nord et de l'est, 1949, 13.

La Préfecture Apostolique de Ta-Ming Fu [J]. Trait d'Union. À nos Chrétiens, 1936, 20 :1.

Le Trait-d'Union. Bulletin des anciens élèves du Collège Français de Taming / 法中期刊。大名法文学同学季刊. 1-20, July 1922 to January 1936.

Arrington Aminta. Recasting the Image: Celso Costantini and the Role of Sacred Art and Architecture in the indigenization of the Chinese Catholic Church, 1922-1933 [J]. Missiology: An International Review, 2013, 41: 438-451.

Aubin Françoise. Un cahier de vocabulaire technique du R. P. A. De Moerloose CICM, missionnaire de Scheut (Gansu septentrional, fin du XIXe siècle) [J]. Cahiers de linguistique. Asie orientale, 1983, 2 :103-117. Original (Leuven, KADOC, Archives CICM, Z. II. h.8).

Aubin Françoise. Christian Art and Architecture [A] // R. Gary Tiedemann. Handbook of Christianity in China. Volume Two: 1800-Present [C]. Leiden-Boston: Brill, 2010, 733-741.

Barberot Étienne. Traité de constructions civiles travaux préparatoires et connaissance du sol, maçonnerie, pavages divers, accessoires de maçonnerie, béton de ciment armé, marbrerie, vitrerie, vitraux, charpente de bois, charpente métallique, couverture, menuiserie et ferrures, escaliers, monte-plats, monte-charges et ascenseurs, plomberie d'eau et sanitaire, épuration biologique, chauffage et ventilation, décoration, éclairage au gaz et à l'électricité, acoustique, matériaux de construction, résistance des matériaux, statique graphique, renseignements généraux (4th edition) [M]. Paris, 1912.

Benedict XV. Maximum illud: Apostolic Letter on the Propagation of the Faith throughout the World [M]. Vatican, 1919.

Bernard Henri. La Compagnie de Jésus en Chine. L'ancien vicariat apostolique du Tchéli sud-est, ses filiales, ses annexes [M]. Tianjin, 1940.

Bornet Paul. Mission de Chine. Le Tche-Li S. E., 1857-1928. Mission de Sienhsien. La troisième étape de 25 ans. 1907-1932 [M]. (typed manuscript of an unpublished book, p. 15) (Vanves, ASJ France, GMC 7).

Bornet Paul. Notes sur l'évangélisation du Tcheli et de la Tartarie aux XVIIe et XVIIIe siècles. Notes sur l'origine de quelques chrétientés du Tchéli et de la Tartarie aux XVIIe et XVIIIe siècles [J]. Bulletin catholique de Paris, 1937 (offprint with personal notes of the author: Vanves, ASJ France, GMC 5).

Bremner Alex. Imperial Gothic: Religious Architecture and High Anglican Culture in the British Empire, 1840-1870 [M]. City of New Haven : Yale University Press, 2013.

Brittain-Catlin Timothy, Maeyer Jan De, Bressani Martin. A. W.N. Pugin's Gothic Revival: The International Style (KADOC Artes) [C]. Leuven: Leuven University Press, 2016 (forthcoming).

Brou Alexandre. Cent ans de missions, 1815-1934: les Jésuites missionnaires aux XIXe et XXe siècles [M]. Paris, 1934; Marianne Monestier. Les Jésuites et l'Extrême-Orient [M]. Paris, 1956.

Cannepin Maurice. Comment on bâtit une église (Esquisses jaunes) [J]. Les Missions catholiques, 1918, 50: 416-418, 429-431.

Carbonneau E. Robert. The Catholic Church in China 1900-1949 [A] // R. Gary Tiedemann. Handbook of Christianity in China. Volume Two: 1800 to Present [C]. Leiden-Boston, 2010, 519.

Chen Alexandre Tsung-ming. Les réactions des autorités chinoises face au protectorat religieux français au cours du XIXe siècle [A] // Alexandre Chen Tsung-ming. Le Christianisme en Chine aux XIXe et XXe siècles. Évangélisation et Conflits [C]. Leuven, 2014, 125-171.

Chen Jianyu. Yingzao Fayuan: Two Editions by the Carpenters and the Architects in 1900s [A] // Austin Williams, TheodorosDounas. Masterplanning the Future. Modernism: East, West & Across the World [C]. Suzhou, 2012, 185-193.

Cluzel Jean-Sébastien, Masatsugu Nishida (eds). Le sanctuaire d'Ise. Récit de la 62e reconstruction [C]. Brussels, 2015.

Cody W. Jeffrey. Striking a Harmonious Chord: Foreign Missionaries and Chinese-style Buildings, 1911-1949 [J]. Architronic, 1996, V5n3.

Cody W. Jeffrey. Building in China. Henry K. Murphy's 'Adaptive Architecture' 1914-1935 [M]. Hong Kong, 2001.

Cody W. Jeffrey. American Geometries and the Architecture of Christian Campuses in China [A] // Daniel H. Bays, Ellen Widmer. China's Christian Colleges. Cross-Cultural Connections, 1900-

1950 [C]. Stanford, 2009, 27-56.

Compagnon P. M. Le culte de Notre-Dame de Lourdes dans la Société des Missions-Étrangères [M]. Paris, 1910, 111-156.

Coomans Thomas, Luo Wei. Exporting Flemish Gothic Architecture to China: Meaning and Context of the Churches of Shebiya (Inner Mongolia) and Xuanhua (Hebei) built by Missionary-Architect Alphonse De Moerloose in 1903-1906 [J]. Relicta. Heritage Research in Flanders, 2012, 9: 219-262.

Coomans Thomas. Sint-Lucasneogotiek in Noord-China: Alphonse De Moerloose, missionaris en architect [J]. M&L. Monumenten, landschappen en archeologie, 2013,32(5): 6-33.

Coomans Thomas. La création d'un style architectural sino-chrétien. L'œuvre d'Adelbert Gresnigt, moine-artiste bénédictin en Chine (1927-1932) [J]. Revue Bénédictine, 2013, 123: 128-170.

Coomans Thomas. Die Kunstlandschaft der Gotik in China: eine Enzyklopädie von importierten, hybridisierten und postmodernen Zitaten [A] // Heiko Brandl, Andreas Ranft, Andreas Waschbüsch. Architektur als Zitat. Formen, Motive und Strategien der Vergegenwärtigung [C]. Regensburg, 2014, 133-161.

Coomans Thomas. Indigenizing Catholic Architecture in China: From Western-Gothic to Sino-Christian Design, 1900-1940 [A] // Cindy Yik-yi Chu. Catholicism in China, 1900-Present. The Development of the Chinese Church [C]. New York: Palgrave and Macmillan, 2014, 125-144.

Coomans Thomas. China Papers: The architecture archives of the building company Crédit Foncier d'Extrême-Orient (1907-1959) [J]. ABE Journal. European architecture beyond Europe, 2014, 5:689, http:// dev. abejournal. eu/index. php?id=689.

Coomans Thomas. A pragmatic approach to church construction in northern China at the time of Christian inculturation: The handbook Le Missionnaire Constructeur, 1926 [J]. Frontiers of Architectural Research, 2014, 3(2): 89-107.

Coomans Thomas, Xu Yitao. Gothic Churches in Early 20th-Century China: Adapting Western Building Techniques to Chinese Construction Tradition [A] // Brian Bowden, Donald Friedman, Thomas Leslie, et al. Construction History Society of America (vol. 1) [C]. 2015, 523-530.

Coomans Thomas. Une utopie missionnaire? Construire des églises, des séminaires et des écoles catholiques dans la Chine en pleine tourmente (1941) [A] // Alexandre Chen Tsung-ming. Le Christianisme en Chine aux XIXe et XXe siècles. Figures, événements et missions-œuvres (Leuven Chinese Studies, 31) [C]. Leuven, 2015, 45-79.

Coomans Thomas. Gothique ou chinoise, missionnaire ou inculturée? Les paradoxes de l'architecture catholique française en Chine au XXe siècle [J]. Revue de l'Art, 2015,189(3): 9-19.

Coomans Thomas, Luo Wei. Mimesis, Nostalgia and Ideology: The Scheut Fathers and Home-Country-Based Church Design in China [A] // History of the Catholic Church in China. From its beginning to the Scheut Fathers and 20th Century (Leuven Chinese Studies, 29) [C]. Leuven, 2015, 495-522.

Coomans Thomas. The "Sino-Christian Style": A Major Tool for Architectural Indigenisation [A] // ZhengYangwen. Sinicising Christianity [C]. Leiden-Boston: Brill, 2016 (forthcoming).

Coomans Thomas. Pugin Worldwide: From Les VraisPrincipes and the Belgian St Luke Schools to Northern China and Inner Mongolia [A] // Timothy Brittain-Catlin, Jan De Maeyer, Martin Bressani. A. W.N. Pugin's Gothic Revival: The International Style (KADOC Artes) [C]. Leuven: Leuven University Press, 2016 (forthcoming).

Coomans Thomas, Luo Wei. Missionary Builders: Scheut Fathers as Designers and Constructors of Gothic, Chinese and Modern Churches in Northern China [A] // Chinese Catholicism and Missionaries from the Low Countries from the 17th to the 20th Century in China (Leuven Chinese Studies) [C]. Leuven: Ferdinand Verbiest Institute (forthcoming).

Costantini Celso. The Need of Developing a Sino-Christian Architecture for our Catholic Missions [J]. Bulletin of the Catholic University of Peking, 1927, 3: 7-15.

Costantini Celso. The Church and Chinese Culture [M]. New York, 1931.

Costantini Celso. L'universalité de l'art chrétien [J]. Dossiers de la Commission synodale, 1932, 5: 410-417.

Costantini Celso. Le problème de l'art en pays de missions [J]. L'artisan liturgique, 1932, 24: 816-819.

Costantini Celso. L'arte Cristiana nelle Missioni (Urbaniana, 2) [M]. Rome: Tipografia poliglotta Vaticana, 1940.

D. G. 1924. Quelques idées sur un art chinois [J]. Bulletin catholique de Pékin, 1924, 128-130: 127-130, 177-183, 218-221.

De Moerloose Alphonse. Construction, arts et métiers, au Kan-sou et en Chine [J]. Missions en Chine et au Congo, 1891, 34: 532-538.

De Moerloose Alphonse. Arts et métiers en Chine [J]. Missions en Chine et au Congo, 1892, 37 :3-8.

De Vogue Melchior, Neufville Jean. Glossaire des termes techniques (Introduction à la Nuit des Temps, 1) (4threvised edition) [M]. La Pierre-qui-Vire, 1989.

Demanet A. Mémoire sur l'architecture des églises [M]. Brussels, 1847.

Doutreligne Denis-Donat. Causerie sur l'art chinois. Peut-on adopter indifféremment dans nos Églises certains détails d'ornementation? [J]. Bulletin de la Société des Missions-Étrangères de Paris, 1925, 42: 341-356.

Duret Donatien. Notions élémentaires d'architecture religieuse [M]. Chavagnes-en-Paillers, 1926.

Garric Jean-Philippe, Nègre Valérie, Thomine-Berrada Alice (eds). La construction savante. Les avatars de la littérature technique [C]. Paris: Picard, 2008.

Gerbes chinoises. Les Jésuites dans la mission de Sien-Hsien depuis 1856 [M]. Lille, 1934.

Ghesquières Albert, Muller Paul. Comment bâtirons nous dispensaires, écoles, missions catholiques, chapelles, séminaires, communautés religieuses en Chine? [J]. Collectanea commissionis synodalis, 1941, 14:1-81.

Goi Paolo (ed.). Il Cardinale Celso Costantini e la Cina. Un protagonista nella Chiesa e nel mondi del XX secolo [C].Pordenone: Diocesi di Concordia-Pordenone, 2008.

Goudallier Léon. Une imprimerie en Chine [J].Cosmos. Revue des sciences et de leurs applications, 1908, 57(9): 524-527.

Gresnigt Adelbert. Chinese architecture [J]. Bulletin of the Catholic University of Peking, 1928, 4:41.

Huc Évariste. L'Empire Chinois [M]. Paris, 1850.

Huc Évariste. Souvenirs d'un voyage dans la Tartarie, le Thibet, et la Chine pendant les années 1844, 1845 et 1846 [M]. Paris, 1850; and L'Empire Chinois, Paris, 1854.

Joüon René. Atlas la Chine [M]. Shanghai-Zikawei: T'ou-sè-wè Press, 1930.

Jung Paul. Fleurs du Céleste empire. Touchante histoire de Stanislas Wang [J]. Les missions catholiques, 1918, 317-319.

Jung Paul. La joie dans la souffrance. Histoire d'une petite Chinoise malade [J]. Les missions catholiques, 1918, 534-535, 546-548.

Jung Paul. From the Mission-field of Taming-Fu [J]. The Far East, 1919, 6-7; 1919, 6-7; 1919, 5-6 (Vanves, ASJ France, GMC, 21/1, 87-91).

Jung Paul. Une grande ville chinoise qui s'ouvre à la foi [J]. Les missions catholiques, 1919, 235 248-249.

Jung Paul. Robinson-Cycliste. Guide du missionnaire vélocipédiste débutant [M]. Xianxian, 1927.

King Stanislas P'ei-yuan. Taming d'aujourd'hui [J]. Le Trait d'Unio, 1922, 1: 10-13.

Korigan Pol. Le manuel pratique de Robinson (2nd edition) [M]. Shanghai, 1912.

Koschorke Klaus. Indigenization [A] // Hans Dieter Betz. Religion Pas & Present. Encyclopedia of Theology and Religion(vol. 6) [C]. Leiden-Boston, 2009, 459-460.

Lamasse Henri. 新國文 Sin Kouo Wen ou nouveau manuel de langue chinoise écrite, traduit et expliqué en français et romanisé selon les principaux dialects [M]. Hong Kong, 1920 (several editions).

Langlais Jacques. Les Jésuites au Québec en Chine (1918-1955) [M]. Québec: Presses de l'Université Laval, 1979.

Launay Marcel, Moussay Gérard (eds). Les Missions étrangères. Trois siècles et demi d'histoire et d'aventure en Asie [C]. Paris, 2008.

Le Pichon Alain. Bethanie and Nazareth: French Secrets from a British Colony [M]. Hong Kong, 2007.

Leboucq François Xavier. Monseigneur Edouard Dubar de la Compagnie de Jésus: évêque de Canathe et la mission catholique du Tche-Ly-Sud-Est, en Chine [M]. Paris, 1880

Leroy Henri-Joseph. En Chine au Tché-ly S.-E. Une mission d'après les missionnaires [M]. Lille-Tournai, 1900, 187-237.

Luo Wei. Transmission and Transformation of European Church Types in China: The Churches of the Scheut Missions beyond the Great Wall, 1865-1955 [D]. University of Leuven, Faculty of Engineering: Architecture, 2013, 120-195.

Magrill Barry. 'Commerce of Taste': Church Architecture in Canada, 1867-1914 [M]. Montreal-Kingston, 2012.

Mertens Pierre. Une grande ville chinoise qui s'ouvre à la foi. Épilogue: achèvement de la cathédrale de Tai-ming-fou [J]. Les Missions Catholiques, 1920, 356-357, 367-369.

Monestier Marianne. Les Jésuites et l'Extrême-Orient [M]. Paris, 1956.

Nuyts Josef. En tournée à travers le vicariat [J]. Missions de Scheut, 1938, 218-219.

Olsborn L. C. The China Story: The Church of the Nazarene in North China, South China and Taiwan [M]. Kansas City, 1969.

Pius XI. Rerum ecclesiae: Encyclical on Catholic Missions, Vatican [M]. 1926. http:// www. svdcuria. org/public/mission/docs/encycl/mi-en. htm.

Planchet Jean-Marie. Les missions de Chine et du Japon. 1925. Sixième année [M]. Beijing: Imprimerie des Lazaristes, 1925.

Planchet Jean-Marie. Les missions de Chine et du Japon. 1929. Huitième année [M]. Beijing: Imprimerie des Lazaristes, 1929.

Renaud Rosario. Le diocèse de Süchow (Chine), champ apostolique des Jésuites canadiens de 1918 à 1954 [M]. Montréal: Éditions Bellarmin, 1982.

Schmit Jean-Philippe. Nouveau manuel complet de l'architecture des monuments religieux ou traité d'application pratique de l'archéologie chrétienne à la construction, à l'entretien, à la restauration et à la décoration des églises, à l'usage du clergé, des fabriques, des municipalités et des artistes [M]. Paris: Manuels Roret, 1859.

Soetens Claude. L'Église catholique en Chine au XXe siècle [M]. Paris: Beauchesne, 1997.

Soulié de Morant George. L'épopée des Jésuites en Chine, 1534-1928 [M]. Paris: Grasset, 1928.

Standaert Nicolas. Inculturation. The Gospel and Cultures [M]. Philippines, 1990.

Standaert Nicolas (ed). Handbook of Christianity in China, Volume One: 635-1800 [M]. Leiden-Boston: Brill, 2001; R. Gary Tiedemann (ed.). Handbook of Christianity in China, Volume Two: 1800-Present [M]. Leiden-Boston: Brill, 2010.

Streit Robert, Johannes Dindinger. Bibliotheca missionum, 13, Chinesische Missions literatur [M]. Rome-Freiburg-Vienna, 1959, 311-312.

Svagr Ian. Le béton chinois [J]. La Cité. Architecture, Urbanisme, 1923, 4(5): 91-92.

Ticozzi Sergio. Celso Costantini's Contribution to the Localization and Inculturation of the Church in China [J]. Tripod (鼎), 2008, 28(148).

Tiedemann R. Gary. Reference Guide to Christian Missionary Societies in China from the Sixteenth to the Twentieth Century [M]. Armonk-London: ME Sharpe, 2009.

Tiedemann R. Gary (ed.), Handbook of Christianity in China. Volume Two: 1800 to Present [C]. Leiden-Boston: Brill, 2010.

Vámos Péter. Hungarian Missionaries in China [A] // Stephen Uhalley, Wu Xiaoxin. China and Christianity: Burdened Past, Hopeful Future [C]. London-New York: Routledge, 2001, 218-231

Van Hecken Joseph. Alphonse Frédéric De Moerloose C. I.C. M. (1858-1932) et son œuvre d'architecte en Chine// Neue Zeitschrift für Missions wissen schaft / Nouvelle Revue de science missionnaire. 24(3):161-178.

Verhelst Daniël, Pycke Nestor (eds). C. I.C. M. Missionaries, Past and Present 1862-1987 (Verbistiana, 4) [C]. Leuven, 1995.

Victoir Laura, Zatsepine Victor (eds). Harbin to Hanoi: The Colonial Built Environment in Asia, 1840 to 1940 [C]. Hong Kong: Hong Kong University Press, 2013

Viollet-le-Duc Eugène-Emmanuel. Dictionnaire raisonné de l'architecture française du XIe au XVIe siècle (10 vol.). Paris: Bance & Morel, 1854-1868.

Wiest Jean-Paul. The Building of the Cathedral of Canton: Political, Cultural and Religious Clashes [A] // Religion and Culture: Past Approaches, Present Globalisation, Futures Challenges (International Symposium on Religion and Culture, 2002: Macau) [C]. Macau: Instituto Ricci de Macau, 2004.

5. 网站/Websites

http://www. catholic-hierarchy. org/bishop/bszar. html.

http://archives. mepasie. org/notices/notices-biographiques/lamasse.

http://www. patrimoine-culturel. gouv. qc. ca/rpcq/detail. do?methode=consulter&id=9892&type=pge#.VkNLTvkvehc.

http://www. svdcuria. org/public/mission/docs/encycl/mi-en. htm.

http://www. vatican. va/archive/cdc/index. htm.

http://dev. abejournal. eu/index. php?id=689.

索 引
Index

人名 Persons

Adrien Dufresne, 7
Donatien Duret, 24
Étienne Barberot, 24
Gagnon, 350
Henri Maquet, 312
Pierre Mertens, 30, 167, 313, 318
阿巴贡 Harpagon, 122, 372
阿道夫·鲁斯 Adolf Loos, 209
奥古斯都·威尔比·诺斯摩尔·普金 Augustus Welby Northmore Pugin, 20, 40, 83, 213, 439
巴鸿勋 Jules Bataille, 29
白玉清 Pierre Perard, 29
慈禧 Cixi, 204, 431
慈禧 empress 即 慈禧 Cixi
丹尼尔·迪福 Daniel Defoe, 233
方济民 Nicolas Vagner, 29
刚恒毅 Celso Costantini, 41, 86, 89-92, 94, 96, 99, 105, 106, 108, 288, 436, 440, 441
葛利斯 Adelbert Gresnigt, XXII, 92-94, 107, 438
郝司铎 Charles Héraulle, 312, 313
和羹柏 Alphonse De Moerloose, XVIII, XIX, XXI, XXII, XXIII, 24, 38-44, 69, 75, 76, 83, 84, 94-96, 99, 105, 106, 108, 109, 111, 114, 136, 346, 438
贺拉斯 Horace, 209
胡克 Évariste Huc, 133
教皇本笃十五世 Benedict XV, pope, 88, 312, 437
孔子 Confucius, 161, 205, 388, 410
郎守仁 Philippe Leurent, 29
雷振声 Alphonse Litzler, 29, 33, 35-37, 80, 81, 96, 109, 111, 284, 297, 306, 315, 323, 342, 436
李诫 Li Jie, 19, 436
梁神父 Henry Lamasse, XIX, XXII, XXVIII, 33-37, 82, 96, 284, 306, 313, 315, 316
刘钦明 Henri Lécroart, 25, 99, 118, 307, 312, 323
刘钦明 liu Qinming, 25, 99, 118, 307, 312, 323
隆其化 Lischerong Gáspár, 289
鲁滨逊·克鲁索 Robinson Crusoe, 22, 29, 233, 423, 440
莫里哀 Molière, 122
沈锦标 Shen Jinbiao, 24, 25, 436
圣母玛利亚 Our Lady 即 圣母玛利亚 Virgin Mary

圣母玛利亚 Virgin Mary, XVIII, XXI, XXVIII, 12, 43, 75, 94, 122, 218, 292, 307, 308, 350

圣约瑟 saint Joseph, 81, 122

陶德民 Charles Taranzano, 29

天德 Petrus De Boeck, XVIII, 4

王亚海 Wang Yahai, XXIII, 103

维利奥雷-勒-杜 Eugène-Emmanuel Viollet-le-Duc, 98, 99, 185, 400, 442

维特鲁威 Vitruvius, 19

卫秉仁 Frans Leineweber, 29

小主 Little Master 即 小主 Xiao Zhu

小主 Xiao Zhu, 323-325

杨·德·拉·方登 Jean de La Fontaine, 185, 209

雍居敬 Paul Jung, XIX, 28-33, 99, 107, 284, 288, 294, 297, 298, 305, 306, 313, 314, 319, 323, 324

地名 Places Names

阿讷西 Annecy, 31

安国 Anguo, XXII, 95

安庆 Anqing, 289

安特卫普 Antwerp, 43

澳门 Macao, 285

巴黎 Paris, XII, XIX, XXI, XXII, XXVIII, 15, 19, 21, 24-26, 33, 35, 74, 82, 85, 87, 94, 96, 99, 101, 133-135, 137, 284, 285, 287, 292, 293, 307, 316, 378, 435, 437-442

蚌埠 Bengbu, 289

保定 Baoding, 106

北京 Beijing, X, XI, XIII, XV, XVIII, XXI, XXII, XXVIII, XXXI, XXXII, 10, 11, 13, 16, 19, 25, 26, 29, 32, 38, 46, 59, 76, 86, 87, 90, 92, 93, 106, 111, 135, 188, 194, 204, 284, 300, 350, 352-356, 358, 359, 361, 364-367, 436, 441

北京 Peking 即 北京 Beijing

波摩莱 Anchialus, 118

布洛涅 Boulogne, 36

长春 Changchun, XXI, 74

朝阳 Chaoyang, XIX, 45

大名 Daming, VIII, XI, XIII, XVII, XVIII, XIX, XX, XXII, XXVII, XXVIII, XXIX, XXX, XXXI, XXXII, 6, 9, 11-13, 15, 16, 25, 26, 29-33, 35-37, 67, 78, 80, 81, 99, 104, 106, 107, 111, 118, 122, 167, 179, 180, 187, 194, 284, 285, 287-293, 295-297, 299-308, 311-323, 325, 326, 330, 331, 334-345, 347-353, 355-359, 361, 364-367, 436

大同 Datong, XXI, 73, 354

迪尤 Thieu, 36

高家营子 Gaojiayingzi, XVIII, 5

根特 Ghen, 38

固阳县 Guyangxian, XX, 67

广平 Guangping, 66, 194, 287, 340, 427

广州 Guangzhou, XVIII, 23, 284, 318

海岛营子 Haidaoyingzi, XX, 60

海门 Haimen, 289

壕赖山 Haolaishan, XIX, 52

河间 Hejian, 36, 287, 307

呼和浩特 Hohhot 即 呼和浩特 Huhehaote

呼和浩特 Huhehaote, 91

吉林 Jilin, XXI, XXII, XXIII, XXVIII, 33, 74, 81, 82, 113, 315, 316, 325

冀州 Jizhou, 287

胶州 Jiaozhou, 307

景县 Jingxian, 288

君士坦丁堡 Constantinople, 140, 380

开封 Kaifeng, XXI, XXII, 69, 88, 92

开州 Kaizhou, 33, 36

拉花营子 Lahuayingzi, XVIII, 4

兰斯 Reims, 29, 215, 216, 415, 435, 436

兰州 Lanzhou, XXI, 71

里昂 Lyon, XVIII, 7, 8, 435

里尔 Lille, 25, 26, 33, 36, 287, 288, 307, 322, 324, 439, 440

连云港 Lianyungang, 289

辽阳 Liaoyang, XXII, 85

鲁汶 Leuven, IX, X, XI, XII, XIII, XV, XXII, 7, 10, 11, 20, 21, 39, 40, 42, 43, 87, 89, 90, 94, 106, 350, 435-439, 442

罗马 Rome, 9, 28-30, 47, 91, 106, 134, 137, 140, 141, 161, 209, 232, 379, 435, 439, 441

洛恩斯特罗 Launstroff, 29

洛杉矶 Los Angeles, 7

玫瑰营子 Meiguiyingzi, XX, 60

磨子山 Mozishan, XXI, 69

南乐 Nanle, 289

南泉子 Nantsuantse, XXIII, 113

尼费尔恩市 Niefern, 36

平地泉 Pingdiquan, XXIII, 109

濮阳 Puyang, 33, 289, 292

青岛 Qingdao, 91

清丰 Qingfeng, 289

曲周 Quzhou, XX, 54

如皋 Rugao, XIX, 51

上海 Shanghai, XIX, XXII, 13, 22, 25, 26, 43, 44, 63, 75-77, 79, 83, 84, 89, 94, 111, 185, 229, 233, 289, 292, 346, 436, 440

舍必崖 Shebiya, XXIII, 39, 114, 438

深州 Shenzhou, 287

沈阳 Mukden 即 沈阳 Shenyang

沈阳 Shenyang, XIX, 33, 35, 142

仕拉乌素壕 Shilawusuhao, XX, XXIII, 58, 64, 65, 103

双树子 Shuangshuzi, XXIII, 108

水东 Shuidong, XIX, 53

苏州 Suzhou, 25, 289, 437

天津 Tianjin, XI, XXI, 31, 32, 43, 55, 58, 63, 75, 79, 86, 87, 94, 111, 143, 162, 168, 185, 218, 229, 230, 284, 285, 287, 288, 292, 350, 435, 437

维也纳 Vienna, 30, 350, 441

芜湖 Wuhu, 289

五号 Wuhao, XXIII, 110

武邑 Wuyi, 287

西昌 Xichang, XXII, 101

西湾子 Xiwanzi, XIX, XXI, 38, 40, 76, 91

希尔德 Schilde, 43

献县 Xianxian, XIX, 6, 9, 18, 25, 26, 28, 29, 32, 36-38, 59, 80, 96, 99-104, 110, 111, 118, 284, 287-289, 307, 316, 324, 325, 337, 338,

340-342, 346, 347, 351, 435, 440

香港 Hong Kong, XI, 26, 33, 34, 92, 113, 133, 285, 350, 437, 440, 442

徐州 Xuzhou, 7, 289

宣化 Xuanhua, XIX, 38, 39, 42, 92, 438

雅典 Athens, 137

兖州 Yanzhou, 66, 194

扬州 Yangzhou, XIX, 53

杨家坪 Yangjiaping, 38, 42, 106

伊势 Ise, 136, 232, 233, 378, 437

伊斯坦布尔 Istanbul, 140

永年 Yongnian, 288

张官屯 Zhangguantun, XIX, 52

张家口 Kalgan 即 张家口 Zhangjiakou

张家口 Zhangjiakou, 106

赵家庄 Zhaojiazhuang village, 32

芝加哥 Chicago, 13

朱家河 Zhujiahe, 287

一般名词 General Terms

北京大学 Peking University, X, XI, XIII, XV, XXVIII, XXXI, XXXII, 10, 11, 284, 300, 355, 358, 359, 361, 364-367

北堂 North Cathedral Xishiku, XXI, 76

大名教堂 Daming church, XVIII, XIX, XXII, XXVIII, 20, 23, 26, 33, 35, 38, 44, 66, 85, 88, 95, 101, 104, 121, 140, 142, 180, 185, 194, 207, 209, 210, 215-217, 284, 307, 313, 315, 316, 318, 323, 325

大名天主堂 The church of Daming, VIII, XIII, XVII, XVIII, XXVIII, XXIX, XXX, XXXI, XXXII, 12, 13, 15, 16, 26, 30, 31, 33, 36, 80, 99, 104, 106, 118, 284, 306, 307, 313-322, 325, 326, 330, 331, 334-337, 339-345, 347-353, 355, 356, 359, 361, 364-367

辅仁大学 Catholic University of Peking, XXII, 89, 91-93, 107, 439, 440

辅仁大学 Fu Jen University 即 辅仁大学 Catholic University of Peking

公立学校 public school, 293

观音菩萨 Guanyin Pusa, 138, 139

管风琴 organ, XXX, 326, 345, 349-351

邯郸教区圣心修院 Handan Parish Sacred Heart Christian College, 298

建造历史国际研讨会 International Congress on Construction History, 13

建筑遗产保护国际研讨会 Architectural Heritage Preservation International, 13

考洛乔圣母修女会 school of the Sisters of Our Lady of Kalocsa, 292

拉瓦尔大学 Université Laval, XII, 7

礼拜堂 chape, XX, XXI, XXVII, XXVIII, XXIX, 67, 69, 76, 81, 127, 128, 179, 180, 277, 297, 300, 302, 305, 306, 325, 326, 336

鲁汶大学 University of Leuven, X, XI, XII, XV, 7, 10, 11, 38, 353, 435, 440

马德莱娜教堂 Madeleine church, 137

南堂 Beijing South Church, 284

女修 convent of Chinese Virgins, 297
女子学校 girls' school, 292, 297
帕特农神庙 Parthenon, 136, 137
牌楼 archway 即 牌楼 pailou
牌楼 pailou, XXVII, 300, 316, 317
圣地特雷耶圣母院 shrine of Notre-Dame de la Treille, 307
圣鲁克学校 StLuke's School, 38
圣米厄尔及圣本笃礼拜堂 St Michael and St Benedict chapel, XIX, 40
圣母得胜堂 Lady of the Victory church 即 望海楼教堂 Wanghailu church
圣母圣心会修院礼拜堂 Chapel of the seminary, XXI, 76
圣索菲亚大教堂 Hagia Sophia, 140
圣约瑟礼拜堂 StJoseph chapel, XIX, 37
圣约瑟神学院 St Joseph seminary, 81
圣约瑟堂 St Joseph church, XXII, 77
市图书馆 Bibliothèque municipale, 7
天津工商学院 Tianjin Commercial University, 32, 292
天坛 Temple of Heaven, 204, 205
天主教学院 Catholic Institute, 134
望海楼教堂 Lady of the Victory church 即 望海楼教堂 Wanghailu church
望海楼教堂 Wanghailu church, XXI, 75, 284
西什库教堂 Cathedral Xishiku, XXI, 76
熙笃会修道院 Trappist abbey, 38
学校 school, X, XVII, XIX, XXVII, XXVIII, XXIX, XXXI, 28-31, 33, 36, 38, 44, 48, 78, 88, 91, 105, 106, 116, 121, 125, 196, 208, 225, 287, 291-298, 300-305, 313, 339, 352, 358, 425
学院 college, IX, X, XI, XIII, XXII, XXVIII, XXXI, XXXII, 7, 11, 21, 29, 31, 32, 36, 81, 82, 233, 234, 284, 290-292, 294, 295, 297, 300, 302, 313, 325, 353, 355, 358, 359, 361, 364-367
杨树浦教堂 Yangtzepoo church, XXII, 84
耶稣 Christ, XII, XVIII, XIX, XX, XXII, XXVII, XXVIII, XXXI, 6, 7, 12, 15, 25-29, 31, 33, 36-38, 43, 44, 51, 53, 54, 77, 81, 83, 94, 98, 99, 111, 118, 122, 132, 134, 135, 208, 284, 285, 287-289, 292-295, 297, 298, 300, 304-307, 311, 313, 315, 324, 339, 346, 347, 358, 378
耶稣 Jesus Christ 即 耶稣 Christ
耶稣会文献 Jesuit Archives, 28, 284
伊势神宫 Grand Shrine, 136, 232
颐和园 Summer Palace, 204
印刷厂 printing press, XIX, 26, 28, 46, 83
震旦大学 Aurora University, 292
主教座堂 cathedral, XVIII, XIX, XXII, XXVIII, 20, 23, 26, 33, 35, 38, 44, 66, 85, 88, 95, 101, 104, 106, 121, 139, 140, 142, 180, 185, 194, 207, 209, 210, 215-217, 284, 307, 313, 315, 316, 318, 323, 325
主徒会修院 seminary of the Disciples of the Lord, 92

住所 residence, XVIII, XX, XXI, XXII, XXVIII, 5, 58, 64, 65, 68, 71, 78, 102, 218, 297, 302, 304

宗教代牧区 apostolic vicariate, XXVII, 6, 86, 286, 287, 291, 307

专门名词 Specialized Terms

保护铺地 Protecting the flooring, 224

玻璃 Glass, XXV, XXX, 22, 73, 75, 76, 79, 142, 204, 217-219, 229, 230, 232, 246, 346, 347, 429

彩色玻璃窗 Stained glass windows, XXI, XXII, XXX, 73, 75-77, 79, 217, 219, 229, 230, 345, 346, 348

打夯 Ramming, 61, 102, 169-172, 430

滴水石 Dripstones, XXIV, XXV, 72, 211, 212, 236, 241, 242, 340

地基 The Foundations, 2, 61, 62, 100, 102, 131, 145, 156, 169, 170, 172-174, 181, 184, 185, 192, 229, 231, 427

杆子 Scaffolding poles 即 脚手架 Scaffolding poles

隔碱层 Saltpetre, 63, 156, 183-187, 224

工程预算 Estimate, XVI, 8, 47, 54, 130, 131

工程预算 Estimates 即 工程预算 Estimate

拱顶 Vaults, XXI, XXV, XXX, 2, 41, 73-76, 79, 80, 82, 97, 131, 139, 189, 209, 213-216, 231, 243, 244, 323, 342-345

勾缝 Grouting of masonry, 77, 100, 193, 221, 222, 426

和泥 Mix earthmortar, 77, 222, 223, 425

建筑布局 Layout, XVI, 8, 47, 49, 126

建筑材料 Building Materials, XVI, 8, 18, 19, 36, 41, 47, 55, 56, 98, 109, 111, 123, 140, 153, 202, 222, 319, 323, 338

建筑朝向 Orientation of the Buildings, 109

建筑细节 Miscellaneous, XVI, 8, 47, 72, 210

焦油 Tar, 79, 186, 229, 430

脚手架（杆子）Scaffolding poles, 168

脚手架 Scaffolding poles, 54, 59, 77, 131, 167, 168, 222, 323, 325, 426

廊子 Veranda, 78, 79, 225, 230, 427

楼梯 Staircase, XXV, 2, 79, 209, 219, 227, 228, 244, 300, 302, 303, 326, 427

麻绳 Rope, 59, 166, 167, 223, 424

磨砖 Polish the bricks, 78, 225, 428

抹泥 Roughcast or coating, 77, 222, 428

木料 Timber, XXIII, 50, 55, 58, 136, 148, 154, 162, 163, 197, 201, 428

欧式屋架 The European Roof Frame, 68, 197-200, 202

排雨水 Flow of rainwater, 229

墙角砖砌合 Clamping, 78, 226

设计方案 Plans, VII, XV, XVI, 6, 8, 10, 18, 22, 24, 26, 33, 35, 36, 47, 54, 55, 104, 105, 111, 118, 130, 131, 159, 189, 227, 284, 315, 323, 370

施工合同 Contracts, XVI, 8, 47, 54, 130, 143

石灰 Lime, XXIII, XXX, 55, 56, 58,

59, 61, 62, 66, 73, 77, 112, 141, 147, 148, 163-167, 172, 173, 177, 178, 182, 193, 195, 213-215, 220, 222, 223, 342, 345, 424, 425, 430

土坯墙的梁下柱 Columns under the beams, in adobe walls, 231

瓦 Tiles, XX, XXIII, XXIV, XXIX, 55-58, 64, 66-68, 70, 141, 145, 147, 148, 153, 160-162, 179, 182, 192-197, 202, 203, 205, 206, 212, 220, 222, 223, 226, 233, 239-241, 339, 340, 425, 429, 430, 432

屋顶 Roofs, XVI, XX, XXI, XXIV, XXV, XXX, 8, 18, 47, 50, 57, 58, 64, 66, 68-70, 78, 80, 81, 83, 97, 100, 112, 128, 136, 138, 139, 143, 145, 160, 161, 163, 179, 180, 182, 190, 192-212, 226, 231, 232, 241, 300, 315, 338-342

屋顶 The Roof 即 屋顶 Roofs

屋面坡度 Roof Pitch, 68, 202, 203

屋面装饰 Roof Ornamentation 即 屋面装饰 Roof Ornaments

屋面装饰 Roof Ornaments, XVI, 8, 47, 70, 110, 158, 203, 205

选址 Choice of the Location, XVI, 3, 8, 28, 47, 48, 109, 111, 123-125

油漆 To oil, 79, 189, 230, 231, 433

灶囱 Fireplaces, 227, 431

中式屋顶 The Chinese Roof, 66, 192

中式屋架 Chinese roof trussing 即 中式屋架 The Chinese truss

中式屋架 The Chinese truss, XXIV, 197-199, 201, 240

钟架子 Belfry, XXV, XXXI, 79, 228, 245, 351, 431

钟楼尖顶 The spire of a bell tower, XXV, 76, 219, 242

注释 Style, VIII, XVI, 14, 55, 96, 116, 117, 133, 187, 377, 401

砖 Bricks, XX, XXI, XXIII, XXIV, XXV, XXVI, XXIX, XXX, 2, 6, 10, 15, 20, 55-58, 61-66, 69, 73, 76-78, 80, 81, 100, 103, 112, 131, 144, 147, 148, 153-157, 159, 160, 164, 166, 170, 172, 174-191, 193, 195-197, 201, 202, 205, 206, 211-213, 216, 217, 219-222, 224-226, 229, 232, 235-239, 244, 246, 248, 249, 251, 300, 302, 313, 315, 316, 318, 323, 325, 335, 338, 339, 341, 342, 345, 354, 426-429, 431, 432

砖拱 Brick arches, XXIX, 73, 209, 216, 217, 335

砖石工程 Masonry, XVI, 8, 18, 47, 61, 169, 173

装窗框和门框 Installing Door and Window Frames, 187